Alfred North Whitehead:

The Man and His Work

The Johns Hopkins University Press Baltimore and London

Victor Lowe

Whitehead

The Man and His Work

Volume I: 1861–1910

Printed in the United States of America

The Johns Hopkins University Press,
Baltimore, Maryland 21218
The Johns Hopkins Press Ltd., London

Illustration Credits

Photographs of the Whitehead family, courtesy of Mrs. T. North Whitehead.
Photograph of Bertrand Russell, by the London Stereoscopic & Photographic Co. Ltd.
Engraving of Ramsgate, courtesy of Ramsgate Library, Kent County Library. Photo by Tony Bryan ARPS.
Map of the Isle of Thanet, courtesy of Ramsgate Library, Kent County Library.

The paper in this book is acid-free and meets the guidelines for permanence and durability of the Committee on Production Guidelines for Book Longevity of the Council on Library Resources.

Library of Congress Cataloging in Publication Data

Lowe, Victor, 1907–
Alfred North Whitehead : the man and his work.

Bibliography: v. 1, p.
Includes index.
1. Whitehead, Alfred North, 1861–1947.
2. Philosophers—England—Biography.
3. Mathematicians—England—Biography. I. Title.
B1674.W354L57 1985 192 [B] 84-15467
ISBN 0-8018-2488-5 (v. 1 : alk. paper)

To the memory of

T. North Whitehead

Jessie Marie Whitehead

Victoria Lincoln

Contents

Preface

Chapter 1 will explain my purpose in writing a biography of Alfred North Whitehead, the plan of the work, and the circumstances under which it was written. I wish at once, however, to express my gratitude to everyone who helped me in the preparation of this volume. As the work was carried on over almost twenty years, I am indebted to a great many persons, more than I can now remember. My thanks extend to all of them.

My deepest gratitude belongs to the three persons to whose memory the volume is dedicated: T. North Whitehead, Jessie Marie Whitehead, and Victoria Lincoln. I am also grateful to Mrs. T. North Whitehead for her constant moral support and infinite patience.

Mr. C. E. Busson, branch librarian at Ramsgate, Kent, supplied the excellent photograph of Whitehead's remarkable grandfather, Thomas Whitehead. Mr. Busson earlier aided me greatly in research at Ramsgate. In Broadstairs, Canon T. E. Prichard was most kind. At Whitehead's public school, Sherborne, the headmaster and the librarian gave me every possible assistance; I owe special thanks to Mr. A. B. Gourlay for consultations as well as for his published history of the school.

At Trinity College, Cambridge, many persons helped me on many occasions: the former Chief Clerk of the College, Mr. I. G. Camps; the former Junior Bursar, Dr. R. H. Glauert; the former Senior Bursar, Dr. T. C. Nicholas; the staff of the College Library, and particularly the former Librarian, Mr. C. R. Dodwell and Sub-Librarian, Mr. A. Halcrow. Among the Fellows of Trinity College, Dr. Robert Robson was unfailing in responding to countless questions about the history of the College in Whitehead's time. For memories of Whitehead himself, I am indebted to several Fellows who are now deceased, and especially to Henry Hollond, to Professors J. E. Littlewood, Sir Geoffrey Taylor, and A. V. Hill. I thank Mr. and Mrs. Arthur Owens for their services in the Cambridge University Library. At Girton College essential help came from the Mistress, Dr. Muriel Bradbrook, and the Librarian, Mrs. M. Gaskell.

The chapter on the Cambridge "Apostles" could not have been written without extraordinary help from Professor R. B. Braithwaite.

My understanding of Whitehead's earliest mathematical papers is due to Professor Bruce Marsh of the Johns Hopkins University. For reading my pages on Whitehead's first book, the *Treatise on Universal Algebra,* I owe thanks to two mathematicians, Dr. David Leep of the University of California at Berkeley and Dr. Blair Kinsman; and to Professor Stephen Barker of Johns Hopkins.

Bertrand Russell was as helpful as he could be when I visited him in 1965 and 1967, and in letters to me. In connection with the chapter on him and those on *Principia Mathematica* I must thank several persons. Year after year, Dr. Kenneth Blackwell, Archivist in charge of the Bertrand Russell Archives at McMaster University, provided information that I wanted; and he was kind enough to read a late draft of most of this book. I owe thanks also to the staff at the Russell Archives and to the Assistant Archivist, Dr. Carl Spadoni. For many discussions of matters concerning Russell, I am indebted to the historian of mathematics Dr. Ivor Grattan-Guinness, to Mr. Ronald W. Clark (Russell's biographer), and to Professor John Slater of the University of Toronto.

I owe exceptional thanks to Mrs. Norah Nicholls, M.B., for information about the health of Whitehead's wife, for reading the chapter on Russell, and for her hospitality.

Thanks are also due to the owner of the Mill House in Grantchester and the owners of the Whiteheads' homes in Cambridge for allowing me to see these houses.

Indispensable help has been provided by two secretaries: in the earlier years, Mrs. Archibald Hart; recently, Mrs. Bennett Gold. I thank the Johns Hopkins University for leaves of absence and for grants that defrayed some expenses.

In 1968 the Secretary to the Syndics of the Cambridge University Press, R. W. David, very kindly sent me extracts from the minutes of the Syndicate and other information about its dealings with Whitehead. The Council of Trinity College, Cambridge, has permitted me to quote from the minutes of the Council.

Letters to and from Bertrand Russell were made available to me by the Bertrand Russell Archives, McMaster University, Hamilton, Ontario. A little use is made of Russell's letters to Lady Ottoline Morrell, with the permission of the Harry Ransom Humanities Research Center at the University of Texas at Austin.

Letters to Whitehead from the Cambridge University Press and from the Royal Society of London were made available by those institutions, which retain the copyrights.

Thanks are due to the *Cambridge Review* for permission to quote (in the notes to Chapter VII) a poem by J. K. Stephen.

Some research in England was done under a 1967 grant from the American Philosophical Society, and some, in 1967–68, as a Senior Fellow of the National Endowment for the Humanities.

Parts of Chapter XIV and of Chapter XV, section i, are drawn from Chapter 7 of my *Understanding Whitehead,* published by the Johns Hopkins Press in 1962.

Alfred North Whitehead:

The Man and His Work

CHAPTER I

Introduction

i/Why a biography of Whitehead.

ii/Whitehead's philosophical system.

iii/Process theology. Neglect of Whitehead's philosophical system.

iv/Genesis of this biography. Paucity of extant documents.

v/In what sense this biography is authorized. Whitehead's reticence.

vi/Qualifications for Whitehead's biographer. The division between Volume I and Volume II. The dedication.

Alfred North Whitehead was an English mathematician and philosopher. For reasons that I shall mention in section iv, no professional biographer in his right mind would touch him. I write out of the conviction that someone who knew him and his work ought to write as much about him as the unvoluminous extant materials allow; not a memoir, but an objective biography.

Whitehead thought biographies of great men were useful, but never thought of himself as great. So said his son, T. North Whitehead, when I first talked biography with him. A brief sketch of my subject's life and work will show how much he deserves a biography.

He was born in Ramsgate, Kent, in 1861. After a career at the Universities of Cambridge and London, Whitehead accepted a call to Harvard in 1924 and taught there until 1937. He died in Cambridge, Massachusetts, ten years later.

"In England, Whitehead was regarded only as a mathematician, and it was left to America to discover him as a philosopher." So wrote Bertrand Russell, his pupil and then his collaborator on *Principia Mathematica.*[1] The remark was edged, and not entirely correct: it was because Whitehead had recently earned high reputation as a philosopher of physics that Harvard invited him to become a professor of philosophy. But it is quite true that he was educated to be a mathematician and held no other professional position until, at an age when he could have retired, he came to America. In the next nine years he wrote books which showed that Harvard had acquired the most powerful systematic thinker philosophy had seen since Hegel.

Whitehead had great personal influence on his American students. He became known as an incredibly benign sage. Benign sages, especially if they are private rather than public figures, become lost in a gentle mist soon after their deaths. This one is worth recovering. A British philosopher who professed inability to understand what Whitehead wrote in America said, "He was equally great in intellect and in character."[2]

Russell disagreed with Whitehead in philosophy, and completely in attitudes toward World War I, but he concluded his "Portrait from Memory" by writing, "I think that in all the abler young men with whom he came in contact he inspired, as he did in me, a very real and lasting affection." Affection for Whitehead was not confined to men nor to the young nor to the able. He was almost universally beloved.

How can the life of so *good* a man be interesting? To make the problem more acute, this good man was a Victorian.* The events that most excite us in biographies of Victorians are those in which the disciplined goodness breaks down in a dramatic self-conflict and we can say, Ah, here is where his diary and letters show him to be a mere human. I do not doubt that Whitehead was human, but he was not enough interested in himself to keep a diary, and he did not write letters to friends about his personal problems, for no friend was a confidant on them.

A life of Whitehead is bound to be lacking in spice. My reading of what others wrote about him has turned up no record of his ever breaking off a friendship, having a liaison, or carrying on a public quarrel with the government. His adventures were adventures of the mind. I think they are well worth following.

ii

Whitehead was a good mathematician but probably not a great one. Certainly mathematicians do not consider him great; this judgment may reflect the fact that his writings were outside the mainstream of the history of mathematics. It is as a philosopher that he stands large in the history of thought. But the philosophy is so complex that only specialists grapple with it. It might as well be mathematics!—might be, but for the once-a-page (on average) appearance of some sentence that someone writing on education or natural science or social philosophy or religion or history (all subjects on which Whitehead read widely and thought deeply) has found to be a gem for his purpose. Whitehead is one of the most quoted philosophers of our century—and one of the least understood.

In his first year at Harvard, Whitehead wrote *Science and the Modern World*. It was immediately hailed, for there was a fair expectation among educated, thoughtful people that a philosophically minded

*See the beginning of Chapter II. Part of the contrast between Whitehead's character and Russell's might be summed up by noting that Russell (eleven years Whitehead's junior) was more of an Edwardian than a Victorian in his ways.

scientist could overcome the dualism between the scientific world view and the values of human experience. Whitehead seemed to be the man. But the systematic statement of his ideas, which appeared in *Process and Reality* (1929), was too intricate and many-faceted to be popular. There ought to be some easier way of doing the job! Why not simply apply the scientist's experimental *method* of thinking to questions about values? John Dewey urged this, and acquired much greater influence than Whitehead. Then logical positivism, existentialism, Wittgenstein, Heidegger, other men and other movements, took over. Whitehead's philosophical system was pretty much left out in the cold. Why not take a fresh look at it, and at the man who wrote it?

In *Process and Reality* Whitehead laid out a metaphysical system in a semimathematical form and discussed its use at length. Just as every particular fact of spatial relatedness exemplifies the general theory of space—though the correct statement of that theory was not settled by Euclid nor by Riemann or Einstein—so it is with Whitehead's metaphysical theory. Its objective is a scheme of ideas such that everything of which we can in any way become conscious can be interpreted—not directly, to be sure—as a particular instance of the general scheme. That is the ideal; as Whitehead said, "weakness of insight and deficiencies of language" prevent philosophers from ever reaching it, from making more than successive approximations to it.[3] When the system fails to accommodate some recurrent experience, the system must be revised. Whitehead made none of the claims to have proved the truth of his system, that were made for the great metaphysical systems of the past. He offered a bold, complex hypothesis and said, "Take it from here." The methodology of his "speculative philosophy," as he called it, was cautious and sophisticated. The system he offered was original, and larger in scope than any that is actively entertained today.

The vision behind the system was that of a universe ever in process. This does not mean simply that events happen one after another, endlessly. They happen, and continue to happen, because of a universal force that underlies every event and its supersession by another; Whitehead named this force "creativity" and wrote that it and the notions "many" and "one" comprise the "Category of the Ultimate" which his thirty-six axioms presuppose. He saw the universe as a cumulative nexus of individual processes of becoming, each an organic unity. (He called his philosophy "the philosophy of organism.") One example of such an individual process is a moment of human experience, which comes into being as an integration of slightly earlier processes in the human body; Whitehead generalized from this example.

The system was pluralistic, yet unified in accordance with his maxim that nothing exists without essential dependence on other things. Like Leibniz, Whitehead produced a monadology, but his was far superior to Leibniz's. In *Process and Reality* he used the depositions of several great philosophers of the past as tests of the power of his system, but he was more concerned that it should accommodate scientific theory and practice, and our social, aesthetic, and religious experience.

Whitehead was raised an Anglican. A flirtation with Rome was followed, in the mid-1890s, by his becoming an agnostic; he remained one for a quarter of a century. After World War I, in which many of his pupils and one of his two sons were killed, he gradually developed his own philosophical theism. He conceived of God, the supreme monad, not as omnipotent Creator but as eternally *with* the temporal world, systematically interacting with it in two ways. God's "primordial nature" is the source of all possible values; it is, he wrote, "the acquirement by creativity of a primordial character."[4] God's "consequent nature" receives, transforms, and keeps forever the monads that have perished in the temporal world; thus God grows with the world, is always in process. Whitehead did not include his highly original concept of God in the categoreal scheme of *Process and Reality,* but fitted the concept nicely onto the scheme. I think the scheme logically required this addition.

iii

A few natural and social scientists have made use of Whitehead's philosophy, but theology is now the main area of its influence. This trend effectively began at the University of Chicago, where several theologians and the philosopher Charles Hartshorne were attracted by Whitehead's conception of God. In 1965 Hartshorne's pupil John B. Cobb, Jr., published *A Christian Natural Theology Based on the Thought of Alfred North Whitehead.* Since then, the movement known as process theology has burgeoned. Although it is centered in southern California (the journal *Process Studies* is published in Claremont), the movement is worldwide and includes theologians who are Anglican, Roman Catholic, and Methodist, among other denominations. Many illuminating studies have been published. Indeed, because of the industry of process theologians, Whitehead's influence, in comparison with the breadth of his interests and his work, is now lopsided.

Professional philosophers of secular persuasion have paid too little attention to Whitehead. Even before he retired in 1937, many at Harvard took up logical positivism and Wittgenstein's philosophy. Yale was more hospitable to his ideas. The universities of his native England were least so. The typical academic philosopher in an English-speaking country has had little interest in general theory, and none in a theory like Whitehead's, which excludes nothing from its scope; he has insisted on discussing detached, bite-sized philosophical problems. The limitation of his interest to specifics that he could argue about with colleagues and students was partly a by-product of excessive democracy in Ph.D. programs. In the sciences a Ph.D. did not get far up the ladder without producing tangible results. In philosophy, talk, so long as it was clear and avoided logical blunders, was thought sufficient. Of course clarity and consistency are virtues, no matter what one's interests are. My point is that taking these virtues as sufficient for progress in philosophy cut down interest in Whitehead's large, unified system among professionals. When they noticed it they called it grandiose, and fostered the myth that Whitehead was deliberately obscure.

The current state of the philosophical world, and of the intellectual world generally, is somewhat different. There are signs of a loosening of the piecemeal method's grip. The alternative to it that was most popular in Europe, existentialism, has lost its novelty. (Whitehead never discussed existentialism, but his philosophy agrees with its insistence upon considering the whole experience of living.) Among philosophers who are read by the public, there is some backing and filling as new directions are sought. I do not know whether the time for Whitehead's philosophy has finally come, but it may be here.

iv

The origins of this biography go back a long time. Between 1929 and 1945 I had many conversations with Whitehead and got to know his wife. I was first called a Whiteheadian more than forty years ago, after perpetrating a long essay on the development of his philosophy in the Whitehead volume of Paul A. Schilpp's *Library of Living Philosophers*. After spending twenty years mostly on other subjects, I wrote *Understanding Whitehead*.[5] In 1965 a need to find out whether any manuscripts of his books existed led me to visit his son North. Upon reading my book, he urged me to attempt a biography.

So I went ahead and sent out countless inquiries, to which the most usual answer was, "I wish I could help you, but I have no letters, only a few memories of times long past." Officials of the public school, colleges, and universities in which Whitehead spent his life gave access to all their records. But a biography cannot be composed of records alone. I soon discovered why, eighteen years after the end of Whitehead's long life, no biography had appeared. The kinds of revealing private documents a biographer wants are almost nonexistent. Whitehead was famous for not writing letters; they took too much time from his work. Natural letter-writers like Bertrand Russell or Virginia Woolf write mainly to express themselves; Whitehead had little of that interest. The only large batch of extant letters consists of those he wrote to Russell during their long collaboration on *Principia Mathematica;* I shall make much use of them in this volume. A smaller batch consists of letters to North written between 1924 and 1929; I intend to print them as an appendix to Volume II of this biography. In his will Whitehead directed that his letters to his wife be destroyed; that makes a great gap, for she is the most important person in his biography. It does not follow, though it may be true, that he wanted to cover up episodes, knowledge of which could embarrass him or his wife. He held an almost fanatical belief in the right to privacy, and thought that the only subject of rightful public interest in him was the work he had published.

Paucity of material limits what can be written not only about Whitehead's life but about his work. As he had requested, his widow destroyed (along with the letters he had received) his unpublished papers and drafts of books, and the manuscripts of the published writings. He idealized youth and wanted young thinkers to develop their own ideas, not spend their best years on a *Nachlass.* His checkered progress on whatever he was writing, with the exception of the *Principia,* is not to be found in letters, for three reasons: first, if he wanted to exchange ideas with someone, he did not write a letter, but tried to talk to that person; second, he worked alone, on problems that he wanted to investigate but that others generally did not; third, he was not given to expressing in letters to friends his feelings about his work in progress. Hence Whitehead's intellectual biography, so far as we can be certain about it, depends mainly on the external facts of his life and the prefaces to his writings. A valiant attempt to discover from the texts of the philosophical books he wrote in America just how they were written and at what points his views must have changed has been made by Professor Lewis S. Ford.*

V

This biography is authorized only in the sense that on May 21, 1965, North Whitehead, having encouraged me to attempt it, gave me written "permission to collect and use all the material you can find, including of course letters written or pertaining to my father and mother, the same to be printed at your discretion with no further appeal to us." Whitehead's only daughter, Miss Jessie Whitehead, also signed this letter.

North and Jessie submitted to endless questionings and did everything they could to help me. When their information consisted of what they had been told by their mother, I had to use my best judgment about its degree of credibility, for their mother tended to dramatize everything. Whitehead did not do so; in using my notes of conversations with him I have had to bear in mind only his reticence, his desire not to hurt anyone, and the fallibility of an old man's memory.

Jessie told me that her father's reticence was general, and strong. He considered his personal problems *his* problems, period. I cannot say what in his early life nourished this reticence, or any other feature of his character. Correspondence with his parents and siblings was not kept, and if there really is some documentary evidence about his emotional experience in the years before he left home for public school, I have not found it. Lack of data puts psychogenesis, always a risky business, out of the question. I shall not try to explain Whitehead's character, but only to narrate its manifestations. His character was so memorable, his intellectual interests, problems, and achievements so singular, that their objective description should itself be worthwhile.

VI

Whitehead's writing of his philosophy was the culmination of his life, but familiarity with it and with him are not the only desiderata for

---

*Lewis S. Ford began this with an examination of *SMW* entitled "Whitehead's First Metaphysical Synthesis," which was published in the *International Philosophical Quarterly* (17 [1977]: 251–64). My skepticism about the project was briefly expressed in the same journal (18 [1978]: 223–26). In 1984 Ford published *The Emergence of Whitehead's Metaphysics, 1925–1929* (Albany: State University of New York Press).

his biographer. Would that I were also a professional historian of mathematics, a historian of modern science, a historian of British education, and a scholar in the humanities! Would that I were English as well as American! When I began to collect material for the present work in 1965, I had a strong sense of my temerity in undertaking a task that calls for a team. That sense is still with me.

As interest in Whitehead is greater in America than anywhere else, I have had American readers primarily in mind. Where necessary to provide context for them concerning his life in England, I have not hesitated to explain things that English readers might be presumed to know. I beg the indulgence of the English reader.

Because of Whitehead's almost pathological dislike of publicity, it was out of the question for anyone to write about his life as long as he lived. I think that his widow, who lived until 1961, would have approved of a biography of him, but only if she had been able to ensure that the picture to be presented was close to the one she wanted to have presented. A biography of a man like Whitehead, however, is no good unless it is informed by a persistent effort at objectivity.

Volume I ends with the first of the two major movements in Whitehead's life: his move from Cambridge to London in 1910, when he was forty-nine. Knowledge of his life and work to 1910 is the one thing that students of his thought ought to but do not yet have. It should be of value even to those who are interested only in his final philosophy, for it makes it easier to understand how a mathematical logician could come to rank beauty above truth and speculation above argument. No biographer can decide questions about the truth of Whitehead's views, since truth is independent of persons; but a full understanding of his views, and even their trustworthiness, can be affected by knowledge of Whitehead the man. A chief concern about Whitehead's thought, in this work, is to understand it as the thought of a most unusual human being.

This volume is dedicated to the memory of the three persons who were of greatest help to me. T. North Whitehead started it all. I found it a privilege to become a friend of that courageous, lovable eccentric, Jessie Whitehead. Victoria Lincoln, my late wife, spurred me on and took time from the writing of her biography of St. Teresa of Avila to participate a bit in my research and writing. My great regret is that because of inefficiency in biographical research and slowness in writing I did not finish Volume I while these three were still alive.

CHAPTER II

The Whiteheads

i/Antecedents.

ii/Grandfather Whitehead.

iii/Grandfather's sons.

iv/Uncle Henry.

v/Whitehead *père*.

vi/Whitehead's mother's family, the Buckmasters.

Alfred North Whitehead always claimed that he was an ordinary Victorian Englishman. Ordinary? Well, in some ways. But he was certainly Victorian, and middle class to boot. He was born in 1861—the year that ended with the death of Prince Albert—and he was forty when the Queen died. Soon after Waterloo, his grandfather Whitehead established himself in the middle class of successful professional men by starting an excellent boys' school in Ramsgate, Kent.

The Whiteheads had lived for centuries in or near the Isle of Sheppey, which hugs the southern shore in the mouth of the Thames. They came originally from Jutland, where *Wichtraed* is a family name. They may have come over with Hengist and Horsa or soon thereafter, as Whitehead's daughter, Jessie, thought likely. Bertrand Russell, in the well-known BBC talk on Whitehead, said, "His family came from Kent and had been clergymen ever since about the time of the landing of St. Augustine in that county." This must be taken with almost as many grains of salt as Russell's conversational remark, "He comes from a long line of bishops." Russell, that unclerical aristocrat, could not refrain from extravagant thrusts at what he felt was a residue of vicarage atmosphere in the philosophy of his mathematical teacher (later his collaborator) and philosophical opposite. There really was a bishop in the Whitehead family, but he was Alfred North's brother, not an ancestor. Whitehead himself made one, and only one, statement in print about his ancestry: "My grandfather born of yeoman stock in Isle of Sheppey, was probably a descendant of the Quaker George Whitehead, whom George Fox in his *Journal* mentions as living there in the year 1670."[1] The conjecture of a Quaker ancestor will not hold up, but there is no doubt about the yeoman stock.*

*Whitehead had no evidence other than the passage his daughter, Jessie, had just called to his attention in Fox's *Journal*. He knew that he was making a surmise, but he liked it; in the manuscript of "Autobiographical Notes," "probably" is replaced by "perhaps" and then restored. Jessie came to disbelieve this theory. George Whitehead came from Westmorland (the name is much more common in the northern counties), he lived all over England, and he had no legitimate children.

Whitehead's son North told me that these yeomen were "fishermen—farmers." In an unpublished autobiography he mentioned other activities:

> The Whiteheads seem to have led a fairly quiet life for a number of generations earning their livings almost indifferently as yeoman farmers (as distinct from gentlemen farmers who did not follow the plough), school teachers and Church of England clergymen, with an occasional doctor thrown in.[2]

Presumably the Whitehead farmers fished for some part of the year, and the sons who had not inherited any land to speak of concentrated on fishing, or took holy orders, or taught in a local school, or, in a couple of cases, became doctors in the locality. We have to remember that higher education was not yet required for any of these occupations.

Farmers prospered in the Napoleonic Wars. (Readers of Jane Austen's letters will remember her complaints about the price of food.) It was possible for Thomas Whitehead, born in the autumn of 1790* to John Whitehead's third wife (Mary Brown), to make more of his life than any Whitehead before him had. His schooling was good, he was a keen reader of ancient and modern history, he had himself mastered quadratic equations and spherical trigonometry, and he had acquired an enthusiasm for teaching. When he was twenty-one he went to London to teach in a charity school in the East End. But he wanted to establish his own school for boys. To do this he came in August 1815 to Ramsgate, the busiest town in the northeast corner of Kent, which is called the Isle of Thanet. He earned a good competency from his school, gave his sons a university education, and in his old age served the town as its chief administrator. By then he had turned Chatham House Academy over to his son Alfred, to whom Alfred North was born on February 15, 1861, in the headmaster's quarters on Chatham Street. (The building no longer exists.) In his "Autobiographical Notes" Alfred North Whitehead remembered: "They were, both of them, most successful schoolmasters, though my grandfather was by far the more remarkable man." Beyond this, he said almost nothing about either one, for he had little interest in his family. He did not even mention his mother's family, the Buckmasters. They too were

*Not in 1794, as Whitehead implies in the first paragraph of "Autobiographical Notes." The registry of East Church parish, Sheppey, shows that Thomas Whitehead was baptized on November 15, 1790.

Victorian middle class. They had a good deal of money—from trade—but could claim no special achievements. The Buckmasters were a varied lot, and only one lived in Ramsgate. Grandfather Buckmaster died before our subject was born, but Grandfather Whitehead lived nearby until Alfred North was sixteen. Grandfather Whitehead and his sons provided the model of a useful life and of moral character that was before Alfred North Whitehead when he was a boy. A short account of these Whiteheads will sufficiently fill out the model for us.

ii

When Thomas Whitehead arrived in Ramsgate and announced that he would take pupils, he set up his school in a single room at the rear of what had been the post office. There was a need for a good private school in the town, for one called Chatham House had been abandoned in 1810, but almost no pupils came. He was tempted to give up, but instead moved his school into better premises, and two years later seized an opportunity to buy Chatham House. Within ten years he made it one of the best schools in England. Its enrollment was largely, though not entirely, local. The great public schools that drew pupils from all over England came later.

Thomas Whitehead's portrait shows a downward smile on the right side of his face, which is set and grim; an upward smile on the left side, which is benign. Outside school he was tender and kind; within, a stern disciplinarian. In his dedication to the cause of education he set down anything that was second-rate in a school as utterly worthless, and he did not tolerate it in either boys or teachers. His ability to size up a situation was uncanny.* His enthusiasm for education was coupled with a genius for teaching and forming the character of boys; Samuel Taylor Coleridge called him a heaven-born schoolmaster. And what training he gave for a life of hard work! Even in winter, Tim, as the boys named him, began the day with his ninety-nine pupils—he would not have a hundred—at 5:40 A.M., and did not let the day-boys go home until 8:30 P.M. (after prayers). A long school day was then customary, but not usually this long. One pupil remembered that during the bathing season the day began earlier, "so that the visit to the sands should not diminish aught from the tale of bricks to

*One alumnus of the school who had been a pupil from 1836 to 1843 wrote that the headmaster never asked which boy had done the mischief, but unerringly singled out the culprit and asked why.

be delivered before breakfast." Another said that none of the boys he knew broke down under this regimen; but he added, "It is true that no weakly boy was received, Mr. Whitehead saying that Chatham House was not a Hospital."

Thomas Whitehead took great pains with the teaching. He himself wrote the text from which mathematics was taught. At a time when most schools ignored science, he bought the best apparatus, gave a carefully prepared lecture weekly on some scientific subject, illustrating it with experiments, and hired outside experts to give some others. On English history he had strong convictions, which he did not soft-pedal in his lectures. Men who stood up for liberty, like Stephen Langton, were praised without stint; so were rulers who knew when and how to give way.[3]

Grandfather's patriotism was unqualified. "I can forgive a man anything, if he loves England, and rejoices in its success."[4] He particularly rejoiced in the Reform Act of 1832 and was all his life a strong believer in Whig-Liberal principles. He imparted this faith to his sons.

This strong man's recreation was simply an hour's ride on his horse before he dined with his pupils. He was a widower of sixty-three when he gave up the school and moved to a house that was an easy stone's throw away. Now, a Whitehead does not retire at sixty-three if there is an opportunity for a new kind of work. The Local Board, composed of the elected Commissioners of Ramsgate, asked Thomas Whitehead to be their chief—in effect, mayor, though his title was Deputy, Ramsgate being still an unincorporated "limb" of Sandwich.* He was not a politician; he was so highly respected that he did not need to be. His son Henry wrote, "No fear of consequences ever deterred him from taking an unpopular line, yet no man in the town was more popular."

Thomas Whitehead was the indispensable chairman for public dinners, noisy gatherings of taxpayers, Ragged School committees, and Bible Society meetings. Always in command, he was noted for a humor that, like his grandson's, was never unkind. It was Deputy Whitehead's habit to give from his own pocket to every needy person who came to him and deserved help. He was as generous as anyone can be who habitually sits in judgment on both himself and others.

After a dozen years as Deputy, Grandfather Whitehead retired; the

---

*The Local Board also saw that no one was better qualified to be a justice of the peace; Thomas Whitehead was appointed one, both for the Cinque Ports and for the county. Until 1888, justices of the peace performed administrative as well as judicial functions.

columns of the *Thanet Advertiser* show that the conduct of business at meetings of the Local Board then went steeply downhill. He lived until June 1877. His funeral was among the simplest the town had witnessed, but most of Ramsgate walked to his grave. Ten years later, a portrait in oils had been painted by public subscription and hung in the Town Hall.[5]

Grandfather's religious views were those of an old-fashioned Evangelical Churchman. Henry wrote: "He did not *take to* religion. He could not help being religious." But as a man of action, he abominated the habit of beginning business meetings with any but the shortest possible prayer. Untroubled by religious or philosophical perplexities, Thomas Whitehead lived out the Victorian ideal of work, integrity, and benevolence. In just one respect may we smile at him. His only published writing that I have found is a long poem of 1817; an elegy on the death of Princess Charlotte Augusta, its heroic couplets are unmarked by any gift for poetry. And according to Bertrand Russell (who presumably had it from Alfred North Whitehead), his friends said that he composed a hymn containing this sublime manifestation of regional pride:

Lord of the Lambkin and the Lion,
Lord of Jerusalem and Mount Zion,
Lord of the Comet and the Planet,
Lord of Sheppey and the Isle of Thanet!

Grandfather's wife was Mary Williams, a London-born woman of Welsh family whom he married when he was starting his school. She died a few years before Alfred North Whitehead was born, but her Welsh blood is said to have shown itself more in him than in his brothers, who were quite different in appearance. She was a gentle, quiet person who counseled patience in all circumstances; and unlike her husband she had sympathy even for the undeserving.

iii

Thomas and Mary had ten children, and three died in infancy. Our Whitehead's father, Alfred, was the youngest of the four sons who grew up. John, seven years older, lit out for Australia many years before Alfred North was born. I do not know what became of John, only that his father's will gave him an equal share of the estate, which was valued at just under £25,000. All the other sons were then clergymen at various places in England. Perhaps John turned away from the ministry because of his father's insistence that Sundays be devoted entirely to church services, catechisms, prayers, and brisk walks.

The paternal uncles who helped shape Alfred North's mind and character were as religious as their father, but as we should expect, they shed his severity. Neither Uncle Tom nor Uncle Henry had Grandfather Whitehead's genius, and neither one advanced far in the Church. Without noticeable ambition to speak to the nation, they were content to do all that they could for other people nearby, in school or parish. Both when young taught for a few years in Chatham House Academy. The oldest, Uncle Tom, served as a parish priest for five years in Ramsgate, followed by fifteen in the Buckingham village of Gawcott. In the year of Alfred North's birth he was in Gawcott, and had finished a small book, *Village Sketches, descriptive of Club and School Festivals and other Village Gatherings and Institutions.*[6] One sketch tells how he started a boys' school; others describe such things as the Penny Bank, a library for children, tea parties, and an evening at what this parson called his "Night Club." The book was successful; enough Londoners wanted to read about village parishes to justify a second edition in 1862 and a third edition in 1870; in the third edition the original ten sketches were increased to twenty-one. All are mildly charming, written with a little cleverness by a loving hand. As depictions of the local social history that British historians neglected until recently, they have some value today, but they make dull reading. The witty satire or righteous indignation that would interest us are absent. The author was not stolid, but he was a *good* man.

After Gawcott, Uncle Tom spent the last nine years of his life as headmaster of Christ's College, Finchley, a preparatory school outside London. He himself had attended Wadham College, Oxford, studying classics and mathematics, but without achieving any distinction. He was not a scholar, but he blossomed out in middle age as a good school administrator. And he had thought about the philosophy of education, expressing his views in a lecture entitled "System: Its Use and Abuse," which was published in 1860.* By "system" he meant "any settled plan for the carrying out of an object or end." He contrasted systems, so understood, with truths, which are necessarily eternal: "Systems, which are their dress, change with every variety of time, place, and circumstance." The moral he drew for the education of every generation was, "It is never essential to perpetuate mere system—nor possible, without injury to the mind." Very good! This

*See page 21 and footnote.

Whitehead also expressed the conviction that the first business of an educator is "to beget power of self-action." Such things had been said before—one thinks of Rousseau—but the point for us lies ahead. Half a century later, when Alfred North Whitehead talked about the aims of education, some of the views he advocated with the force of genius came naturally to him.

iv

Uncle Henry's career was entirely in the Church of England. Alfred North Whitehead had in his library two memorial tributes to this uncle: a long pamphlet by a Nonconformist minister who considered Henry Whitehead his "ideal man,"[7] and a book by an Anglican canon.[8] The book makes generous use of its subject's written and reported words. Often they express views similar to Alfred North's, though in the manner of the preceding generation: Uncle Henry was born in 1825 and died in 1896. These two Whiteheads also shared many traits of character.

The course of Henry Whitehead's life would have disappointed anyone interested in self-advancement. As a student at Lincoln College, Oxford, he missed the First Class that his tutor thought he would get. (The tutor was Mark Pattison, one of the great teachers at Oxford; they became close friends.) He was a good scholar, but a better talker, with an abundant store of mother wit, a merry laugh, and an ever-ready sense of the ridiculous. He was known as the College parodist, equally deft in Latin and English verse. But there was no malice in him, any more than there would be in his nephew. He was a friend to the College bores as well as to the wits. Nobody could provoke him to anger, and nothing could embitter him. In short, he was something of a saint.

Soon after graduating from Oxford, he was ordained and started work as a curate at St. Luke's, Berwick Street, in the slums of Soho. When the yellow flag went up at the top of Berwick Street in the terrible cholera epidemic of 1854, he was a tireless helper to all who were afflicted. As soon as the epidemic abated, he visited all the families who lived or had lived in his parish, and in his friendly way readily got full information about their living habits. His collation of this evidence corroborated Dr. John Snow's hypothesis—about which he had been skeptical—that the disease is spread by drinking contaminated water, and that in his neighborhood the Broad Street pump was responsible. When cholera struck the East End in 1866, he left his family in order to live with the sufferers and work with the inspector of public health, John N. Radcliffe, who said that next to Snow, Henry Whitehead did more than any other Englishman to control the cholera.

From Soho, Uncle Henry went to a parish in Westminster that was known as the "Devil's acre." A visitor in its streets could say that he had seen the same people in Hogarth. Henry Whitehead was a reformer—a quiet, loving reformer, not a grim one. He enjoyed life, but his pleasures were strictly limited by an overriding consideration for other people. He was fond of playing the cornet, but set it aside for good when he first took up lodgings in London, because, he said, "many poor people might object and not like to tell me."[9] The first vicarage he accepted—in 1871, after twenty years as a curate—was in Limehouse. The more miserable the district—and in those days a dock hand's job was at the mercy of the wind—the more good there was for him to do.

Needless to say, he did not encourage class antagonism. The only kind of socialism he respected—perhaps the only kind he knew—was the Christian Socialism of F. D. Maurice, founder of the Workingmen's College in London. Henry Whitehead did his best to bring out the good potentialities that he saw in everyone, but especially in poor boys. In 1860 he told clergymen that talk about "the whole question of elevating, as it is called, the working classes" was *no good* "unless from the very depths of your soul you hold the grand doctrine of the essential equality of mankind." The wrong attitude toward workingmen was expressed by clergymen who asked each other, "How shall we *get hold* of them?" instead of trying with all their hearts to know and understand. What Henry Whitehead had was a sympathy, inspired by the New Testament, with each human being. He was tagged as a disciple of Maurice, who had preached the centrality of love in Christian doctrine and had been dismissed (in 1853) from his professorship at King's College when the Bishop of London could no longer stomach his disbelief in the everlastingness of Hell. Maurice, the great Victorian foe of dogmatism in the Church of England, advocated "a free and open church," and Henry Whitehead followed him in that. Henry was a Broad Churchman whose tolerance of diverse religious views—*if* they were sincerely held—was absolute. "Charity," he said, "would induce us to believe that there are Christians who are not of our faith and name." In his late years the place of honor in his study was given to a portrait of George Fox. Although he deplored the trend toward sacerdotalism in the Church of England, one of the reasons he gave for declining a wealthy benefice was that it would be hard on the members of that congregation to be saddled with a Broad Churchman after two generations of High Anglican teaching. With him, people counted more than movements or institutions.

Henry Whitehead had not enjoyed the Sundays imposed upon him by his father. "Don't drive men to church," he said, "but lead them there. Don't think them bad because you do not find them in the pew." He had no tricks for leading people to church, and his sermons were more thoughtful than appealing. He simply did not have it in him to be a popular preacher. Nor did he in other respects try to adapt himself to the world or consult his own self-interest. He was always in the world—besides his pastoral work, he concerned himself with schools, free libraries, reformatories, sanitation, and so on—and never of it. He would not argue about religion, but he reasoned with people about these other things, persistently and well. "It is worth taking any trouble," he said, "to get people to do right of their own accord." He knew what was right, to be sure; he was a Victorian with firm convictions. But imposing those convictions on other persons was, to his mind, contrary to the Christian ideal. He encouraged innovations but would not force them on anyone. Uncle Henry really believed in individual freedom, and in the power of the Holy Spirit to move the hearts of human beings, however slowly. In this he reminds us of the philosopher Whitehead, who maintained that the right name for God's agency in the world is "the Divine persuasion." Tolerance in religion, helpfulness to all, and an unusual regard for people's feelings, are other traits the two men shared. One difference is that Uncle Henry is said to have had an almost Quixotic reliance on the ways of the Holy Spirit as against the ways of the world. His position among the clergy was almost one of splendid isolation. I have become quite fond of Uncle Henry.

During Alfred North Whitehead's boyhood in Thanet he and his uncle Henry saw each other at least a few times a year; much less after 1874. In that year Henry Whitehead left London for the border county of Cumberland, where he served in the Carlisle diocese for the remaining twenty-two years of his life, first at Brampton and finally at Lanercost. He loved the country people of the border as much as he had loved Londoners, and took great pleasure in learning their language, customs, and traditions. Local history became a passion; he made the past live again in accurate studies of old buildings, parish registers, and church bells.[10] He was also the kind of country vicar who sat down on a heap of stones to smoke a pipe with a tramp.

Along with a love of people, he possessed a shrewdness and a tact which ensured that if there was much dissension in a parish when he came to it, there would be none when he left. His nephew, too, was a reconciler.

Henry Whitehead did not have the intellectual curiosity of a philosopher. He was a practical man, with a simple faith, to whom the battles between bishops and Darwinians were unimportant. What difference could they make in his dedication to the Christian duty of neighborly love? As for the higher criticism, that too could be ignored.

Uncle Henry was the first member of the Whitehead family to include writing among his regular activities. He published a fair number of articles (beginning with several on the cholera in London), three books of sermons, and the antiquarian studies already mentioned.[11] His writing is clear, but its pace is too slow to make his pieces attractive so long after the Victorian age. Parish magazines were of course important to him, and he was a devoted editor of them. It was he who got his brother Tom to write *Village Sketches,* and he was responsible for the 1860 book of lectures in which Tom's thoughtful philosophy of education appeared. He wrote nine of the thirteen pieces that made up the book, which was entitled *Lectures, chiefly on subjects relating to Literary and Scientific and Mechanics' Institutes.** Such institutes, associated with free libraries, were often cursed by a patronizing attitude on the part of the clergymen and gentry who sponsored them; Henry Whitehead set himself against that. From his lectures I draw two final illustrations of his views. In "Reading Rooms" he wrote that libraries for poor people (he had a part in starting more than one) should include what they already enjoy reading as well as what will educate them; they must have a free choice. The short lecture called "Divide! Divide!" is a warning against the tendency of discussion to turn into debate and voting. His nephew's attitude would be the same.

V

Alfred North Whitehead's father was a horse of another color; to me, a less colorful horse. This is partly, but by no means solely,

*Three were by William Driver, a close associate of Henry Whitehead's who had been a cobbler and devoted himself to helping poor boys and teaching in Ragged Schools. Driver, Henry Whitehead, and Thomas Whitehead are named as coauthors on the title page. Henry did the editing, though he is not there called editor. All but one of the lectures were given in Clapham, where Henry was a curate for eight years.

because his was a completely local life. He was born in Ramsgate, got a university degree without leaving home for long, and died at St. Peter's (Broadstairs), three miles away, in his seventy-first year. His adult life was spent in those two towns of Thanet as schoolmaster and clergyman, the role of clergyman being uppermost. The Old Testament was one of the two books he read and reread. The other was Gibbon's *Decline and Fall of the Roman Empire.* That is strange only at first sight. Alfred North Whitehead wrote:

> I do not think that any of Gibbon's chapters shocked him; for his robust common sense told him that the people of East Kent, with whom he was quite content, were really very unlike the early Christians. His favorite character in the Bible was Abraham, who exhibits many features to endear him to the East Kent mentality.[12]

The Reverend Alfred Whitehead was an upright man with countless friends. North Whitehead summed up his character in these words: "He never asked a favor of anyone and never shirked what he considered to be a duty, but it cannot be said that he spent more time absorbing the lessons of the New Testament than was necessitated by his calling."[13]

Whereas his brother Henry was one person in a million, Alfred Whitehead was one of a type. Trollope drew him perfectly. "Reading him," Whitehead told Lucien Price, "I can hear my father and his clerical friends talking." North Whitehead wrote of his grandfather:

> I often think of him when I come across Archdeacon Grantly in *Barchester Towers;* they had much in common. My grandfather kept a hunter or two and a well-appointed carriage and pair; when an unwise investment necessitated some economy he did not keep a carriage and a single horse (which he could well have afforded), he used his legs instead. He never did anything by halves.[14]

North called him "a good example of a full-blooded hunting clergyman."

Alfred Whitehead was born November 9, 1827, and died March 16, 1898. His schooling, of course, was received at Chatham House. Henry had been their father's assistant there for two or three years before going off to college at twenty-three. Alfred did the same; his university, however, was Dublin—Trinity College, which at that time was almost a colony of Cambridge, and especially of its Trinity College. The College records at Dublin show that he was admitted as a pensioner in October 1850, that he passed the Bachelor of Arts examination with a Second Grade in the Michelmas (autumn) term of 1853, and that he received the baccalaureate in 1854, the M.A. in 1857.* They show nothing more—not the subjects he studied† nor the terms of residence he kept.

In a ceremony performed by his brother Tom, Alfred Whitehead was married to Maria Sarah Buckmaster of London on December 20, 1851. He was then twenty-four, his bride nineteen. I do not know how he met her, and no story of their engagement has been handed down in the Whitehead family.

The University of Dublin did not require residence for the Arts degree—only that its examination be passed. Thus Whitehead *père* had but one year of typical undergraduate life at a university. By the fall of 1852 he was in Ramsgate as headmaster of Chatham House Academy, though he probably did not succeed his father until he had his B.A. two years later.

Alfred Whitehead's clerical life began in 1856, as a curate at St. George's, Ramsgate. In the year of Alfred North Whitehead's birth he received a better position, as the minister of St. Mary's Chapel of Ease, also in Ramsgate. In the mid sixties he gave up the headmastership of Chatham House for his clerical work; the school was subsequently sold outside the Whitehead family. But Whitehead's father did not drop—he never dropped—active interest in schools. His clerical duties included superintendence of the parochial ones. Alfred North Whitehead remembered that as a small boy he often accompanied him on the daily visits to his three parochial schools—for boys, for girls, and for infants.

In 1871, when Alfred North was ten, Archbishop Tait appointed his father to the excellent benefice of St. Peter-in-Thanet. The parish was large and mostly rural. The church—a fine one begun in the eleventh century—was in the village of St. Peter, a mile inland from Broadstairs. The vicarage house, of brick, was large, dark, and drafty. (It has been razed and replaced by a much more modest one with modern conveniences.) Seventy to eighty persons could meet in its hall; forty could be seated in its dining room. Such things happened,

---

*A "pensioner" at Dublin, as at Cambridge, was one who paid for his room and commons. The M.A. degree required no graduate study, only the payment of a fee three years after earning the B.A. By becoming a pensioner Alfred Whitehead was also thereupon incorporated a member of Exeter College, Oxford, without undertaking to study there; this was simply a courtesy arrangement between the two institutions.

†The customary subjects were classics, ethics, astronomy, and natural philosophy.

for Alfred Whitehead was active and influential among the laity and clergy throughout East Kent.

This "thoroughly countrified vicar," as Whitehead called him, was a man of quite ordinary brain power; fortunately, he wrote no books. What he had was common sense, kindness, force of character, and a commanding presence. He was described—in one section of a delightful article, "England and the Narrow Seas," which Whitehead wrote for the *Atlantic Monthly* in 1927—as the last example of the clergymen who were the leaders of East Kent, on secular as well as religious occasions, until the twentieth century arrived. "He had all the habits of thought of a man who had always taken the lead, not because he thought about it, but because it was the natural thing to do." It is hard for us to realize that community leadership in small English towns and rural areas was not in those days something for which extraverts from several walks of life would compete. In big things and little, the clergy took the lead. In the Isle of Thanet it was chiefly Whitehead's father who persuaded wholly uneducated, obstinate villagers to send their children to school. When Bertrand Russell, as a child of five visiting Broadstairs in 1877, refused to believe that the earth, as he had been told, was round, it was the vicar of the parish, Whitehead's father, who was called in to persuade him.[15]

This vicar was much more concerned with the everyday lives of his parishioners than with saving their souls. In this, he was typical of the Anglican clergy of the region at that time. When Whitehead in 1927 looked back, he rather liked what he saw:

> These East Kent clergy of the old school . . . were simple and direct in their methods, and yet they got at the heart of the people in a way denied to the present generation. As they walked through their villages, or across country footpaths, they stopped and chatted with every man, woman or child whom they met. They knew all about them—whether their patch of vegetable garden was good or bad, whether they were sober or whether they drank, what their fathers were like, and how their sons had turned out. They had homely advice and kindly sympathy to give. Above all, they saw to it that every child in the village went to school and had an education according to the lights of those days. . . . It was a humanizing, kindly influence, which trusted mainly to the mercy of God to save the souls of men.[16]

This was religion-in-the-world, as the future philosopher first observed it. It had little to do with the essence of religion, toward which his thoughts would turn much later.

When Whitehead's older brother Bishop Henry Whitehead read "England and the Narrow Seas," he wrote to him, confirming the description:

> What you said about Father was so very true. There was more humanity and fellow-feeling with the parishioners in him and in the parsons of his type than in the modern brand. On the whole the clergy do more and better work than they did 60 or 70 years ago; but something of value has been lost.[17]

Alfred North Whitehead called his father "homogeneous" with his people. In the 1870s and 1880s they, like him, were natives of the Isle of Thanet, not Londoners; the relations between people of different social and economic classes were friendly (their differences would not be exploited until later); and there was not much parochial apparatus between the parson and his flock. Such a happy simplicity could not last. While it did, the Reverend Alfred Whitehead was the man for the time and place.

He was quite a preacher, too, and was sometimes invited to preach in London. "My father," Whitehead wrote, "was a natural orator, equally at home in the pulpit or at a mass meeting either of townspeople or of countrymen. His church was always crammed with villagers, and with townspeople who had walked some miles, and with Londoners spending their holidays in the district."[18] He had a powerful voice, and a sort of sturdy eloquence. There was nothing unusual about what he had to say. Obituaries call him a faithful and doughty standard-bearer of the cross that the Apostle of the English had planted in Thanet thirteen centuries earlier. Alfred North Whitehead called him "an equal mixture of a High Churchman and a Broad-Churchman." Cambridge undergraduates—the honor students, at least—of the 1880s and 1890s would have found his sermons and his conversation boring. Of course he was a very plain man, without subtlety, but equally without pretense or deviousness. The dubious false magic with which popular preachers in America now call their radio hearers and television watchers to Christ was foreign to him. Nor did he emotionalize Christianity. To understand the man this passage from "England and the Narrow Seas" is indispensable:

> He was entirely devoid of any artificial tone of "uplift." In fact he hated it, and expressed his opinion of "cant" with direct Saxon vigor. But in his generation a tenderness of tone had crept in, and he was an example of it. When the Baptist minister of the village was dying, my father was the only minister whom he would see.

> Despite all the differences between their churches, they were both East Kent men; and when they read the Bible together they understood each other without many words.[19]

In 1892 Alfred Whitehead was elected a Proctor in Convocation for the diocese of Canterbury; he was later appointed an Honorary Canon of the Cathedral. His influence among the clergy was based on his popularity with the general mass of the population of Thanet. And on his character. An obituary letter to the *Times* reports as follows on his activity in one of his Anglican offices, that of Rural Dean of Westbere (consisting largely of the Isle of Thanet):

> He exercised for many years a commanding influence over laity as well as clergy, often solving by sheer force of character obstinate problems which had defied the more measured procedure of ecclesiastical authority. It was a surprise to no one that he was elected chairman of the parish council, or that his supremacy over it was exercised without friction and without dispute.[20]

Such exercise of authority was not a trait of either his brother Henry or his son Alfred North; their way was that of tactful, often shrewd reconciliation.

We may be tempted to assume that the Reverend Alfred Whitehead was in a similar degree an authority figure to his children. The assumption would not be warranted. When he spoke, there was no doubt about his meaning. But he was an affectionate, sensible father; he doted most on his baby, Alfred North. So the philosopher's children said. No family letters, alas, have been kept.

There are only stories, most of them told by Alfred North Whitehead's wife. One, passed on to me by Jessie Whitehead, is that when her father, a freshman settling in at Cambridge in 1880, wrote his father that he was pleased to have met some Darwins, the response was, "Don't ever let me hear that name again!" In the stories Jessie's mother told, nobody made qualified remarks. All we can say about this episode is that Alfred Whitehead would have been less likely to express strong pro-Darwinian feeling. He was an eighteenth-century man whose interests were local, not intellectual.

When the son went off to Cambridge, the obviousness of his father's limitations made it easy for his own intellectual interests to develop. Although his character was quite different, he was on the whole fortunate to have had as his father this plain old-fashioned parson who loved him.

vi

What shall we say about Alfred North Whitehead's mother, née Buckmaster? In his writings there is no counterpart to his memories of his father; she is not even mentioned. Do not guess that she died when he was a babe; she lived into her nineties. There are some men—at least there were in Victorian and Edwardian times—in whose lives there is only one woman of importance. Alfred North Whitehead was such a man; the woman was his wife. She had a low opinion of her mother-in-law and of her family. Whitehead would defend his mother in private, saying, "She always did her duty as she saw it," but he would not bring her into what he wrote for publication. His son and daughter, unfortunately, could tell me very little about their grandmother Whitehead, and that was mostly negative. They did not see a great deal of her; for most of her long widowhood she lived with her daughter, sometimes with a son, but not with the married son who is our subject. His children had only a general impression of an unimaginative, rather small-minded woman with some wit but no sense of humor. The bit of available testimony outside the family is to the same effect. The family photographs suggest a shrewd, fairly bright woman who was used to getting her way.

We must remember that the English middle-class world in Whitehead's youth was a man's world. In his family background the important persons were the Whitehead men I have described, his older brothers, and certain servants. He also saw a good bit of one maternal uncle. There is no family tradition of his mother's having been an influence in any positive way; if she was, the fact has been submerged, and in the absence of family letters is not recoverable.

The Buckmasters of London, however, have a part in the story of Whitehead's life. They were not connected with the unpropertied Buckmaster family from which Whitehead's contemporary, the jurist who was Lord Chancellor in 1915–16 and later Viscount Buckmaster, came. The family into which Alfred Whitehead married was rich. The bride's father, William Buckmaster, had a very prosperous business as a military tailor.

When Alfred North Whitehead was a small boy, William Buckmaster's widow, Sarah, lived at 81 Piccadilly, overlooking Green Park.[21] He spent two months of several years there, and from the windows often saw Queen Victoria go by. As an old man, he gave Lucien Price one other vivid memory from his grandmother's house;

it was not of her, but of her housemaid.* Price reports just two remarks of Whitehead's about the Buckmasters. One: "My grandmother was a wealthy woman. She made the mistake of having thirteen children, and by the time even a large fortune is divided among thirteen it is no longer large, and when it came to the grandchildren, not much came to me."[22] The other remark was, "My mother's family, while socially somewhat better placed,[23] hadn't the intellectual distinction of my father's, and they were very contentious people." There was not enough devotion in their affluent lives to let them be models for him, as his father's people were. We can see him more clearly if we have in mind a little more about some Buckmasters. He was always friendly with them. But what a contrast they pose!

Whitehead's mother was the oldest of the thirteen children. I have identified seven brothers. For them, a public school, then Cambridge—except that John North Buckmaster took an Oxford degree, and Uncle Harry Buckmaster took no degree. After being educated, none of these uncles did anything memorable. To Alfred North Whitehead they represented the middle-class Englishmen who led comfortable lives on inherited money, outside the academic world that was his world, and outside of public life.† Horses and sports, however, were absorbing to most of the Buckmaster men.

The uncle Alfred North often saw was John North Buckmaster, who was four or five years younger than Whitehead's mother. He was the young curate at St. Mary's Chapel of Ease in Ramsgate when Alfred Whitehead was the minister. His younger son, Herbert, twenty years younger than Alfred North Whitehead, is the most interesting of the Buckmasters. In his fifties he wrote an autobiography entitled *Buck's Book: Ventures, Adventures, and Misadventures.*[24] In it he calls his father "more of a country squire than a clergyman." "I still have a photograph of my father, my uncle Harry Buckmaster (who was one of the finest judges of a horse in England), and Linley Sambourne, the famous *Punch* artist, hunting with the Thanet Harriers."[25] He does not name any other of his father's hunting companions, but Whitehead's father was surely one of them.

Herbert Buckmaster asserts that Alfred North Whitehead's middle

---

*See Chapter III, section i.

†One uncle was admitted to the Inner Temple and called to the Bar, but never practiced law. Another spent his life in the Church, most of it as Vicar of All Saints' in Epping, Essex; he is remembered for his generous donations to the church building fund.

name, North, was given him in honor of John North Buckmaster. That seems likely, though there is not now any way to confirm it. In any case, the name was considered a good one. Whitehead named his eldest son Thomas North. North, as he was called, told me that the name was first used by the Buckmasters in the belief—of which he was skeptical—that they were of the same family as Lord North.

John North Buckmaster acted as curate for only a few years, until he inherited enough money to retire. He stayed in Ramsgate, and when he died, about forty years later in his villa overlooking the sea, left an estate of £80,000.

His son Herbert represents him as a stuffy person who preached a negative kind of Victorian morality to his sons: do your duty, whatever it may be, avoid evil companions, have the courage to say no, become a total abstainer as I now am. Also, "Never write a letter you would be ashamed to have published in the *Times*."[26] "Buck" wrote: "When I got into trouble, which was often, I had only one thought, and that was to keep it from father."[27]

Herbert Buckmaster naturally grew up to be a man about town in London. (He is one of two among Alfred North Whitehead's maternal cousins who became newsworthy. The other was Walter Buckmaster, Cambridge man and a stockbroker, who for many years was the best polo player in England; his "Old Cantabs" honored the University in a way that no Whitehead did.) Herbert's father wanted him to go to Cambridge, but at public school (Whitehead's school, Sherborne) he never reached the Sixth Form. (He liked to say that "Cousin Alty," whom he admired, had taken his share of brains.) After fighting with the Imperial Light Horse in the Boer War, he tried to be a gentleman-farmer, but gave it up to become a professional backer of horses. Gladys Cooper ran away from home to marry him; his second wife, Nellie Taylor, also was an actress. In the First World War he was a Captain in His Majesty's Household Guards. It had been vainly hoped that he would find a career in the Army, but it was as the founder of Buck's Club that he became known. Its nucleus consisted of his fellow Household Guard officers who survived Ypres in 1918. He reports the answer of a man who was asked why he did not join: "Oh, a very tiresome club. You have to practically ride up the staircase on horseback in jodhpurs, and the only conversation you ever hear at the centre table is about horses."[28] The founder was also given to talk about physical culture and how England could be restored to its former luster by a patriotic, disinterested National Party founded on Boys' Clubs.

We have no further interest in *Buck's Book* except for one episode at about the turn of the century. At five o'clock one morning, at the bar of Prince's, in Piccadilly, Herbert knocked down a stranger who had called him a ruddy Jew. "I never knew I looked like a Jew before; at any rate it was the wrong hour of the morning to call me one."[29] There may be some truth in the opinion, conveyed by Whitehead's wife to their daughter, that the wealthy military tailor came from Jewish stock.

CHAPTER III

Childhood

i/Early childhood.

ii/The Isle of Thanet. Ramsgate. Canterbury.

iii/A link with the past: Archbishop Tait.

iv/Love of England. Early instruction at home.

When Alfred North Whitehead was born on February 15, 1861, his father was thirty-three, his mother twenty-eight. He was the fourth—and the last—of their children. The first, Charles Selby, was eight years old; Henry was seven; his sister Shirley, two. He was the baby of the family and was so treated.

In size he never caught up with his father, sister, or brothers. To his parents Alfred looked like a child in delicate health. Although measles was his only disease of any importance, he was treated as a boy who was not robust; he might not live to grow up if he were allowed to do the things that healthy boys do. As a small child he once cut his hand in cutting up the meat on his plate. On Mother's insistence, his meat thereafter was cut up for him—until he was nine. His brothers had been taught to ride and to drive the family horses; Alfred wasn't. Nor was he sent as a day boy to Chatham House; he was not sent to any school. Its demands, and the rough and tumble with schoolmates, might be too much for him. Until he went away to Sherborne in 1876 he learned his lessons at home, from his father.

As Alfred was a quick learner, he had a good deal of free time. Much of it was spent outdoors. He always remembered the family's old gardener as a fine man and a teacher of the names of plants and how things grew. "Servants have always been my friends," he told Lucien Price near the end of his life; this is confirmed by persons who served him in America—at home or at Harvard. Another servant he remembered gratefully from his childhood was Grandmother Buckmaster's housemaid in London, Jane Wychelow, who read *David Copperfield* and the *Pickwick Papers* to him by the hour before the grate fire. "She made them tremendously alive," he said to Price. That first gave him his taste for literature; "and it was *Pickwick* which gave me my first insight into the English social system." In Ramsgate his parents, being fairly well off, had quite a few servants, by our standards. There was a nursemaid for him, but he was not wholly in charge of a nanny; that would have been unusual in a provincial family without social ambitions. His mother could not lavish attention on him, since being a minister's wife is almost a full-time occupation.

He had more time than a schoolboy would have had for solitary wondering about everything he saw. (His eyes were good, and his legs too.) The habits of wonder and reflection stuck, and gradually became necessities. As a philosopher of seventy, he would say that he was no good unless he had a couple of hours each day in which he could get off by himself and think.

By and large, his childhood was not one of suffering. Mother was the sterner, and not the pleasanter, parent. Probably she found nothing amiss in her brother John North Buckmaster's attitudes. But it is said that little Alfred was never beaten. He was the apple of his father's eye, and father could take bad behavior in stride. Alfred North Whitehead liked to tell his American friends of an incident from his fourth year. This is how Lucien Price heard it: In a Swiss railroad restaurant, very thirsty, he drank one glass of water after another,

> till an old gentleman sitting opposite said, "Little boy, you shouldn't drink so much water."—Whereupon I picked up a spoon, threw it, and hit him in the mouth! My father, quite sensibly, did not punish me. Partly he was amused, and partly, I suspect, he thought it served the old gentleman right.[1]

The boy Whitehead, partly because he was the baby of the family, partly because he was bright, partly because he looked frail, received a good deal of petting. Its effect was nowise unfortunate, as it so often is in other situations, where the boy is mother's darling and most of the people in the house are women. Alfred was especially lucky in his two older brothers. As I have said, they adored him. Charlie and Harry were active, kindly, intelligent, and knowledgeable. Their affection for him lasted. When he won a scholarship to Trinity College, Cambridge, they celebrated the achievement by taking him on a trip to Switzerland. When he married in 1890, Harry performed the ceremony.

ii

Whitehead had roots in the land. I do not mean that he belonged to the landed gentry—he didn't. I mean that in the formation of his mind the land in which, the shores on which, he grew up played a firm and lasting part. Of course this is not unusual—to put it mildly—in the life of an Englishman. It *is* something quite absent from my makeup, and from that of almost all Americans who may read Whitehead or read about him. We all carry in us something from the physical character of our childhood surroundings, but for us those surroundings are not steeped in history. (Forgive the hackneyed phrase; it is exactly right.) When I was born in Cleveland, Ohio, the city had existed for little more than a hundred years. In the eyes of my family and those of most of my schoolmates, it had no history. The country near it, and Lake Erie, meant nothing to us; Cleveland was simply an industrial center to which our people had come from Europe in the second half of the nineteenth century. But Whitehead grew up surrounded by landmarks of ancient and medieval history, and beside waters whose dangerous character was part of the history of Western civilization.

Bertrand Russell told his BBC audience that the Isle of Thanet was a much more intimate part of Whitehead's makeup than Cambridge was. Whether or not it is intrinsically interesting to us, we must learn the character of this "island" before we go any farther with the man Whitehead.

Thanet is small, about one-thirtieth of Kent, the rural northeast corner of the county. It runs up to a dozen miles from east to west, and is four to six miles wide. On an Elizabethan map it is marked *Tenet Insula,* and the broad blue line along its southern and western edges shows that it is an island by rights. When St. Augustine the missionary landed in Thanet, in 597, he had to cross almost half a mile of water to reach the British mainland. For centuries, ships for London used that channel, the Wantsum, but it had not been navigable since about 1500. The tides and currents inexorably closed it, so that the safe passage became mere marshland, and Ramsgate changed from a hamlet to a port of entry.*

Ramsgate is on the curve of the Thanet shore from Broadstairs (which faces due east) to a large, shallow bay famous in history, Pegwell Bay. The town faces south and southeast; on an exceptionally clear day Whitehead could see the coast of France by looking straight out from the West Cliff. If we visit Ramsgate today, we are likely to come in the summer; we shall enjoy the water and the firm sandy beaches, and we shall see, besides the foreign tourists, what George Eliot called a strip of London come out for an airing. When Whitehead was a child, the steam packets and trains had been bringing Londoners

*Its own fishing fleet in 1870 numbered about 300 vessels.

on summer weekends for decades. But the seaside resort was the least important part of Whitehead's Ramsgate.

If we are invited onto a sailboat that is to go outside the harbor, our sailor should be first-rate. In "England and the Narrow Seas," Whitehead told his American readers how those seas, north of Dover Straits, "put up almost every form of difficulty known to sailors—tides, fogs, winds, dangerous headlands, sunken shoals."[2] He never became a sailor himself; I doubt that he ever had anything more than a rowboat, and do not believe that his mother let him have a rowboat for many years. What is important is what he saw and heard. From the nursery windows on winter evenings he saw the flashes from the lightships on the Goodwin Sands. (We can find these shoals southeast of Ramsgate on any good map.)

> Sometimes during a fog the boom of a gun would be heard at slow intervals across the sea. It was a ship ashore on the Goodwin Sands. At other times we saw rockets rise mysteriously from the dark waters. It was the Gull lightship signaling a wreck. Next day we were taken down to the harbor, and there was the lifeboat decked with flags: during the night it had been out and had saved the crew of some vessel slowly sinking in the merciless quicksands.[3]

There were also quicksands in the marshes south and west of Ramsgate; in his childhood Whitehead once watched a horse sucked down into them, "the rider barely escaping."

The harbor to which he was taken had been built in the mid eighteenth century after Ramsgate was selected as a harbor of refuge from the Downs (the fine anchorage off Deal, vital until steam replaced sails). Most of the vessels coming into Ramsgate were blown, or ran, from the Downs in bad weather. In Whitehead's boyhood, reliance on sails had by no means ended: in a squadron of the British fleet that he saw anchored in the Downs during the Franco-Prussian War, all but one of the ships had sails as well as engines.

In a conversation with me, Bertrand Russell remarked that he always sensed that his friend felt most at home in Thanet. Yes: at home breathing that live North Sea air, remembering the excitement and the dangers of those waters and knowing that he was safe. His memories show that there was much more to this than the feeling of safety that everyone whose childhood was happy has in the place where it was spent.

Men who did not know and respect the force of this sea came to grief or, like King Philip II, saw their plans ruined. Whitehead liked to dwell on its part in preventing the Invincible Armada from landing. And he believed that the dangers of those waters had an effect, generation after generation, on the coastal population. They bred "obstinacy and a tendency to lonely thought." When I visited Ramsgate I asked the town librarian and his deputy whether their experience of the native population agreed with this characterization; both answered that it did, adding, "they're introspective." Whitehead gave some illustrations of obstinacy among the East Kent people he knew, like "Old Saxby," the fisherman who rowed the Whitehead children out to bathe from his boat. In the war of 1914–18, Whitehead wrote,

> a bomb fell on the house where I was born, another in the garden where I played, and a third blew up a powder magazine on the quay where Old Saxby used to embark us for bathing. I do not think anyone left the town by reason of these little incidents. People repaired their windowpanes and stuck it out with East Kent obstinacy.[4]

I should count Whitehead himself a splendid example of the tendency to solitary thought; he was also obstinate—on important matters, never on minor ones.

In 1927, Whitehead could also observe that all the free governments of the world were historically connected with the coasts of the Narrow Seas. "The Pilgrim Fathers were their offspring." Today we have got in the habit of attributing the push for freedom to mass movements that attack established institutions and favored classes. We ought not to forget the impulse to freedom that lies in quiet obstinacy and solitary thought, in a personal love of independence.

There were two striking ruins for the boy Whitehead to see. Richborough Castle was once the fort of Rutupiae, built by the Romans at the eastern end of the Wantsum channel. During their long occupation of Britain it was their chief supply base, and the usual debarkation point of legions coming from Gaul. The boy saw tall, thick walls of flint, stones, and red brick. More than half the length of the walls still stood, a reminder of Rome's third- and fourth-century defiance of ever-increasing raids. There was also a large platform with a cruciform mass of rubble on top, where once the Romans had a watchtower and in happier times a large, decorated triumphal gate celebrating the conquest of Britain. These acres had intrigued antiquarians since Elizabethan times. Portus Rutupinus had been the strongest of the nine forts between the Solent and the Wash.

The other end of the Wantsum, at the northwest corner of Thanet, was then guarded by the fort of Rugulbium. Only parts of its foundations were to be seen; here the tides and currents had not added land but taken it away. This place, Reculver, was settled long before the Romans came. After they left, in the seventh century, King Ethelbert's priest built a fine church, St. Mary's Abbey, in the middle of the castrum. Early in the nineteenth century a foolish vicar took it down. Not entirely. The west façade, added in the twelfth century, had twin towers, "the two Sisters"; because they were marks for sailors, that facade was bought from the parish. Those towers, their spires replaced by pyramids of ironwork, marked the desolate spot for young Whitehead's eye. How much you could learn in this corner of England about what nature does and the things that silly men do!

Besides the stark visible monuments, a curious boy could find any number of vestiges of the distant past. The most common were coins. Nature was known to take a hand: when a cliff at Broadstairs fell into the sea early in the eighteenth century, old coins appeared in quantity. Whitehead was seven when Roman potsherds were found in a pit near the church, soon to be his father's, of St. Peter's in Thanet. When the railway to Ramsgate was cut, a fifth-century Saxon burial ground was unearthed a mile west of the town. As land-draining, road-making, and building went on during Victoria's reign, more and more remains of variously civilized or barbarous times were accidentally discovered. What would not be revealed by deliberate systematic excavations! These expectations were copiously fulfilled during and soon after Whitehead's lifetime. What he saw or was told, however partial and inaccurate it may appear today, had a massive effect on him. And the sense of the undiscovered, waiting for the sustained effort or the lucky chance, was early on a part of his experience. So was a sense of the reality of legends in the minds of people.

The air of Thanet was full of spirits. According to an undoubted tradition (probably true), it was at Ebbsfleet, on the shore of Pegwell Bay, that Hengist and Horsa landed in 449, and St. Augustine with his monks in 597. Whitehead remarked to one of his Harvard classes that his grandfather was alive when the oak tree under which Augustine was said to have first preached to King Ethelbert was cut down. At Sandwich, just south of Pegwell Bay, Canute had landed his army. St. Clement's, there, was a glorious Norman church, but the town had been almost a ghost town since the seventeenth century. The sea had silted up the harbor of what in medieval days was the most active of the original Cinque Ports. Dutch and Flemish buildings told of the time when Queen Elizabeth encouraged weavers from the Low Countries to settle in what had become an inland town.

The land of Thanet was dotted with beautiful Norman churches. There were none in Ramsgate itself; it was not old enough. Safety from piratical raids had dictated an inland location for the old villages and their churches, which would hold all their people. A mile back of the opening in the shore cliffs at Ramsgate, in the mother parish of St. Laurence, stood the large church of that name, dating from about 1200. Back of Broadstairs, in the village of St. Peter's, was the church that Whitehead, naturally, knew best, with its big Early Perpendicular tower of blackest flint. (That tower was fitted up as a naval signaling station in 1813.) The church he admired most was the mother-church of the island, at Minster; here a famous monastery was founded in the seventh century. In the eleventh century the Canterbury monks built the splendid cruciform church—a monumental rebuilding that used some of the pre-Norman parts. When Whitehead wrote about Thanet (in "England and the Narrow Seas") he praised these churches and could have praised more of them. What concerns us now is his early education in art. The emphasis on the aesthetic in his philosophy is mainly a result of his marriage. He grew up in a family devoid of taste, and as a boy learned nothing of value about music or painting. But the old churches were there to be seen; they were the things of beauty in his youth. The historic aspect was an inseparable part, and possibly the larger part, of the appeal they made.

But there were also, in Ramsgate itself, two fine nineteenth-century churches. St. George's was only a step from Chatham House, and Whitehead's father was curate there when he was born; it was easily the most splendid church in Kent of the unsplendid Regency period. And on the West Cliff stood A. W. N. Pugin's masterpiece, St. Augustine's Roman Catholic Church. Architecture was the only art form, aside from poetry, for which Whitehead ever had immediate appreciation.

When we describe the land in which Whitehead grew up, we must bring in one place outside Thanet. Canterbury was sixteen miles from Ramsgate, but part of the terrain; the marshes that bounded Thanet also ran westward to its edge, and you could see the cathedral towers on any clear day. Outside the east wall of the city stood St. Martin's Church, where Augustine and his followers first worshiped. And the Venerable Bede had written that it was built before the Romans left. In any case, some authorities considered it the longest-used church in England, the mother-church of the kingdom.

In the garden of Canterbury Cathedral Whitehead could see two Saxon columns that had been saved from the destroyed church of St. Mary's at Reculver.* When he visited the Cathedral, he saw English history. There was St. Augustine's marble chair, in which the archbishops had been enthroned since the thirteenth century; the tomb and armor of the Black Prince; and much more. Becket was one of his heroes. When Whitehead was eighty he said in his "Autobiographical Notes," "As I now write I can visualize the very spot where Becket fell A.D. 1170, and can recall my reconstruction of the incident in my youthful imagination." In his mind's eye he could go with his father into a medieval building, the Archbishop's palace, which had been plundered in the peasants' rising of 1381.

iii

Among Whitehead's early acquaintances there was a magnificent link with the past: Archbishop A. C. Tait. Tait's summer residence, Stonehouse, was in St. Peter's parish, and he was a close friend of Whitehead's father. In an article entitled "Memories," which Whitehead wrote at seventy-five, there is just one memory of a scene in the old vicarage house. It is of the primate of all England, "tall, commanding, stately," standing with his back to the fire and laughing heartily as Whitehead's father tells of "the great comfort which his leading parishioner, a silly old gentleman, took in the doctrine of eternal damnation."

Whitehead called Tait "one of the few great men whom I have met":

> Archbishop Tait ought to have been a prime minister. Fate made him Archbishop of Canterbury. I have always been grateful for my glimpse of him during half-a-dozen years, and for the family tradition of him during a longer period. To have seen Tait was worth shelves of volumes of medieval history.[5]

This gratitude was partly that of a man seeking a picture of the cultural life of England from Augustine to the beginning of the twentieth century. Historians are likely to say, Ah yes, Tait was the last of the medieval archbishops. For Whitehead, there was more to it than that. For over a century before Tait's appointment in 1868, the primacy had been given to safe, retiring men. Tait was not one; he was forward-

---

*These Saxon columns are now in the Cathedral crypt.

looking, and more influential in Parliament than any bishop for three hundred years. James Bryce wrote that his preeminence was not in oratory but in statesmanship and informed common sense. For anyone who, like Whitehead, develops a point of view on the history of a civilized nation, it is a great advantage to have known in his youth a leader with the quality of statesmanship. Tait was a leader in the social and civic life of England; in his evangelical and compassionate work as Bishop of London, in his hospitable archiepiscopal policy toward Dissenters, and in his friendliness to the march of science, he tried to make the Anglican Church of the 1870s once more the prime source of spiritual leadership for the nation.

The Archbishop was energetic and practical, tolerant and generous. But the forces against him were too many and too strong; Whitehead came to see him as a magnificent failure. His constitutional authority in the Church was small; most of its lay and clerical leaders were narrowly dogmatic, and too interested in fighting each other over rituals. He could laugh with a friend at a silly parishioner's pleasure in the thought of eternal damnation; but when he said in Convocation that nobody in the Church accepted the damnatory clauses of the Athanasian Creed in the plain, literal sense, 315 Anglican clergymen declared in writing that they so believed them.

Against the rising tide of unbelief, Tait could do little. When in his last year he called a meeting of scientists to discuss the relations between science and religion, few came, and those who did could not agree on what, if anything, they should say to help the cause of religious faith. Of course Darwin had been invited, and of course he declined to attend. This was in 1881, when Whitehead was a freshman at Trinity College, Cambridge.

iv

Back to Whitehead's boyhood. Tait's appearance was "commanding," he looked every inch the primate, but his manner was the very opposite of dictatorial. Inevitably, the boy saw that the national Church was right and good. His father and brothers were good, his grandfather an outstanding man. The ways of Thanet were right, it was good to grow up a "man of Kent." ("Kentishmen," Whitehead wrote later, "are an inferior brand who live west of the River Medway.") Because of his absorption of historical tradition from his physical surroundings, he knew that Kent was the first county of England. It was good to be an Englishman. Whitehead never became chauvinistic, and he honestly decried nationalism; but he loved his country, would never think of changing his citizenship during his twenty-three-year stay in the United States, and could never be taken for anything but an Englishman. I do not know of any violent commotions in his boyhood that could have evoked a spirit of rebellion against his family or against the fairly prosperous professional middle class to which it belonged.

It happens that we know the first poem Alfred North Whitehead learned as a child, for he taught it to his children, and as an old man recited it at a meeting of the Harvard Society of Fellows.[6] It was from the immensely popular collection of poems, or songs, composed by the Scottish journalist Charles Mackay and published in the mid fifties. (Others were "Cheer, Boys! Cheer!" and "Tubal Cain.") The boy's reciting of it might remind us that Ramsgate was on the sea, that Thanet was rural, and that there was class structure and a work ethic.

> To the West! to the West! to the land of the free,
> Where the mighty Missouri rolls down to the sea,
> Where a man is a man if he's willing to be,
> Where a man is a man if he's willing to toil,
> And all can enjoy the fruits of the soil.

His father rewarded the recitation with a penny.

The schooling Whitehead's father gave him was normal at that time when Greek was a condition of matriculation at Oxford and Cambridge. The study of Latin began when he was ten, Greek—from a Latin grammar—at twelve. And then:

> Holidays excepted, my recollection is that daily up to the age of nineteen and a half years, some pages of Latin and Greek authors were construed, and their grammar examined. Before going to school pages of rules of Latin grammar could be repeated, all in Latin, and exemplified by quotations.[7]

Even so, Whitehead was no more than competent in Latin and Greek at public school. (He never studied English grammar, but absorbed it from his reading.) The mathematics that his father taught him was normal too: from the multiplication table, through simple fractions and decimal notation, to elementary plane geometry. Somehow, he soon thereafter developed an interest in mathematical ideas, so that he shone in "Maths" at Sherborne.

CHAPTER IV

Sherborne

i/H. D. Harper's Sherborne School. The town and the country. King Arthur and King Alfred.

ii/The English public school. The boys' life at Sherborne.

iii/Whitehead's studies at Sherborne.

iv/Sports. Whitehead, Captain of the Games.

v/The Debating Society at Sherborne.

vi/Head Boy.

vii/Sherborne's headmasters and mathematics teachers. Commencement.

In September 1875 Alfred left the Vicarage and his father's tutoring to spend five years at a public school that had recently attained a very good reputation—Sherborne. There are no letters or diaries to tell us how he felt, away from home. If so protected a boy had been sent off to school at eleven or twelve, as many were, the first year would have been rough. But Alfred was almost fifteen. And life would be easier for him at Sherborne than it would have been at other public schools. In 1870 both his brothers had gone there to finish their school education. Charles, the older by a year, had left it in 1872 with a scholarship to Magdalene College, Cambridge; Henry, in 1873, with a Sherborne School scholarship to Trinity College, Oxford. Henry had won school prizes in Latin verse, history, and modern languages, and had been captain of both the Eleven (cricket) and the Fifteen (football). Charles's school days were less distinguished, but he was to return to Sherborne as a housemaster in 1876, staying until 1892. One of young Alfred's brothers, then, was teaching at Sherborne during all but his first year there. And both brothers had been in the largest and most prestigious of the houses, the School House (over which the headmaster presided); to it Alfred in turn was assigned.

Sherborne School was in the town of that name, near the Somerset border of West Dorset. There were several good schools in Kent, such as Tonbridge and the King's School at Canterbury, to which the Reverend Alfred Whitehead could have sent his sons. The boys would have been a little out of place at Eton or Harrow, perhaps at any of the "illustrious seven" that were commonly considered the great public schools of England (Eton, Harrow, Rugby, Winchester, Westminster, Charterhouse, and Shrewsbury). But in the 1850s and 1860s the number of good public schools had much increased. Several great headmasters had changed small grammar schools, which taught Latin, Greek, and divinity to an almost entirely local enrollment, into boarding schools with modern curricula, buildings, and playing fields, where hundreds of boys lived that peculiar kind of disciplined life, first set up by Thomas Arnold at Rugby in the 1830s, which had come to be the essential meaning of "public school." The newcomers undertook to provide as good an education as Eton or Harrow, "but without the adventitious social price."[1] In this they largely succeeded. Sherborne under H. D. Harper must have seemed as attractive as any. Its fees were among the lowest. It also differed from most of the good public schools in imposing no restrictions other than age and merit. Probably the Whitehead boys were admitted on testimonials from their father.

When the Reverend Mr. Harper came to Sherborne in 1850 it was a very ancient, impoverished little grammar school without living traditions, in an out-of-the-way place. In twenty years he converted it into a public school with a tradition, greatly increased its size (the arrival of the railway in 1860 helped him in that), and pretty well realized his aim "to give my boys *every* advantage (except social) which they can get anywhere."[2] In 1870 Harper was not yet fifty years old. I have no evidence that Whitehead's father knew him personally; if he did, he could not but have liked the energetic, forthright, warm-hearted extrovert with the loud cheery voice, like himself an unsophisticated Old Testament man, who had done so much for his students while carrying out the School's stated purpose of supplying "a liberal Education in accordance with the principles of the Church of England."[3] Certainly the former head of Chatham House Academy knew Harper's reputation. Only a minority of Sherborne's graduates went to Oxford or Cambridge, but their records there were generally good. In 1875 the fair success of Charles in mathematics at Cambridge, and Henry's outstanding success at Oxford (he won a First Class in Classics Moderations in 1874), spoke well for Sherborne. In all probability no other school was considered for Alfred. He was granted one of the three annual scholarships to Sherborne.

The Sherborne *Admission Register* shows that Alfred North Whitehead was admitted in September 1875 as one of thirty-two boys in Form IV A (there were two Fourth Forms). He was in good health, having had measles but not scarlet fever or "other diseases." But the most striking thing is the entry in the column for the boy's proposed pursuit in life. It consists of the single word, "Cambridge."

What were his new surroundings like? He described them in the most delightful of his short pieces, "The Education of an Englishman," written for the *Atlantic Monthly* in 1926.[4] In the present chapter I shall quote from it and enlarge on it, for Whitehead's move from Thanet to Dorset and Sherborne was crucial to his development—as much so as his later moves from Cambridge to London and from London to the American Cambridge.

Sherborne was, and is, a rather sleepy little market town (six thousand people) on a long hill, with not much in it except the School at the bottom, a very little light industry, the usual small shops, a Tudor castle to the east (part of it built by Sir Walter Raleigh in 1594), the ruins of its Norman predecessor—and the big Abbey Church. This, a very fine example of fifteenth-century Perpendicular Gothic, includes some recognizably Norman work and even a bit of pre-Norman wall. Visitors are most likely to remember its wonderfully rich fan vaulting and the superbly beautiful oak choir stalls. Whitehead, who admired the Abbey and loved to dwell on the presence of the past, also remembered its magnificent bells. They were the gift of Henry VIII, he says, brought from Tournai when he returned from the Field of the Cloth of Gold. That is doubtful, but the bells do date from the sixteenth or seventeenth centuries. Schoolboy Whitehead was then partly right in hearing them as "the living voices of past centuries." He later came to believe that they "were a great factor in the moulding of the school character."

The town is good to look at, for it is built not of ugly brick but, like the Abbey, of stone from Ham Hill quarry. The country roundabout is lovely. Whitehead recalled Dorset as "a rich agricultural district, with apple orchards, and woodland, and ferns, and rolling grass downs." (The North Dorset Downs lie on the farther, or south, side of the River Yeo.) He also referred to "intimate scenery such as, in all the world, only the West of England can provide." That is a lover's pardonable exaggeration. Whitehead came to Dorset from the East Kent coast at the very age when a boy first becomes keenly aware of the beauty of nature. As an old man remembering that new environment, he as usual emphasized its human significance:

> Geography is half of character. The soil there is rich, loamy, and gravelly. The climate is formed by warm currents and warm moist winds from the South Atlantic. My own home was in the Southeast of England, where we are formed by the polar currents and Siberian winds which come down the North Sea, with interludes of South Atlantic weather from the English Channel. But the interludes in the East were the habitual climate in the West. . . . The peasantry . . . were a kindly folk; if a schoolboy on a country walk asked for water, he was given cider and no payment taken.[5]

The lovely country is also the melancholy country of Hardy's Wessex Novels. (Sherborne is his *Sherton Abbas*.)

As in Thanet, Whitehead at Sherborne saw English history as a main ingredient of life. The notion that King Arthur held court nearby was embodied in the school song, but as Whitehead says, "Every respectable district in the West of England claims King Arthur." Historians consider the town to be of late-seventh-century Saxon origin. It is first mentioned in 705, when St. Aldhelm was made bishop of the new diocese of West Wessex. In Sherborne he built his cathedral church and founded a college of secular clergy. It is believed that along with this he established a grammar school for boys. In 1905 the town and Sherborne School together celebrated their twelve hundredth anniversary with a great pageant.*

King Alfred was a boy in Sherborne in the 860s, when the town was for a short time the seat of government; of course the school claims him as a pupil. Whitehead wrote, "His connection with the school was mythical, but undoubted."[6] At the end of the tenth century Benedictine monks were established at the Abbey, and for five and a half centuries usually maintained a high level of learning at Sherborne. Shortly after the dissolution of the monasteries the school was given a new foundation by one of the earliest of several similar acts of Edward VI. "The King's School" was intended, as a free grammar school, to ensure the hold of the Church of England on the minds of the young of that region. Thereafter the school, under local governors, had its ups and downs but did not become really important until Harper's time.

The town never regained its pre-Norman importance. In 1875 England was changing rapidly, but Sherborne was not. That was a great year for modern social legislation: Disraeli pioneered in legislation for workers, set up a charter of trade union liberties, and passed acts on public health. Some bad omens for the future were the great coal strike and lockout in Wales and the beginnings of the great agricultural depression that was to hurt England for a quarter of a century—not least the Oxford and Cambridge colleges, whose incomes went down with farm rents. In his mature life Whitehead would maintain a lively interest—as a responsible citizen, never as an agitator—in such matters, but in 1875 the people and boys at Sherborne knew little of them. Political interest was concentrated overseas. In Whitehead's first term Disraeli was negotiating the purchase of the Suez Canal.

*Some misinformation must be responsible for Whitehead's statement in his "Autobiographical Notes" (1941) that the School was to celebrate this anniversary in that year (*LLP-W*, p. 5).

ii

The English public school has been defined as "an independent, non-local, predominantly boarding school for the upper and upper-middle classes."[7] In Whitehead's time the public schools had recently become *the* system for their education. In 1873 Leslie Stephen wrote: "Neither the British jury, nor the House of Lords, nor the Church of England, nay, scarcely the monarchy itself, seems to be so deeply enshrined in the bosoms of our countrymen as our public School."

Whitehead's schoolmates in 1875–80 were, he remembers, sons of families of the kind Anthony Trollope wrote about:

> The fathers of the boys were archdeacons, canons, rectors in the Established Church, or officers in the Army, or small squires in the Southwest of England, or lawyers, or doctors. There was a sprinkling of boys from large commercial families.

Whitehead was very much a public school man. His happened to be Sherborne, but the description that follows was true of all the schools. Each had several boardinghouses; at Sherborne there were (in addition to the School House) seven of these, averaging thirty boys each and run by housemasters on their own money. Outside the classroom there was not much contact between the boys and their teachers. Instead there was the system of prefects and fags. A fag's duties were usually not heavy, but I do not know how it was in Whitehead's case, nor how often—if ever—he needed the fagmaster's protection, to which he was entitled. When young Whitehead came to Sherborne, he was made some senior boy's fag, to black his boots, tidy his study, make toast and tea, and run errands. Two years later, when he entered the top, or Sixth, Form, the situation was reversed.

In each of the school houses one of the senior boys was appointed prefect, or "head," and the headmaster appointed each year a Head Prefect, or Head Boy, for the entire school. The masters ruled the school through the prefects, who ran the house activities, helped make rules, and were responsible for seeing that the rules were obeyed. They judged most offenses and, themselves exempt from punishment, did much of the punishing. This was the system that Arnold had devised at Rugby. It was intended—in conjunction with the fagging system—to diminish the prevalent brutality, end the natural enmity between boys and masters, instill a sense of responsibility that would prepare the top boys for important places in the hierarchy of British life, and help turn every boy into a Christian gentleman. These goals were in the main achieved. It is easy to call the system undemocratic, but in fact many prefects developed an ear for consensus among boys. The light touch in giving orders, habits of courtesy and moderation, and appeals to fair play, "good form," loyalty, and the good of the group became the most effective characteristics of prefectorial regimes. Putting the prefects on their honor as disciplinarians worked so well that Arnold's system was soon copied in all the public schools and approved by public opinion. Along with the etiquette, dress, and speech distinctive of each school, the school's customs in fagging and its form of prefectorial discipline became in a very short time the venerable and inviolable constitution of life at that school. Although the fagging system could be abused in some schools, it was not in Whitehead's time at Sherborne. As we shall see, he participated most fully in this system of duties and privileges. It was his closest approach to military life. (There was no Cadet Corps at Sherborne until much later.)

All this could happen only in England. Those Nazis who studied the English public schools to learn how to produce leaders were, for their purposes, misguided.

Whitehead recalls that at Sherborne, unfortunately, the prefects "were chosen merely according to their standing in the intellectual life of the school."

> If the prefects were also athletic and of high character, the system worked very well; otherwise it worked very badly. In my own schooldays for about half the time it worked badly* and for the other half extremely well. There was some teasing, but no gross bullying.

The roughest life was the communal life of the Junior Boys' Common Room, where bullying was least reportable. Arthur Waugh,† who was in it a few years later, pictured it as hell in his autobiography, *One Man's Road*. If Whitehead had any of this life, it would have been when he first came to Sherborne. It would have been especially hard to take because he had never before been a student in any kind of school—and he was not a big boy. The School historian, A. B.

---

*Notice the absence of the emphasizing adverb here, and its inclusion in the next clause. Five of the six head prefects between 1875 and 1881 were athletic (Register of Sherborne School). I conclude that on the whole the system at Sherborne in Whitehead's time worked fairly well.

†Arthur was a very different person from his son Alec, who offended Sherborne by his sensational first novel, *The Loom of Youth* (London, 1917).

Gourlay, ventured to me the opinion that as a very clever small boy Whitehead was probably assigned very soon to a private study. The fact that he was a scholarship holder, that he was the younger, less robust brother of Charles and Henry Whitehead, may at once have secured a study for him. (About seventy School House boys were in studies, twenty in the Common Room.) Of course he was rather lost in his first months at school but he can not have been there very long before he and the other boys discovered that he could play football; that made things much easier.

The private studies, each for one or two School House boys, had been constructed by Harper, by rebuilding the fifteenth-century residence—once sumptuous but long since gutted—of the monastery's Abbot. Although this was done in the 1850s, the notion arose that the boys' studies had actually been monks' cells. Whitehead writes that the study he occupied during his last two years at the School (when he was one of the leaders of the Sixth Form) was commonly believed to have been the Abbot's cell. In it Whitehead could indulge in private reading; the boys were not overworked. It was at this time that "poetry, more especially Wordsworth and Shelley, became a major interest, and also history"; so Whitehead wrote in "Autobiographical Notes."

What were his days like? "We rose—nominally—at 6:30 A.M. and were in chapel at 7 A.M., if our state of dress, or undress, enabled us to pass the prefect at the chapel door. If not, we had to write out some lines in Greek." Morning Prayers were short and always the same: Te Deum, Collects, Lord's Prayer, Litany, School Prayer. The last, first set out in statutes of 1679, was read by a different boy each day. It reminded them that they were being trained "to serve God in Church and State." Whitehead, remembering this, adds, "and we never conceived life in any other terms."

The hours spent in classrooms were often easy; that depended on how lazy and how permissive the master was. (A few years later, John Cowper Powys was unexcited by Charles Whitehead's teaching of Euclid but happy in his class because he left the boys alone as much as he could without allowing anarchy. None of the Whiteheads had any of the drill sergeant in them.) Afternoons were mostly for sport. An important part of the public school system was the channeling of aggression and competitive spirit into games, run entirely by the boys. Participation was then almost entirely voluntary, though it had become the thing to do. There was also time for country walks, which Whitehead enjoyed very much. Evening study was displaced once a week, for senior boys, by meetings of the Debating Society. Evensong consisted simply of Psalm VIII, the Creed, and Collects.

Harper saw to it that his boys were properly fed. But comfort was not a feature of public school life. The School historian writes:

> Rooms, classrooms especially, furnished with primitive heating pipes or none at all, could be glacial. Illumination was provided by . . . dim fishtail burners . . . the supply was cut off at bedtime and if light was required later, it had to be from candles. . . . Hot water from a tap was unheard of.[8]

Whitehead recalls that in the School House

> there were ninety boys and four baths . . . I am recording and not defending. Of course there were washbasins in our bedrooms, the water being put there in jugs. Labor was cheap in those days, and plumbing was barely in its infancy.

The then general belief of public school authorities was that a shared Spartan existence made men of boys. I dare say it saved the many whose work took them to the colonies from being prone to grumble. Also, Harper's physical expansion of the School had been costly; it would scarcely have been possible without the generosity of the squire of the district, Earl Digby. The school's governors, whose chief work concerned its scattered income-producing lands, would not spend a penny on comforts for boys and were sometimes careless about necessities. Whitehead missed an epidemic of typhoid by two years.

At the close of "The Education of an Englishman," Whitehead wrote, "Altogether we were a happy set of boys, receiving a deplorably narrow education to fit us for the modern world." Before turning to their education, let me note that their fun was not limited to the playing fields. Under Harper something like a big happy family feeling had grown up at Sherborne School. His daughter described the feasting and singing the night before the Christmas holiday began, and the Sherborne Ball, which began with a quadrille and ended with a gallop down the long hall, to "Post Horn" or "John Bull."[9] It was probably on these occasions that Whitehead learned to try to sing and dance, neither of which came naturally to him.

iii

A Practical Handbook to the Principal Schools of England, edited by C. E. Pascoe and published in 1877, shows the relative prominence of the various subjects of secular instruction at Sherborne by the following numbers: English, Latin, and Greek languages and literatures, 7; mathematics, 2; geography and modern history, 1; some branch of physical science, 1; a foreign European language (French or German), 1; drawing and vocal music, 1. This was the basic curriculum for all boys; Sherborne did not offer, as some public schools had decided to offer, two curricula, one classical and one modern; nor was there a "military side" for boys who planned to enter the Army or Navy or engineering. What was new at Sherborne was a policy of modifying the basic curriculum in particular cases so as to give a Sixth Form boy more time for what he was best at. Whitehead was excused in the composition of Latin verse and the reading of some Latin poetry, to allow more time for mathematics. This, he recalls, was "very well taught." Indeed, he more than once said to his daughter that his Sixth Form maths master was the best teacher of mathematics he had ever seen, "bar none." In the summer of 1880 the visiting examiners noted in their report Whitehead's special excellence in every branch of "Maths." By the time he went to Cambridge, he told Lucien Price, "For a boy of nineteen I was a good mathematician."

The work at Sherborne was more heavily weighted in classics and in history than Pascoe's figures suggest. Whitehead described the work in four golden pages of "The Education of an Englishman." Latin and Greek, he wrote,

> were not foreign languages; they were just Latin and Greek; nothing of importance in the way of ideas could be presented in any other way. Thus we read the New Testament in Greek. At school—except in chapel, which did not count—I never heard of anyone reading it in English. It would suggest an uncultivated religious state of mind. We were very religious, but with that moderation natural to people who take their religion in Greek.
>
> The difficulty as to the Old Testament was surmounted by reading the Septuagint in class on Sunday afternoons, though the lower forms had to descend to the vulgarity of the King James Bible.

Because Greek and Latin literature included the study of Herodotus, Xenophon, Thucydides (a favorite of Whitehead's), Sallust, Livy, and Tacitus, much ancient history was taught "very thoroughly so far as it went," Whitehead recalls. He adds:

> But it was characteristically limited according to the prejudices shared equally by the Liberal schoolmasters, the Tory parents, and their children who were the scholars. Our reading was closely

> limited to those periods of history which, if we might trust our national pride, were closely analogous to our own. We did not want to explain the origin of anything. We wanted to read about people like ourselves, and to imbibe their ideals. . . . For example, in Roman history we stopped short at the death of Julius Caesar. Freedom was over then. A gentleman could no longer say what he liked in the House of Lords or in the House of Commons—that is to say, in the Roman Senate or to the citizens in the Forum. . . . Cicero needed no explanation—he was the Roman substitute for a Lord Chancellor.
>
> But Athens was the ideal city, which for two centuries had shown the world what life could be. I do not affirm that our image of Athens was true to the facts. It was something much better; it was alive. The Athenian navy and the British navy together ruled the seas of our imaginations.

Whitehead was convinced that the classical education was superb "as a training in political imagination." And it was quite relevant to the future lives of many of the boys.

Like other English boys of his class, Whitehead had begun to study Latin when he was ten and Greek at twelve.

In examinations at the end of his first year at Sherborne, Whitehead stood high in mathematics, but merely above average in classics. Three years later, among thirty-two boys examined under the auspices of the Oxford and Cambridge Schools Examination Board, he stood first in mathematics and twenty-first in classics. In his last term at Sherborne he again took the Board's examinations. This time the chips were down; success would exempt him from an examination at Cambridge that was prerequisite to reading for honors in mathematics. He came out first in mathematics and eighteenth in classics, of thirty-five boys examined.

Whitehead's Sixth Form teacher in classics was the new headmaster. Harper left Sherborne at the end of 1877 to be head of his old college at Oxford. His successor, Edward Mallet Young, was a fine scholar. A former Fellow of Trinity College, Cambridge, he brought a reputation as a brilliant classics teacher. Matthew Arnold, whose two sons were his pupils at Harrow, warmly recommended him to Sherborne. Whitehead's only reference to Young notes his "passion for accurate philology":

> I remember the headmaster stopping a boy who, when translating into English before the assembled class, reeled off the familiar phrase, "Alas, alas, the glory of Israel hath departed," with "No, no, laddie: The glory of Israel has gone away as a colonist."[10]

As a liberal in educational policy, Young in 1880 ceased to require that all Sherborne boys study Greek. Whitehead, who had just left, had got what he needed from his study of the classics; he maintained all his life a fine appreciation of Greek and Roman civilization, and he never lost the ability to read the Greek as well as the English in the Loeb Library editions of his favorite authors.

The formal study of English literature concentrated on Shakespeare, and was not successful. "To this day I cannot read *King Lear,* having had the advantage of studying it accurately at school." There was no drama society at Sherborne. Scenes from Shakespeare were presented as parts of long programs on the annual Commemoration and Speech Day. So it was that on July 3, 1879, when the long-desired New School Room was opened, Whitehead played Bassanio.

Reminiscing with Lucien Price in 1943, Whitehead said: "Of the modern languages, German we took seriously; but the two subjects which were beneath our serious consideration were French and physics."[11] The records of Whitehead's examinations at Sherborne include French, but neither science nor German. Probably one language was all he attempted until, after a few years at Cambridge, he found an exciting new mathematical literature in German. And how did he do in French? After one year he stood at the bottom of his class. He was still at the bottom in 1878, and a year later he had advanced only to thirtieth among the thirty-two examined under university auspices. He was plainly the kind of schoolboy who would not work much at what was of no real interest to him. In French, as in Classics, he could and would do better when he had to. Of the thirty-five boys who took that examination in 1880, he made fifteenth place in French.

Young Whitehead's forte was mathematics, including the elements of physics. It is unfortunately impossible to know how interested he was in experimental and observational science. The School had recently opened a laboratory, a workshop, and a museum, but not much science was taught. The *Shirburnian Magazine* records Whitehead's election in 1878 to senior membership in the Natural History Society.

All his life Whitehead delighted in the nonprofessional observation of nature. I have no evidence that he ever conducted experiments with apparatus; his curiosity found vent in experiments with ideas. At Sherborne he probably spent some time in the laboratory; the absence of an examination score primarily reflects the fact that only mathematics was needed as preparation for the university. The teaching of physics and chemistry to boys, which the Public Schools Commission had advocated in 1865, was a new thing, newer than the teaching of natural history. When he entered Sherborne it was one of but thirteen public schools that had laboratories. Its first science master, H. W. Lloyd Tanner (appointed in 1873), constructed telephones that his pupils and colleagues were sure would never be more than amusing toys. Although Tanner left Sherborne in 1876 and so may not have taught Whitehead, the boy must have known about these telephones. A story in the Whitehead family says that his brother Charles was interested in improving them, and that young Alfred did some relevant mathematical calculations with him. When, sixty-odd years later, the philosopher was talking to me about the advance of civilization in his lifetime, his first illustration was to point with admiration to the instrument on the table at his side.

Every year Sherborne School awarded its Classical Medal and its Mathematical Medal to the best Sixth Form students. Whitehead was eligible in '78, '79, and '80, and won the Mathematical Medal each time. He also won the Digby Prize for Mathematics and Science three years running. He won none of the other prizes that Sherborne gave—for Latin and Greek prose and verse, English verse, or the study of Divinity ("and especially of the Greek Testament"); nor yet the English essay prize of the Digby for History and Modern Languages. His preeminence in mathematics, which was the most important subject after classics, caused him to be recognized as one of the intellectual leaders of the Sixth Form.

Apart from this, it was in games and as prefect that Whitehead's experience as a senior boy was highly important. Since most of my readers are probably Americans, I shall describe these activities of his in a little detail.

My information comes mainly from the *Shirburnian Magazine,* which the senior boys published. There is no specific information about the activities of Fourth and Fifth Form boys; their life, as at every public school, was a routine preparation for the Sixth Form. Hence I can say nothing about Whitehead's life in the academic years 1875–77.

iv

In the autumn of 1877 it was discovered that Whitehead could not only play football, but play it so well as to make the first Fifteen, that is, the boys who represented the School in games against teams of old Shirburnians, or of town boys, or against mixed groups arranged ad hoc. Most of Whitehead's teammates were a year or two older. He was also a member of the first Fifteen the next year, and its Captain in 1879–80. Henry Whitehead, it is true, had accomplished precisely the same thing in 1870–73. But Alfred was the son who was considered frail by his parents, too frail to be put on a horse or to undergo schooling at Chatham House Academy. And it is Alfred whom the *Shirburnian Magazine,* describing "The Characters of the Fifteen" in April 1880, called "the best forward the School has ever had." This is something that only a master who had seen many Sherborne Fifteens could say.*

American readers must not suppose that to play varsity football these English boys of the 1870s had to come up to anything like the size and weight of today's high-school players. Still, their football was rough. It was the Rugby game, not Association football (i.e., soccer), which was played at Sherborne. It involved running with the ball and getting tackled (but not blocking), as well as kicking. It was not only the ball that was kicked. Whitehead, telling his Harvard colleague William Ernest Hocking about this in 1934, said: "I used to play in the middle of the scrum. They used to hack at your shins to make you surrender the ball."[12] "Hacking" was in fact the subject of much official discussion in English schools; at Rugby—though not at Sherborne—it was forbidden in 1877. In his conversation with Hocking in 1934 Whitehead was looking for some paradigm of the Real. He said: "Compulsion—symbolized by the traffic cop—No, this is still too intellectual—*being tackled at Rugby, there is the Real.* Nobody who hasn't been knocked down has the slightest notion of what the real is."[13]

The scrum, a more important part of the game then than it is now, consisted of nine or ten boys (the "forwards") on each side, standing three deep and tight together. To "play in the middle of the scrum" means that from whichever end the ball is tossed in, you will try to hook it, while your teammates behind you are shoving you forward and your opponents are shoving you backward. Once the ball is well off the ground, the boy who gets it in his hands may run with it or pass it; but when a ball-carrier is tackled he must release it immediately for kicking by the first boy who can reach it—play is not stopped by a successful tackle. The fullback is the only player who (because his job is like the safety man's in American football) is not constantly in motion. Substitutions of one player for another were not allowed.

*Although the *Shirburnian Magazine* was written mostly by boys, such items as "The Characters of the Fifteen" were written by sympathetic masters.

Obviously, the game was designed by the Devil to test a boy's stamina and grit. Furthermore, padding was not worn, and injuries were frequent. Whitehead's immediate successor as captain of football, W. J. Harper (son of the former headmaster), suffered a fatal injury in 1881.

The *Shirburnian Magazine* at Christmas 1878 said of Whitehead:

> shows to equal advantage in tight and loose scrimmages. By working the ball through the scrimmages and carrying it on, often gains considerable ground. Collars [i.e., tackles] and passes well, but unfortunately lacks pace [i.e., is not a fast runner].

Sherborne's historian, on reading this with me, said, "He must have been very wiry, what we call a 'piano wire' boy; also very good at not letting the ball get away from him and at advancing the ball."

The importance of his hours on the playing field in fixing Whitehead's character and outlook can be indicated by two quotations. The first is from an undergraduate who played Rugby at Harvard in 1968. "Rugby tradition dictates that sportsmanship and self-control are as necessary to a rugger as are good wind and sturdy legs. Talking in the scrum, profanity, and protesting a referee's decision are not tolerated." At Sherborne ninety years earlier, there were no referees at the games; debatable points were decided by consultation between the two captains, a circumstance that made fairness, and sportsmanship generally, of the very essence. The second quotation is again from Hocking's report of what Whitehead said in 1934:

> They used to hack at your shins to make you surrender the ball, a compulsory element—but the question was *How you took it*—your own self-creation. Freedom lies in summoning up a mentality which transforms the situation, as against letting organic reactions take their course.[14]

Of course Whitehead was reading his own philosophy, written much later, into what happened in the scrum. My point is that his play there was a case of personal response, positive and achieving—of responding to hostile force not by giving in but by "taking it" and finding a way to change the situation into one in which his team had the ball. Three years of this on almost every schoolday afternoon from October to March was bound to have a lasting effect on the way in which Whitehead would conceive of life, namely, as consisting of processes of compulsion and of that reaction to compulsion in which individuality—strong or weak—is created.

In "The Education of an Englishman" Whitehead pointed out another fact about the games at Sherborne: freedom from supervision—positively speaking, spontaneity and initiative:

> We managed the games ourselves, and trained ourselves. We played cricket, and football, and fives, because we enjoyed those games and for no other reason. Efficiency, what crimes are committed in thy name! Today (1926), throughout English schools, the games are supervised by the younger masters. Fifty years ago at Sherborne no master either played at games or interfered with advice, except by the express invitation of the boy who was captain of the games. We were not efficient; we enjoyed ourselves.[15]

The slightest acquaintance with Whitehead's final world view brings to mind his recurrent use of the term *self-enjoyment,* his emphasis on the value of self-determination as opposed to determination from the outside, and his doctrine that the nature of things always permits some degree of self-determination. Also, everyone who knew Whitehead at Harvard immediately recalls his insistence on letting each student pursue just those investigations which appealed to his trained instincts, not those which a professor knew to be efficiently manageable topics for theses. I wonder if he ever urged such a topic upon a student; he never said anything of the sort to me, nor did he give advice when I was writing my Ph.D. thesis. The business of the faculty, when students have reached the thesis stage, is to see that they have ample opportunity to exercise their own powers. So Whitehead insisted: the pursuit of research, like the playing of games at school, is done best when it is done freely. A far cry from present academic practice!

At Sherborne, football was made compulsory in 1893. The Governors of Sherborne decreed in 1874 that games were to remain solely the boys' province. How much anarchy on the playing field this entailed depended on the characters of the team captains. The very rules of football there had been for a quarter-century "of a fluid nature, varied from time to time at the whim of the players." When Alfred began to play, the lesson of the necessity of fixed rules had been learned.

The games were arranged not by an athletic office but by the Games Committee of the first Fifteen, consisting of the Captain and a couple of other boys. Most of the games in which Alfred played, however, were between the School House (in white jerseys with black and magenta stripes) and a team drawn from other houses (such battles usually went on for two or more days, eighty minutes each day!), or between groups that the boys picked out in order to get up a game—for example, A to L versus the rest of the Alphabet, Cornwall and Devon versus The World, Sixth Form versus the School. At any given time there would be ten or eleven Fifteens in the School, for rather more than half the boys played football regularly.[16]

In those days the popularity of football at Sherborne was increasing rapidly, while that of cricket (still the premier sport) and of fives, which ranked just after cricket, was decreasing. Fives was essentially what we call handball, played outdoors. (Sherborne had no gymnasium, and all sports were outdoor sports.) The School had some excellent courts, and crowds would gather to watch two boys playing. Fives was strictly an individual sport; only singles matches were played at this time, and there were no team competitions. Now, Whitehead's superiority is not mentioned in the *Shirburnian Magazine* in connection with *any* individual sports competition nor in the School record of the winners of the cups annually presented. He was not a top athlete.

He was, however, Captain of the Sherborne cricket team in his last year. He was not until then a member of the Eleven, though he played in house matches. Other evidence that he was not outstanding at cricket is the fact that the awards for bowling and for runs went to other boys. I suspect that he was chosen Captain, over boys who had already earned their colors in cricket, only because of his demonstrated excellence as a leader; he was already Head of the School, and had been Captain of football; as Captain of cricket, he would automatically become Captain of the Games (all games) at Sherborne.

Sherborne cricket gave Whitehead his showiest experience of hard-won team triumph. (At Cambridge he would be only a spectator of matches and boat races.) Annual matches with Clifton had begun in 1865, when it was a new public school. "But the infant grew up quicker than was expected," Mr. Gourlay has written. Before Whitehead's time, Clifton's enrollment had passed Sherborne's; it was now more than double Sherborne's. Arthur Waugh wrote about the 1880 match that he, a junior Sherborne boy, watched:

> The impossible is happening; the miracle of miracles has dawned. Sherborne is beating Clifton, . . . a thing that has not been known for years, . . . a thing that, as it turned out, was never to be known again. Those were the great days of Clifton cricket. . . . The best

> we had hoped for was a creditable fight, and here we were winning, with plenty to spare.[17]

After praising the performances of three of Sherborne's players, Waugh continued:

> But a liberal share of the credit must go to the captain, that indomitable leader, A. N. Whitehead, who made a cricketer of himself against his native quality for the sake of the school he had so finely led in football.[18]

Cricket is a game of individual performances, not the team game that rugger is. The outstanding cricketer has to be a natural athlete. Henry Whitehead was probably such a boy. Alfred was not. He loved games; he was superb as a team player and a leader. It was in the character and mentality which success at Rugby called for that he was unequaled at Sherborne.

In 1879, when Whitehead was Captain of football, we are to picture him appearing on the field at the precise hour for games and calling, "Come up, please," stopping the unorganized kicking practice that had been going on and getting the contests of the day started. At the end of this, his last season, the *Shirburnian*'s comment was:

> A. N. Whitehead has proved himself an excellent captain, and to his energy and ability in that capacity the School may attribute much of its success during the past season. He has fully maintained the reputation he so justly earned last year—of being the best forward the School has ever had.

V

The life of a bright and active boy would also include membership in one or more of the societies at the School. In the pages of the *Shirburnian* the activities of the Debating Society are those most fully reported. For about two years Whitehead was its most prominent member. He is first mentioned as one of six who on February 16, 1878, successfully opposed the motion, "That this House would view with satisfaction the introduction of cremation into England." (Whitehead himself came to accept it, and was cremated when he died.)

The vote against cremation had been 10 to 8. On one occasion, a year later, the voting audience split, 8 to 8, after "a very exciting and well contested debate" on a motion proposed by Whitehead, "That we ought not to oppose by force of arms the secession of any of our colonies." The meetings were not as small as these voting figures suggest. Besides the approximately thirty active members of the Society, other people came, and probably found it hard to keep silent: in October 1878 Housemaster Charles Whitehead defended from the floor the proposition, "That the monks of the Middle Ages deserve our gratitude and respect." Presumably he spoke only with the express permission of the boy who was President of the Debating Society.

Whitehead's keen interest in School debates is shown by the fact that in 1878 and 1879 the *Shirburnian* credits him more often than any other boy with proposing subjects. One which he proposed was, "That Public Picture Galleries and Museums ought to be opened on Sundays"; this was still being urged at Cambridge years later. Another, "That the employment of Indian troops in Europe without the previous consent of Parliament is unconstitutional and unnecessary," was twice debated. A kind of question frequently chosen by the boys was the admirableness of various men in England's political history. Most of the questions debated at Sherborne, apart from these moral-historical ones, show a lively awareness of real issues of the present or the future; examples, besides those already mentioned, are the desirability of annexing Cyprus, of women's suffrage, and of compulsory service. Unfortunately the *Shirburnian* says nothing about the debate that took place in February 1878 on the boldest motion of all: "That friendships between Seniors and Juniors in Public Schools are not conducive to the welfare of either."

There was no literary society at Sherborne until the year after Whitehead left; but the *Shirburnian Magazine* had been founded in 1859, and Whitehead in his last year was its Editor—though probably not for the whole year. The editorials (unsigned) are of only transient significance, and it is impossible to identify the authors of the articles published. Socrates is the subject of one, but our future philosopher was not yet a senior when it appeared, and it does not at all resemble his writing for the *Cambridge Review* a few years later. Whitehead was, however, the Editor in March 1880, an issue which included a letter from "Archbishop Whately" protesting the "scraping of violins" at the School, and a tart reply from the Editor. The *Dictionary of National Biography* informs us that Richard Whately, Anglican Archbishop of Dublin from 1831 until his death in 1863, "was without ear for music." To see the circumstances of the joke, let us turn to the place of music at the School.

There had long been a tradition at Sherborne of ignoring music almost entirely. Harper got away from this in the early 1870s by appointing first the energetic J. R. Sterndale-Bennett, "ostensibly to teach mathematics, but in reality to encourage the school music," and very soon thereafter Louis Napoleon Parker as choir master and organist.* Parker worked wonders in the School. He is known chiefly as the father of the modern pageant; but that came later. (It began with the great historical pageant at Sherborne in 1905.) In 1880 Parker was an ardent young Wagnerian. He had started a fine School Orchestral Society in 1876; each year it gave six or more concerts in which the great German composers were favored. The building up of music at Sherborne was continued under Young.

In 1879 and again in 1880 the School House won the prize that Young had offered for the House Glee competition. But the scant records of the burgeoning of music at Sherborne make no mention of young Whitehead. His children did not remember him as musical; in his old age he listened to the radio for news only. He could be moved, and could see that others were moved, by the organ and by bells. His response to other music was, I suspect, a vague sense of emotional meaning, together with recognition that many people enjoyed it more fully and more delicately. Whitehead had a lifelong habit of learning much from conversation. That, I suppose, is how he learned enough about music to appreciate its importance.

Whitehead may have concocted the letter from Whately. In any case, the young Editor's reply is a case of Whitehead taking sides on behalf of something important and relatively new, something championed by the young (by Parker) and blindly opposed by those who are old in spirit. He would do this all his life. The question of whether the opposition in 1880 consisted of more than a handful of old Shirburnians and the many boys who would not be quiet during a concert is less important than the fact that Whitehead could recognize and would automatically defend a valuable new development, even one which he did not know much about and by which he was not personally moved. Whately is a natural symbol for Whitehead's typical target. He had been almost as dead to painting, sculpture, and architecture as to music. He had distributed his contempt impartially upon Wordsworth (already a favorite of Whitehead's), evangelicalism, the Oxford movement, and the higher criticism. J. M. Rigg says in the

*Sir William Sterndale-Bennett, at that time the best-known musician in England, paid visits to the School and sent Parker there.

Dictionary of National Biography: "He ignored metaphysics and dismissed theology. . . . His most characteristic mental trait was strong common-sense." We shall often find Whitehead cudgeling Whatelys.

vi

Before Whitehead began his last year at Sherborne, his future for at least three years thereafter had been assured. In mid April 1879 he had gone to Cambridge to take Trinity College's entrance examinations for scholarships and he had won one.

I do not see how this achievement could have been absent from the mind of the headmaster—himself a Trinity man—when he appointed Whitehead his Head Prefect for 1879–80. Whitehead was nowhere near the top of his Sherborne class in any subject except mathematics. The summer term of 1879, which took into account both term work and his marks in the Oxford and Cambridge examinations, shows him to have been better in daily work than in examinations and overall third highest in his class. In view of his character and qualities of leadership it was natural for Young to choose Whitehead as his Head Boy.

Whitehead's last year at Sherborne was not an anticlimax to his Trinity scholarship examination. He wrote in his "Autobiographical Notes":

> My time was mostly occupied with my duties as Head of the School with its responsibility for discipline outside the classrooms, . . . and as Captain of the Games, . . . very enjoyable but taking time.

His son North wrote to me: "My father then believed that *he* ran the school with some help from the headmaster. . . . My father sometimes told me that he had never been so old and mature as when he was head boy." Such was generally the case in English public schools. Pascoe has this to say about the Head Boy:

> He lives in a miniature principality, where his school-fellows are his subjects, and he reigns supreme. He stands senior of the senior class; . . . his ability finds due meed of recognition as editor of the school-journal; he is confidential adviser of the head master . . . ; he has the honor of numbering the assistant masters among his friends; and in his little world he carries himself with distinction, as if he were the leader or conqueror of battalions.[19]

Naturally some heads became conceited prigs, but almost none became tyrants. The usual effect of the earlier years in the school, and of the whole system, was to make a prefect value himself not as an outstanding individual but as part of the group and of a historical continuum, and so to do an unselfish job. Of the six head boys at Sherborne from 1875 to 1881, Whitehead was the only one who reached high distinction as an individual. I know of no evidence of either conceit or tyranny in Whitehead at any time in his life.

At Sherborne as at other schools the code of behavior was detailed and there was quite a variety of punishable offenses. Young, as fond of boys as Harper was but unable to inspire fear as he did, especially needed a head boy of strong character who could rule with authority outside the classroom. He got such a boy in Whitehead.

In every public school, flogging was the accepted punishment for many offenses. It was not as brutal or as readily administered as it had been fifty or a hundred years earlier, but it was not even beginning to be on the way out. At most, its maintenance was about to become an amusing proposition for college debating societies; in 1881 Whitehead took part in such a debate, easily won by the defending side, at Trinity College, Cambridge. Flogging would be customary for another half century. Whether it was called flogging, thrashing, beating, whipping, or caning, it was a severe whacking of the culprit's buttocks or shoulders (his arms being draped over the back of a chair), and was done in the presence not of any of the masters but of all his classmates, or, for some offenses, before all the boys of the school. Whitehead recalls caning a boy for stealing money from another. He privately consulted the headmaster, who said, "Either you must thrash him in the presence of the school or I must expel him." Expulsion could ruin a boy's career, especially if he hoped to get into a university. Years later—about 1940—Whitehead told some of the Junior Fellows at Harvard that he considered this caning one of the most disagreeable things he had to do.

In "The Education of an Englishman" we read:

> I remember cuffing a big boy over the head because I found him twisting the arm of a small boy; but I apologized afterward, because I found that the small boy had called his elder "a captain of Barbary apes"; this was unpermissible insolence in the school world.

If Whitehead's memory that there was no gross bullying was right, such arm-twisting did not often happen. Some bullying, not so much by the prefects as by the bigger boys just below them, was usual in public schools. Sherborne in Whitehead's time was better than most.

There was something else at Sherborne on which Whitehead found cooperation with the headmaster a disagreeable task, though he would never mention it in writing or in ordinary conversation. His son North, in his unpublished autobiography, noted that headmasters who were steeped in the classics and particularly in the literature and life of the Athenians ought to have known that large numbers of boys, segregated for months at a time from female society, would develop homosexual habits. North Whitehead wrote that homosexuality "was a major curse when my father was head boy at Sherborne School." It was not much talked about; it was what a master had in mind when he said that there was "an evil" among his boys. Although English headmasters (Harper probably included) long maintained an unawareness of it, Edward M. Young did not. He asked the cooperation of his Head Boy in suppressing homosexuality. Whitehead, his son told me, was saddened by the prowling he had to do. I doubt that he did as complete a job of it as he could have. The punishment was flogging or expulsion or both. The idea that homosexuality might be something more than deliberate wickedness in those days occurred to no one in the schools. It was handled just as the boys' stealing, smoking, and drinking were handled.

In the two decades 1870 to 1890, criticism of the public schools was at a lower ebb than at any time before or since. But in 1881 the increase in homosexuality was denounced in the pages of the *Journal of Education,* and before the decade was out, the handling of a case at Sherborne had ruined both Young and (for some decades) the school. Late in 1888, a sharp quarrel broke out between Young and the tutor he had hired to help him supervise the School House—a quarrel over fairness in punishing two boys caught in homosexual practice. The tutor, E. C. Malan, brought suits for slander and libel against the headmaster. The case was heard *in camera,* while masters, Old Shirburnians, and even Harper (now one of the School's governors) took sides. The dispute brought to a head a growing dissatisfaction with Young's Sherborne because it was not Harper's Sherborne. The natural result—augmented, of course, by the fact that the suits were heard *in camera*—was that parents stopped sending their sons to Sherborne. (When judgment was rendered in open court, the judge had to deny rumors that Young and Malan themselves were guilty of sexual immorality.) The unfortunate headmaster, a generous man but a thin-skinned and tactless one who could not be reticent under criticism, was almost entirely cleared of the charges, but the damage was irreparable. Two years later he was asked to resign.

When the time came to send North Whitehead to a boarding school, his parents chose neither Sherborne nor any of the established public schools, but a new coeducational school.

Two traits of Whitehead's character that stand out in his mature life are an unusual reticence and a habit of ignoring personal criticism. When people were rude to him or made personal attacks, it was as if he had not heard. And whereas other witty people (notably his wife) would make devastating remarks about the moral weaknesses of their acquaintances, Whitehead preferred (and on principle if they were young acquaintances) to mention what he liked about them and to keep his feelings of moral disapproval private. Not all his reticence falls under this head; some of it was due to caution. Without pretending to any knowledge of what in his childhood first encouraged these traits in him, let me simply observe that if anything that transpired in his youth or early manhood was needed to convince him of their value, the case of Malan versus Young was that thing. I think Whitehead was also convinced—again, if he needed convincing—that spying is a bad business; there is no need to know every sin of your associates, and there are some things that it is better not to know.

Whitehead learned a good many things in his year as Head Boy at Sherborne. Most important of all, that power is to be exercised with restraint: morality and common sense call for that, and so does a care for the effectiveness of one's rule over others. He surely learned that there are many kinds of boy nature, and so of human nature. The belief that it is good to be kind is one to which he was already receptive by reason of the kindness of his father and brothers (especially Henry); I think it now grew into an axiom for his own conduct. He certainly learned that firmness is good too, and absolutely necessary. (There is no worse misconception of Whitehead's character, among those who knew him slightly at Harvard, than the notion that, being so obviously benign and gentle, he was all milk and water.) He had an opportunity to see that firmness presumes that everyone has a clear understanding of just what the rules are; avoidable vagueness is something Whitehead always disliked. He saw in the school the truth of one of the lessons of the history he had studied—that the use of force can be diminished but not eliminated. He learned not to shrink from the use of force, disagreeable though it is to you, when the alternative is acceptance of much greater evil. You should make sure of your grounds, then go ahead. One of the incidents which Whitehead recalled suggests that he knew well that sometimes you should apologize; we all make mistakes. More generally, the importance of good manners, even of etiquette, was borne in upon him. (I have never seen anyone with better manners than Whitehead.)

By practicing for a year, in his position of highest authority among the boys, that use of the light touch, moderation, and courtesy which I have mentioned as characteristic of the best prefectorial rule, Whitehead made it a settled part of his character. To infuse in yourself the spirit of mannerly dealing and moderation causes you to eschew militancy and extreme positions, and to dislike them in others. In later life Whitehead was always a liberal, never a rebel.

vii

In the great internal reform debates at Cambridge University in the middle and late Victorian years, the opponents of reform included a large percentage of the very numerous resident mathematicians, while the chief reformers were men in what we now call the humanities. The mathematical training, more than one reviewer of that time has observed, easily led men who did not have other interests to look at the university as a closed system of fixed units and fixed relationships. This static view did not tempt the mathematician Whitehead. Some credit for that must be given to his work as Head Boy, dealing not with units but with boys, in his last year at Sherborne.

There is an interesting negative fact about Whitehead's published recollections of Sherborne. He mentions no one by name—no master and no schoolfellow. In order to describe "The Education of an Englishman" for readers of the *Atlantic,* he did not *have* to name anybody after Thomas Arnold; but most writers would naturally have said something about Harper, and named the headmaster who was consulted on the punishment for theft, and inserted a word about one or two of the teachers. In the "Autobiographical Notes" there are only five hundred words on Sherborne, so perhaps the omission of names by a man of eighty whose lasting friendships were formed at the university is natural; but one would expect at least—what his biographer sorely misses—an identification of his superb mathematics teacher. In the only other place where we can read Whitehead on Sherborne School—Lucien Price's *Dialogues,* nobody is mentioned by name, not even in Price's manuscript.

Whitehead's memory of the scandal of 1888 at Sherborne might be a sufficient explanation of this reticence. But I think that a more general cause is involved. When Whitehead writes about his student days at Cambridge he does not mention his mathematics "coach," though every other Wrangler of the time whose recollections I have seen does so. Whitehead names some of his many Cambridge friends, but says nothing *about* any of them except Bertrand Russell. In the "Autobiographical Notes" he does not name a single colleague, student, or friend of his thirteen years in the University of London. His life is amazingly full of friendships, but I have not come across a clear case of a passionate friendship, nor of a broken friendship—two facts that surely have in part the same cause.

I am not forgetting that for an Englishman of Whitehead's generation and social class, personal relations were private, not things to talk or write about. Still, the fact that he does not in his recollections say even the little about them which it would have been proper and, for others, natural to say, suggests that he was not much interested in personalities. His permanent interest was concentrated on ideas, institutions, types of action, and types of people. The people were indeed unique individuals in his eyes: when he wrote or talked history, he showed a wonderful sense for their individual contributions. He was not at all deficient in sensing the feelings of persons in his life; but I think he did this on principle, not because he had—as his wife did have—an urgent emotional interest in their personalities. That is one reason, I surmise, why he was a good head prefect.

A man who does not, except to his wife and one or two very intimate friends, spontaneously talk about persons he has known, makes it very hard for his biographer to say much about personal influences on him. On this head I cannot say anything about his friends among the boys at Sherborne, and only a little about the two headmasters and his Sixth Form Mathematics master. The last is the most important influence.

My efforts to identify the man have narrowed down to two experienced teachers, both trained at what was then the leading mathematical college in Cambridge, St. John's. One is the Reverend Alexander Wood, a Scotsman of striking appearance whom the boys called "the Gaffer." He spent almost no time at the blackboard, but simply set problems for the boys to do at their desks. If Whitehead was his pupil, any special help to the budding mathematician must have been given in private.

The other master is John Blanch, who later was unhappily married to Whitehead's sister, Shirley. Blanch took on extra responsibilities without pay. The Sherborne scholars liked him. He sounds more like the teacher who would go out of his way with a promising student.

Nothing is known about his teaching, but what is known about Wood's tells against the probability that Wood was Whitehead's teacher. If Blanch was the man, family loyalty and embarrassment over his unexplained suicide in 1907 would explain Whitehead's failure to name to Jessie Whitehead the Sixth Form Mathematics master whose teaching he praised.

I come now to the two headmasters. Harper had an influence, positive or negative, on everyone in the School House. As scholar and teacher he was merely competent; he was rather, as his biographer says, "essentially a man of action."[20] He was a simple man with simple tastes in literature (when he took time to read it) and no taste for intellectual subtleties. Of unquestioning, simple Christian faith, he would not have theological books put in the School library.[21] What he continually preached by word and example was the gospel of work, and contempt for loafing. In his last, very characteristic sermon at the School, he said: "Work: get something done, though that something may not be much, and it may at times be dull work doing it." That is a rule which Whitehead always made it his business to follow. To Harper it had a religious basis:

> What is our work, and whatever at any time may be our work, if only we are not conscious that we are seeking our own will in it and not God's, we may fairly believe it to be the work which God would have us do.

It was not until about sixteen years after Whitehead left Sherborne that he ended a long period of reading and thinking about theology by concluding that the rule of work would stand for him entirely on its own feet. That human life is not as simple as it looked to Harper, Whitehead must have realized much sooner.

Harper's example and preaching should probably be given some credit for leading the boy Whitehead, in his first years away from his family, to adopt the habit, which he never dropped, of getting to work early and keeping steadily at it, without worrying about its personal rewards.

What positive values he got from Young it is impossible to say. It was Young who had learning, style, and elegance, but he lacked effectiveness with boys and men.[22] It would, however, be unrealistic to suppose that Young's weakness was much noticed by Whitehead—that he learned an important negative lesson from the contrast between Harper's success and Young's failure—until years later, when Young was on the point of defeat.

"It has often been said that the last two years at a public school are probably the happiest in a boy's life." So wrote Stanley M. Leathes, in agreement with this saying. He was Captain (Head Prefect) at Eton the same year that Whitehead was Head Boy at Sherborne. He became Whitehead's classmate at Trinity College, and later his colleague there, before he became a top civil servant. Some could say that Eton and other illustrious public schools bred caste-conscious boys. Sherborne, especially under Harper, bred love and loyalty to school, country, and God, but it was not an aristocratic public school; a boy from the Whitehead family would not there learn to treat people from the lower classes any differently than his equals. Critics of the public schools had noticed some tendency "to thwart and repress social, moral, and *intellectual individuality,* to mould the individual to one pattern, to laugh down 'oddities,'" even to "discourage genius."[23] The tendency toward one pattern was of course present at Sherborne, but I doubt that intellectual individuality was much repressed, or that the few potential geniuses among its boys were discouraged. Whitehead himself was never given to oddities. His last years at public school were both happy and fruitful.

They formally came to an end on Commencement Day, also called "Commemoration and Speech Day"—June 23, 1880. The Abbey was completely filled. After service the boys, masters, and visitors adjourned to the big New School Room. Then, says the *Shirburnian Magazine:* "A. N. Whitehead (Captain of the School) began proceedings with a Latin address (a translation of which was thoughtfully provided for the uninitiated), setting forth the most important events of the past year in connection with the School." This "Contio," the School historian tells me, was in all probability written for Whitehead by the headmaster. Besides reciting it, Whitehead took part in the program as recipient of the Mathematical Medal and as an actor in the two plays (doubtless much cut) that were presented. In Aristophanes' *The Peace* he was the hero, Trygaeus; in *A Midsummer Night's Dream,* Whitehead played Snout.

On visiting Sherborne School in 1973, an American who had studied Whitehead's philosophy of science found there a club named after Whitehead, with a mathematics master as faculty adviser, and heard one of the boys read a paper on particle physics.[24]

CHAPTER V

Student at Cambridge

i/Late-Victorian Cambridge.

ii/Trinity College. The scholarship examination.

iii/Freshman at Trinity.

iv/The honors program.

v/Friendships: W. R. Sorley, D'Arcy Thompson, Henry Head. The *Cambridge Review*. Chapel.

vi/Scholarship examination in Whitehead's first year.

vii/Friendships with dons and new freshmen. The Magpie and Stump.

viii/Young women at Cambridge.

Whitehead was very much a product of late-Victorian Cambridge. As almost none of the people who have become interested in him or his thought may be presumed to know what Cambridge was like a hundred years ago, I shall include a good deal of description in the story of his student years. Any reader who already knows the setting is advised to skip the paragraphs in which Whitehead is not mentioned.

His life there, which would go on for thirty years, began the second week of October 1880. If you would visualize the university to which he came, you must make a huge subtraction from what you see today. Subtract almost everything except the splendid pre-Victorian buildings, the Fitzwilliam Museum, and the college "backs." West of the Queen's Road (which still went by its old name, "Backs of the Colleges"), there was nothing save a few professors' homes and one hall of Newnham College. Cambridge had an observatory and a couple of small scientific museums, but Maxwell's Cavendish (now directed by Rayleigh) was the only modern laboratory, and engineering existed only in a little workshop where everything but the walls was owned by Professor Stuart. No "research students" from abroad; it would be fifteen years before the first, Rutherford, arrived. Boats on the Cam, yes; but the stream was like an open sewer. That would be remedied, many years later; baths and electric lights in the colleges would come too, here and there. Improvements and expansion were constantly debated, and by fits and starts accomplished. On average, a new cornerstone would be laid every three years. In the streets, horse-drawn trams, promised for the beginning of this term, would appear in a few weeks. In the Great Court of Trinity, you must picture the College's serving-men filling up their buckets at the splendid Gothic fountain.

As the holder of a Trinity College scholarship, Whitehead was entitled to rooms in the College. (Slightly more than half of Trinity's students would live in licensed lodging houses.) Along with a hundred other students, mostly freshmen like himself, he was assigned to

Whewell's Court, across Trinity Street from the Great Gate. It consisted of two narrow courts, together named after the greatest of Trinity's nineteenth-century masters, who had bought the land and found the money to build. Of pure bad Victorian style, Whewell's has been called the dreariest court in Cambridge. What a delight to walk away from its gloom into the spaciousness of the Great Court!

For the next two years Whitehead would occupy C-1, on the north side of the ground floor of the Whewell court that faced Trinity Street: a dark bedroom, and a study with narrow south windows* through which a little sun would shine, come spring. Alongside the north-facing bedroom window, All Saints Passage carried the pedestrian traffic from Trinity Street to Sidney Street. These were not secluded quarters! The view from the bedroom window—a short view—was of a small graveyard. Its flat stones, mostly from the eighteenth century, looked mighty quiet. Off to the side there was a lovely walnut tree; in the center of the graveyard and of Whitehead's view, a brand new, undistinguished twenty-five-foot column surmounted by a cross, erected in memory of the dilapidated medieval Church of All Saints, which once occupied the site of Whewell's Court.

A manservant, or "gyp," and a bedmaker—a woman whom the College statutes *required* to be neither young nor beautiful—served Whitehead and his immediate neighbors in the Court. Like others who lived in college, Whitehead would have to buy his furniture. He did not spend much time on that; until he married Evelyn, a room was a room. He could simply buy what he needed from the furniture left by the previous occupant. Throughout his thirty Cambridge years he would have a room in Trinity College, though after his marriage he never lived in it.

The academic year was divided into three sessions, which Cambridge called the Michaelmas (or October), Lent (beginning in mid January), and Easter terms. The minimum residence for the baccalaureate was nine terms. Whitehead hoped to stay longer; the best honors students did so to take the most advanced examinations and to try for a college fellowship.

Before Whitehead could begin his studies, he had to buy his cap and gown from a University tailor. The Trinity gown was blue, with black facings on the foreparts and sleeves.† He must wear his cap and

---

*The windows were improved half a century later.

†The gowns of the other colleges—except Caius, also blue—were black, with slight variations distinguishing one college from another.

gown when he left his rooms on any academic business, and when in the streets on Sunday and after dark on other days. All college gates were locked at ten; anyone who went out for a leisurely evening had to be a good sprinter or a sure climber. The University's two proctors, in splendid silk, and their four deputies ("bulldogs") patrolled the streets, to apprehend students who were not in academic dress or who were violating any other rule of conduct. As almost all the undergraduates were between nineteen and twenty-three years old, some were irritated by the constant proctoring. There is no evidence that Whitehead was among them. The numerous regulations did not hinder witty and humorous men from forming clubs in which they had a great deal of fun. A little later, some of the most athletic undergraduates defiantly took up the art of roof-climbing; but not Whitehead.

ii

Why was Whitehead now at Trinity instead of one of the sixteen other Cambridge colleges? His brother Charles had gone to Magdalene, a very small college with low standards. Alfred was up to the best. Trinity was the largest (larger, too, than any Oxford college), and none had a better staff; most Englishmen considered it the greatest college in the world. Its biggest Cambridge rival, St. John's, commanded the loyalties of Wood and Blanch at Sherborne, and had graduated more Senior Wranglers than any other. But Isaac Newton's college was catching up in that competition. Always strong in classics, Trinity was from Whewell's mastership (1841–66) onward a leader in mathematics and natural science. Undoubtedly Headmaster Young strongly recommended his own college for Sherborne's outstanding student of mathematics.

There was also a special reason for choosing Trinity: its regulations on entrance scholarships. Everyone who aimed at a professional career tried for a scholarship, whether or not his family needed the financial aid.* The scholarship was a badge of merit and a condition of further advance. Without it, Whitehead could not someday become a Fellow of his college. (His brother Henry was already a Fellow of Trinity College, Oxford.) In the *General Almanack and Register of Cambridge University,* Trinity and St. John's supplied all the information anyone

*If aid was necessary, a boy could compete for a "sizarship," which entitled him to commons as well as a sufficient allowance. Whitehead's father entered him at Trinity as a pensioner, i.e., one who pays for his commons and lodging.

could wish about their scholarships, including the texts of the preceding year's examinations. Whitehead's examination would not be easy. Trinity permitted him to attempt it a year early, St. John's did not. In Easter Week of 1879 he went up to Trinity for six days of examinations. If he failed to win a scholarship, he could try again a year later, there or at St. John's or elsewhere; as it turned out, he did not need to. When faced with an ordeal, Whitehead liked to prepare early and get it over with as soon as possible. I do not know whether this habit was mainly the cause or an effect of his successful effort in 1879.

The competition was fierce. Whitehead did not come out on top, or next to the top, but he did well. This was typical of his competitions. His future preeminence would be of a quite different kind. Trinity College reserved for prospective freshmen six entrance scholarships (also called minor scholarships), each tenable for three years; three of these were valued at £75, and three at £50. There were also eight £40 awards known as exhibitions, and several exhibitions for boys from particular schools, of which Sherborne was not one. It was a large company of contenders that sat down to write at the tables in Hall. Whitehead won a £50 scholarship.*

The ordeal of this examination was a sign of the great difference of late- from mid-Victorian Cambridge. As before, a good part of the student population cared little about academic success. But the ratio was changing; in a few years more men would be reading for honors than for an ordinary degree. The demand for excellence had come to the fore, and the urge to demonstrate superiority was very strong in the public school sons of the professional middle class, who now set the tone. Twenty years earlier a Royal Commission had called for entrance scholarship examinations in order to pick out the very brightest and best-educated boys and make sure that they could go to Oxford or Cambridge. This was one of many steps in the modernization (and declericalization) of Cambridge, a process in which Trinity College was generally in the forefront. Trinity had not at first welcomed this innovation, but as Cambridge's historian, D. A. Winstanley, says, "The appetite for clever schoolboys came with eating." It led Trinity to offer early examinations.

iii

We have nothing in Whitehead's hand about his experience as a freshman—only a few golden memories of his later student years. We

*About half of all the awards that Trinity made to prospective students in 1879 were for the study of mathematics, three for natural science, the rest for classics.

must put out of our minds the best-known autobiographies and biographies of other Cantabrigians. To Whitehead's contemporary Goldsworthy Lowes Dickinson, and later to Bertrand Russell, Leonard Woolf, and E. M. Forster, arrival in late-Victorian Cambridge was a wonderful release from a miserable existence under oppressive, or at least stuffy, private tutors, or schoolmasters and prefects, who could not appreciate one's gifts and would not listen to one's opinions. Whitehead had been his own self at school, not forced into an alien mold. Now, like a frog that has been the biggest in his pond, he was suddenly turned into a tadpole in a vast lake. The Head Prefect in charge of 300 boys became one of 865 freshmen, 164 of them at Trinity. The headmaster's confidential adviser became one of over forty freshmen whom Henry Martyn Taylor, M.A. and Fellow of Trinity, met as a group at the beginning of the term; he was their Tutor—that is, he would give advice on courses of study and stand *in loco parentis* to all of them. The Captain of football at Sherborne found the Trinity playing field preempted by Etonians and Harrovians four days of the week—"a social iniquity," as one Harrovian has confessed; but there it was. The premier sport was rowing, which did not exist at Sherborne, and he was not cut out for it. The athletic career of Alfred North Whitehead was over.

It is quite possible, though of course I cannot assert, that all of his immediate companions were new. Only four of his Sherborne classmates had come to Cambridge,* and none were at Trinity. What college you were in made more difference than it now does, for teaching was then the business of the colleges, not of the university.† Trinity undertook, and even small colleges pretended, to provide everything that an undergraduate needed to pass the university's examinations. Only the bolder honors men chose—and not as freshmen—to attend lectures given by a professor in the university.

Of course there were university-wide organizations that you could join. The big permanent one was the Cambridge Union, which had built its own quarters a stone's throw from Whewell's Court. It held debates weekly. Everyone who hoped some day to sit in Parliament was a frequent debater, but so were many future dons and lawyers. The Union had a good library, and a reading room with the best periodicals and about ninety newspapers from all over the world.

---

*Many more, including one of his best friends, had gone to Oxford.

†The shift of instruction to the university did not begin until 1883 and was not completed until forty years later.

Whitehead joined the Union at once and valued it enough to pay the £3 fee every year. But he was never an active member. The Minutes show that he made just one speech—after two years had passed. Nor did he ever nominate anyone for election as an officer, much less stand for election himself.

There were dozens of smaller university-wide societies—the Pitt Club, the Carlton, the Reform, the Browning Society, the societies for music, chess, Shakespeare, skating, Temperance, what you will. Their activities were regularly reported in the *Cambridge Review.* Whitehead is not mentioned in any of these reports.

But, you may say, everybody who knows anything about him knows that he was a member of the famous, select discussion club called "the Apostles." Yes, but he was not elected to it until he was in his twelfth term at Cambridge, a graduate student.* He was not a quick bloomer.

The change from Sherborne would have been less overwhelming, and Whitehead's social life easier, in one of the smaller colleges. Trinity was not only the largest; its student population was several times that of every other except St. John's. The majority of the freshmen came from the twenty recognized public schools; many, like Whitehead, were sons of country clergymen; but there was a variety of backgrounds.†

The 164 freshmen were required to dine together at the long tables in the magnificent Gothic Hall, all at five o'clock—the first of three seatings. The man of Kent's tendency to solitary thought was only superficially discouraged.‡

Sherborne was not what an American would call an Ivy League prep school, and this was sometimes an obstacle to easy friendship. Whitehead always disliked men who put on airs. He was himself wholly unpretentious, and not a climber; he didn't concern himself with his image. He was reasonably sociable, and so fell in with the custom of spending a couple of evening hours in talk. In the after-

---

*Tradition discouraged the election of freshmen, but many of the "Apostles" were elected in their second undergraduate year.

†Trinity College and Cambridge University records show that Whitehead's memory sixty years later—"We all came from the same sort of school, with the same sort of previous training" (Abiog. Notes, in *ESP,* p. 7)—was a considerable simplification.

‡The American reader just might suppose that Whitehead had a roommate. The practice of assigning more than one student to a set of rooms was not dreamed of in those days; it would have infringed the freedom proper to a gentleman and student.

noons he probably did what was expected that first October and showed up at the river for "tubbing." (A "tub" was a broad, paired-oared boat in which a coach taught freshmen how to row.) My guess is that, like many others who were scholars and good walkers but not rowers, he soon gave it up and spent the hours from two to four walking with a friend or friends, most often along the Granta, as the upper Cam is called. He was a good walker, a good listener, and either was or soon became a good talker (though not a quick or voluble one).

It was this conversation—sometimes with dons—that supplied the nonprofessional part of Whitehead's university education. All the lectures he attended were on mathematics, pure or applied. "I never went inside another lecture room," he wrote in his "Autobiographical Notes." Cambridge *did* have "general education" requirements, ancient and unchanging. The subjects were mathematics, classics (Latin and Greek), and divinity (with emphasis on Paley's *Evidences of Christianity*). The University set an elementary examination in them that was intended for all students in their first year. This was "the Little-go," officially named "the Previous Examination." It was not much of an obstacle to men who had had Whitehead's kind of school education.* In fact, he had earned complete exemption from it when he passed the examination given in his last term at Sherborne under the auspices of the Oxford and Cambridge Schools Examination Board. That is how he came to Cambridge as one of those who, in the words of the *Student's Guide* (an authoritative volume written by several dons), could "enjoy three years of uninterrupted study of the subjects in which they wish to graduate with honors."

iv

The honors program at Cambridge was education for examination's sake. After three years Whitehead would take the Mathematical Tripos.† The colleges did not, as in the United States, offer courses in which grades were given, the student's final standing being deter-

---

*Relaxation of the insistence on the study of Greek was under consideration when Whitehead came to Cambridge, and was periodically voted down throughout his Cambridge years. Greek lasted until 1919; Paley, two years longer.

†"Tripos" was first applied to the three-legged stool on which a Bachelor of Arts sat to dispute with candidates for degrees; then to the man himself; then to the set of verses composed by him; then to the list of successful candidates for honors; then to the examination.

mined by the grades earned over the years. They offered lectures on the subjects that would be covered in the Tripos, and many of these lectures were more like lessons in school, being devoted to drill and the solving of problems by the students. Whitehead's progress would be tested by college examinations near the end of his first and second years, but these did not count. Only the Tripos counted.

Cambridge was the mathematical university *par excellence,* and the Mathematical Tripos was believed to be the reason. Most famous of all examinations, it had been instituted not long after Newton's death. In the nineteenth century it was updated from time to time, without change of its character. What it tested was a man's ability to solve a host of problems—many quite complicated—in minimum time. This went on for six days. The results would be ceremoniously announced on Commencement Day in June 1883, when the students' names would be read out in order of rank. Those who placed in the first of the three divisions—the Wranglers—would be admitted to the concluding part of the Mathematical Tripos the following January, to write for three days on the most advanced subjects. This division of the Tripos would not take effect until 1882, but was announced in 1880. If Whitehead hoped for a mathematical career, he would have to become a Wrangler, one of the first dozen if possible. The chance of a good career, and the chance of a Trinity fellowship, depended on that.

Memoirs of the period remark on the hopelessness of trying to be a top Wrangler without employing a private tutor, or "coach." The *Student's Guide* for 1880 called this "almost a universal practice." It had been so for more than a century. Honors students paid tuition to their colleges for lecture courses, cut as many of these lectures as they legally could, and paid a coach—for three years and (usually) July and August of two summers—to train them for the greatest possible success in the great academic Derby.* The established coaches were dons who took private pupils without regard to college connection. To be the coach most sought after was the highest distinction in Cambridge short of being a professor or head of a college.

The freshmen honors students knew very well what they faced, for the *Student's Guide* in 1880 included some thirty pages on the Mathematical Tripos: the mass of rules governing it, the subjects that would

---

*Reformers wanted to abolish the time-honored order of merit. That had now been accomplished in all the triposes except the Mathematical; there, the battle would not be won until 1907. Whitehead's part in the battle is described in Chapter VI, section ii, below.

be covered, the schedule for each day, advice about books and plans of study, and the advice to get a coach. The examination questions of the preceding year had been published; 98 passed it that year, and 33 of these were Wranglers. Whitehead was beginning the race as one of a large crowd. In fact, more than a third of the honors men were mathematical students. The high proportion reflected the historic dominance of this Tripos at Cambridge.*

Mathematics had been the leading field of study since the early eighteenth century. It was still thought to train the mind well for a career outside mathematics—in the law[1] or the Civil Service, professions that many Wranglers of Whitehead's time entered. We have also to remember that physics at universities had not yet become a department separate from mathematics. Dynamics and statics, astronomy, optics, sound, heat, electricity, and magnetism bulked larger in the Mathematical Tripos than pure mathematics did,† and mastery was not made easier by the absence of laboratory work to do. Mathematical students with experimental interests went to the Cavendish Laboratory only after they had been through the Tripos.‡

Small wonder that Whitehead, up against the best mathematical students from more than half of England's schools (Oxford was not the place for mathematics) and a good sprinkling of graduates of Scottish universities, did not make commitments outside his field. Why should he be a joiner? It was enough to drop in at the Union, and to talk with friends about history, literature, politics, religion, philosophy. He was interested in all of them, and did a good deal of unplanned reading in them. Mathematics was what he read steadily and deliberately. One of the books he bought in his first term has turned up: *A Treatise on the Analytic Geometry of Three Dimensions,* by George Salmon.[2] The *Student's Guide* said that after study of solid geometry in

---

*For a long time it was the only one. Thereby the university of the immortal Newton was kept a mathematical university, however much pain this might cause brilliant undergraduates who, like Macaulay, had no talent for mathematics. It was not until 1857 that classicists and reformers got Cambridge to permit any other Tripos to be taken in place of the Mathematical.

†There was also a Natural Science Tripos, which only 31 men took in 1880; not much physics entered into it.

‡The great physicist J. J. Thomson was one of these. When he came to Cambridge in 1876 from Owens College, Manchester, he had already done experimental work on electricity that would soon be published. The man whose work would inspire his achievements at Cambridge, Clerk Maxwell, was the Cavendish Professor until he died in 1879. Thomson had no contact with him.

an easier book, Salmon's "may be read by those who have time enough to make an elaborate study of the subject." This undergraduate was what they called "a reading man." Some unserious men called him a "smug" (Americans, read "grind").

V

The friends that Whitehead gradually made in his first year were not, most of them, mathematical honors men. In fact, most of those men who had good minds were so intent on working for the Tripos that they were dull company, when they were company at all. But at Cambridge, as Whitehead has said, "Groups of friends were not created by identity of subjects for study."[3] Of the men in other triposes, perhaps the most brilliant, and at the same time most lovable, was Leslie Stephen's nephew, J. K. Stephen, President of the Cambridge Union when Whitehead joined it, and the wittiest speaker since Maitland. He was already one of the "Apostles"; Whitehead would get to know him well in that group, which will be my subject in Chapter VII. Let me say a few words here about the three friends of young Whitehead who were not Apostolic brothers but are mentioned in the "Autobiographical Notes": William Ritchie Sorley, D'Arcy Thompson, and Henry Head.

Thompson and Head entered Trinity with him. Sorley had come two terms earlier to read for the Moral Sciences Tripos, which was the preparation for a career in philosophy. Among Whitehead's fellow-students, Sorley was probably the first with whom he had many conversations that were sure to be mainly about philosophical questions. Whitehead was the learner: Sorley was five years older, and had been a university student since entering Edinburgh at fifteen. After taking his baccalaureate (in philosophy and mathematics) he had spent several years studying theology there and in Germany before he decided to devote himself to philosophy instead. He soon became much concerned to demolish the ethics of naturalism, something to which the theory of evolution had given new life. In 1883, when Whitehead was finishing his undergraduate work, Sorley won a Trinity fellowship with a dissertation on this subject. The philosophy that attracted him was Idealism[4]—but not Absolute Idealism.

That speculative conception of an all-embracing Mind or Spirit in which every bit of finite experience finds its place and fulfillment was the most natural terminus of idealist philosophy. Sorley's mind was simpler and more limited by tradition. His interest, early and late, was in the connection between moral values, persons, and God. In the end, he produced a personalistic metaphysics, commonly classified as personal—rather than Absolute—idealism. Whitehead nowhere refers to it. But the conviction with which Sorley approached metaphysics is another matter. It was that our chief clue to the nature of ultimate reality was to be found not in Nature as studied by science nor in the knowledge-process as analyzed by idealists, but in moral values and ideals. If you look for your clue not so much in moral as in aesthetic values and ideals, you will be following in Whitehead's steps.

Sorley succeeded Henry Sidgwick as the professor of moral philosophy at Cambridge, delivered Gifford Lectures on "Moral Values and the Idea of God," and is still mentioned in encyclopedia articles on moral arguments for the existence of God. Otherwise, he has long since been forgotten. I do not suppose that Sorley's writings were of any importance to Whitehead when he turned to philosophy. It was the conversations of the early 1880s, some years before McTaggart came to Trinity College as a student, that played a role, by giving Whitehead his first personal education in philosophy, and particularly in a species of idealistic philosophy.

D'Arcy Thompson and Harry Head were reading for the Natural Sciences Tripos and concentrating on biology; a great day for it had begun at Trinity, under Michael Foster and young Frank Balfour.* Thompson became a great zoologist, with a keen interest in the greater problems of biology and a novel tendency to think of biological phenomena in mathematical terms. He was a year older than Whitehead, a big, genial, red-headed man from Edinburgh. While an undergraduate he translated Hermann Müller's treatise on the fertilization of flowers by insects. Darwin greatly admired that book, had wanted it translated, and wrote a Preface for Thompson, two months before he died, in 1882.[5] When Whitehead discussed philosophical issues half a century later, he often picked his illustrations from something he had learned about at Cambridge in the eighties. Thus, when he ridiculed the positivist doctrine that we should be content with the observed facts instead of entertaining theories, he did so by applying it to the observed fact that insects visit flowers. "It is a fact of charming simplicity. There is nothing further to be said upon the matter, according to the doctrine of a positivist."[6]

*Francis Maitland Balfour (1851–82), Thompson's mentor, was a younger brother of the future Prime Minister. He enjoyed a laudable reputation among European biologists. When he was thirty Cambridge established a Chair of Animal Morphology for him, but before he could occupy it, he died attempting to scale Mont Blanc.

Head's family were East Anglian Quakers. He came to Trinity from six months at the German University of Prague, and after his baccalaureate would return there for physiological research before beginning medical studies at Cambridge. He would then be, for forty years, both a consulting physician (consulted by the Whiteheads) and a research scientist in neurology. He became, among other things, the authority on aphasia. Harry Head often shocked his companions with wild ideas: his imaginativeness was matched only by his passion for exact knowledge. He was also, even as an undergraduate, a man of wide culture.

When philosophers read Whitehead on sense perception, or on the mind-brain problem, Head is the first scientist of his time that they should think of. Thompson's magnum opus, *On Growth and Form,* is one of the scientific books to have in mind when reading *Process and Reality*.

Whitehead's friendships with Thompson and Head gave him an opportunity to glimpse some coming trends in biological thought. It is unlikely that as an undergraduate he did any reading to speak of in biology. The point is that his education outside the classroom was not limited to the humanities.

Whitehead's time of coming to Cambridge was fortunate in another way. The weekly *Cambridge Review* had been founded the year before. The brightest young men in the humanities, and a few scientists, spoke in its columns. Some of the contributors and editors were advanced undergraduates, more were recent B.A.'s or young dons. The book reviews were numerous and good. Most of the poetry was light verse about life in Cambridge, but the *Review* also printed Karl Pearson's translations of "Sozialdemokratische Lieder." There were letters from India, informed discussions of Spinoza, Hegel, Thoreau, Ruskin, etcetera, and disputes on the definition and on the spread in Cambridge of positivism and atheism. The *Review* performed well as Cambridge's "journal of university life and thought."

The Trinity College Chapel had an essential part in young Whitehead's life. It was not as beautiful as King's, but it was handsome; and there was good music by a paid choir, though the acoustics were abominable. In the antechapel, as Whitehead knew from Book III of *The Prelude*—before his years at Trinity began—stood Roubillac's great statue of Newton. He greeted the "silent face" at least a hundred times a year, for attendance at Chapel was obligatory: not every day, to be sure, but on two days of the week, and twice on Sundays (wearing surplices). Markers were present to ensure the satisfaction of this requirement, from which only students who declared that they were Roman Catholics or Nonconformists were excused. Whitehead did not feel personally oppressed by this rule, but he probably doubted its wisdom. Wordsworth had written in *The Prelude:*

> The witless shepherd who persists to drive
> A flock that thirsts not to a pool disliked.

There had been other protests against compulsory Chapel, but they were ineffectual. The issue would not be hotly debated for two more decades.

vi

An entrance scholarship at a Cambridge college was but a stepping stone toward the real prize, becoming a Scholar on the College Foundation. At Trinity in 1880 there were 74 of these. Each received about £100 annually, not merely until he took his B.A. degree but for three years after, whether or not he stayed in residence. Whitehead could compete for a foundation scholarship (also called a major scholarship) in Easter Week of his first year. Before he did that, he took a crack at a University scholarship that had fallen vacant, the Abbott. It would run for only three years, but it paid a third more than his entrance scholarship, which would expire in two, and it bestowed prestige—there were only two Abbott scholars at a given time. The advantage of standing for the Abbott was that its examination was held five weeks before the Trinity College scholarship examinations; Whitehead could again play it safe. His father had a part in this decision: by the terms of John Abbott's bequest, this scholarship was for "sons or orphans of Clergymen in the Church of England who stand in need of assistance to enable them to obtain the benefit of a University Education." Apparently the family's financial security had slipped enough to enable the Reverend Alfred Whitehead to say that after sending one son through Cambridge and another through Oxford, he qualified for this help in educating his third.

The examination for the Abbott took up five mornings and afternoons in the middle of March. Whitehead was elected Abbott Scholar on April 1, 1881. This did not make him ineligible for the Trinity scholarship examination; indeed, it would have been thought odd if he had been content with the Abbott. And it was not in character for him, or for any Whitehead, to rest on his laurels. The papers he wrote in Easter Week were good enough to secure the conversion of his minor scholarship at Trinity to a major one.

The labor of writing examination papers continued. In the first week of June, five half-days were given over to Trinity's examination of its freshman honor students. Although this examination, as I said, did not count—only the Tripos, two years ahead, would count—it was given all the importance that a milestone can have. The mathematical students were ranked by the marks they made, and grouped into six classes. Each man who made the First Class was awarded a prize of books worth £35. Whitehead ran true to form: a First Class, but not first in the First Class; he ranked fourth among the seven prizemen. The generous gift was especially welcome, for the use of the College and University libraries by undergraduates was quite restricted in those days.

When he began his second year at Cambridge, Whitehead's situation was solid. If his expenses remained average, the two scholarships would take care of about five-sixths of them for the next three years.* Admission to the select company of scholars on the foundation at Trinity also did a lot for him intellectually. At each seating for dinner, the college reserved a table for them; here Whitehead could talk with Sorley, Thompson, Head—in fact, with the brightest young men in all fields of study.[7] Some of them were already B.A.'s, working on their dissertations for fellowships. The Scholars' Table, later discarded, was characteristic of the meritocracy in which he lived.

vii

In this second year, friendships with dons grew. The last quarter of the nineteenth century was the period in Cambridge's history when the student-don relation was happiest. At the heart of the modernization and revitalization of the University that quickened about 1860 was the rise of a new generation of dons who were real teachers. They were professional scholars or scientists who gave a professional education to the honors men. But they also opened their rooms to students at all hours and made them feel important. Many of the dons came close to fulfilling the informal educational ideal of a college.[8] Some with large sets of rooms delighted in evenings when they were

---

*According to the *Student's Guide,* the average cost of a year at Cambridge was £200. The charges to Whitehead entered in the bills books of Trinity College show that his spending was in the average range.

crowded with undergraduates, and not only those of their own college. The fantastic Oscar Browning of King's was one such; "O.B.," though he hated mathematics, was the most overpoweringly genial of those who got young mathematicians and classicists excited about modern history, music, and so on. The top Greek scholar, Henry Jackson of Trinity, was another; "the least donnish of dons," he was the most Socratic character in Cambridge.

The autumn of 1881 brought some remarkable freshmen to Cambridge. No Trinity man kept a bear in his rooms, as Byron had, but Connop Williams once rode a pony up the stairs into his rooms. Williams eventually achieved a fellowship, but could not fulfill the promise he showed; he went mad. Another gifted freshman, who died in early middle age, was Arthur Llewelyn Davies. His father was John Llewelyn Davies, Chaplain to the Queen, translator (with Vaughan) of Plato's *Republic,* admiring friend of Uncle Henry Whitehead, and himself one of the most active reformers among the Broad Churchmen inspired by F. D. Maurice. His sons all came to Trinity College on scholarships, took the Classical Tripos with flying colors, and eschewed holy orders. Arthur did something Whitehead would have loved to do but did not: he played Rugby for Cambridge.* All the children of John Llewelyn Davies came into the lives of Whitehead and his wife—the youngest sons, Crompton and Theodore, most of all. But that is a later story.

Whitehead had more friends at King's College than at any other except Trinity. Two men of genius who were his friends throughout their lives entered King's in October 1881: Walter Raleigh and Goldsworthy Lowes Dickinson. Something will be said about them in Chapter VII.

In every college there was a small debating society. Trinity had two—the prestigious Magpie and Stump,† and the Irregulars. Being in any way irregular was not in character for Whitehead. He was elected to the Magpie and Stump a few weeks after this autumn term began. It met on Friday evenings, in members' rooms. Usually, about ten men spoke to the motion of the evening, which might be about anything *except* religion; the business of the Magpie and Stump was

---

*A solicitor, Arthur Davies married George du Maurier's daughter, and when he died had five sons, whom James Barrie adored. When Arthur's widow died, the boys were practically adopted by Barrie, whose dramatic imagination gave birth to *Peter Pan.*

†The Magpie and Stump had been founded fifteen years earlier. Two histories of it have been written—one by J. D. Duff in 1924, and one by C. L. Ferguson in 1931.

debating for fun.* Among the topics debated in 1881–82 were the Government's Irish policy, abolition of the House of Lords (a favorite in future years), the desirability of maintaining the system of corporal punishment in public schools (carried by a large majority), and the desirability of a gradual separation of England and India (defeated by one vote). But when it was D'Arcy Thompson's turn, he moved "that a liberal indulgence in Alcohol is conducive to Physical and Intellectual improvement," and got six votes out of twenty-six. The motion that Whitehead brought forward this year was less outrageous, but he won only two votes (out of seventeen) for it: "that horse-racing is on the whole beneficial to the nation."†

Whitehead at twenty-one was not a natural debater. For that matter, he never acquired the kind of self-assurance that the superior debater needs. And the frequent hesitation in his voice that some have called a slight stammer, though not a disadvantage in a lecture, was one in debate.

viii

In the foregoing account of Whitehead's first two years at Cambridge, women have not been mentioned. Cambridge University was then as completely a male society as a public school was. Perhaps more so. Schoolmasters were usually married men with families, but Cambridge dons were not permitted to marry. Until the bars were let down by the statutes of 1882, a don vacated his college fellowship upon marriage, *unless*—how specific the law was!—he was the head of a college, or a professor in the University, or the University Librarian, Registrary, or Public Orator. As there were only seventeen heads of college and twenty-seven professors, the number of University families was tiny in comparison with the number of male students. J. J. Thomson reached his estimate, "not more than sixty," by including local clergymen, doctors, and lawyers who were members of the

---

*The Magpie and Stump had a rule that each member must speak at least two minutes in debate each term or pay a fine of one shilling. That was not enough to force a shy man to speak. The penalty for a year's silence, however, was expulsion. A few years later Stanley Baldwin (elected as a freshman) was thus expelled from the Magpie and Stump; he was ceremoniously restored to the privileges of membership upon becoming Prime Minister in 1923.

†The Ordinances of Trinity College prohibited students from riding in a horse race or otherwise promoting horse racing.

university, that is, alumni. Except for them, these sixty families included few young men and women, since they were the result of marriages made rather late in life. And there were not near Cambridge, as there were near Oxford, residential areas inhabited by gentlemen's families. Of course one had nothing to do with the families of tradesmen. One's acquaintance with young women was practically confined to University families, and restricted by the requirement of chaperonage.[9] A town girl on the street with a student could be arrested by a Proctor, tried in the Vice-Chancellor's Court, and imprisoned in the University's jail; this was one of the many legal powers that the University held throughout the town.

It will fill out our sense of the distance between the 1880s and our age if we observe that some dons founded a Cambridge University Association to Promote Purity of Life, that is, continence outside marriage. Whitehead did not need their urging.

The students at Girton and Newnham, strange young women, were carefully kept from dangerous proximity to men. Girton College was first located in Hitchin, 26 miles away. In 1872 the idea of moving it into Cambridge was broached, but abandoned; a site two miles outside the town was close enough. Since then, Newnham College had been started just back of the Queen's Road. Both colleges, however, were self-contained; they provided in their own halls all the lecture courses that their students needed to get a Cambridge type of formal education. In the University precincts, some professors allowed women to come to their lectures; women might also attend some advanced courses that were offered jointly by several colleges. Thus Whitehead in his third year as an undergraduate saw a few female students of mathematics in his classes. They did not sit beside the men; they sat in a back row with their chaperon and her knitting. The "Journal of University Life and Thought" ignored women, and none acted in the excellent performances of the University Dramatic Association.

The women's desire to be allowed to show themselves the intellectual equals of Cambridge men was formally satisfied when the Tripos examinations were opened to them in 1881–82.* But no woman could be a Wrangler; her sole reward was the announcement that her score equaled that of the nth Wrangler. Girton and Newnham were not, like women's colleges in the United States, degree-granting in-

*The date in 1881 on which the Senate of the University acted became a day annually celebrated at Girton.

stitutions, and their students were not permitted to take Cambridge degrees until 1948.*

The custom of "dating" did not arise until much later. The University did not allow a theater to open in Cambridge until 1896. There *were* concerts, very good concerts; respectable mixed parties could be arranged for hearing Joachim's string quartet. And in the second week of June, called May Week, there were a few college balls in the Guildhall, and festivities on the river. You could then enjoy the company of a friend's chaperoned sister who came to Cambridge. But May Week was not as big a thing as it became in the nineties.

Of course, some Cambridge men had homosexual relations; the time when they could be open, even fashionable, was far off. I do not think anyone is going to find a homosexual relationship in Whitehead's life.

---

*Competition for University scholarships and prizes was not opened to women until 1928.

CHAPTER VI

Mathematics at Cambridge

i/Mathematical staff at Trinity: J. W. L. Glaisher, H. M. Taylor, W. D. Niven. Professors Stokes and Cayley. E. J. Routh.

ii/Ordeal of the Mathematical Tripos: Whitehead, Fourth Wrangler.

iii/The advanced part of the Tripos: Whitehead, a First Class.

iv/Fellowship dissertation on Maxwell's *Treatise on Electricity and Magnetism.* Whitehead's election to a Trinity fellowship, 1884.

Whitehead wrote his "Autobiographical Notes" at the behest of an American editor, for American readers interested in him as a philosopher. Hence it is not entirely surprising that in so short an account—only a dozen pages—he said nothing about what he did as an honors student of mathematics at Cambridge in the early 1880s. The coaching system is not mentioned, and the series of competitions that marked his progress, from the entrance scholarship examination to the culminating Tripos, is passed over without a word. He would not likely have said anything about that to any audience: one of the extraordinary things I *like* about Whitehead, hard though it is on a biographer, is that he just did not write about his trials, disappointments, and triumphs. It is a pity, however, that he was silent about who his teachers were and what they gave him. His undergraduate education in mathematics seemed to be something of a merely personal interest that had become irrelevant to him, as it was doubtless irrelevant to the readers of the "Autobiographical Notes." But it is an essential part of the story of a life which, for forty years, was that of a professional mathematician. I shall now try to fill this gap.

The education in mathematics that Whitehead got as an undergraduate was better than we might suppose it could have been in nineteenth-century England. During the first two-thirds of the century (very roughly speaking) little mathematical research was carried on at Cambridge, and few college lecturers thought of a branch of mathematics as much more than a source of problems for their students to solve in examinations. But by 1880 the situation had changed greatly. Anyone who gives himself the pleasure of reading J. J. Thomson's, A. R. Forsyth's, or Karl Pearson's recollections of their student years in Cambridge[1] will see that the state of mathematics, and of its teaching, had become quite creditable.* They took their

*Cambridge was still far behind Göttingen and Berlin, especially for the study of pure mathematics. The cause of Cambridge's inferiority to Continental universities was the Mathematical Tripos (discussed in section ii of this chapter). This is emphasized in recollections written by Bertrand Russell more than once (see *MPD,* pp. 37ff.); G. H.

degrees in 1880, 1881, and 1879, respectively. No Cambridge mathematician whose undergraduate years were identical with Whitehead's seems to have written memories of them, but Trinity's staff was the same as it was for Thomson and Forsyth, both of whom were Trinity men. Both, and Pearson too, were coached for the Tripos by the same man, E. J. Routh of Peterhouse, who (I think) coached Whitehead.

Let us look first at the Trinity staff. It consisted of seven lecturers, at least three of whom were real mathematicians. From each of them Whitehead got something of special value. They were J. W. L. Glaisher, H. M. Taylor, and W. D. Niven.

Glaisher, twelve years older than Whitehead, was the bright light at Trinity. While still an undergraduate he had written a paper that Cayley sent to the Royal Society for publication in its *Transactions*. He was elected F.R.S. a few years later. Would he some day succeed Cayley in the Sadleirian professorship of pure mathematics? He did not; he was more a stimulus to others than a pioneer. In Whitehead's student days he was Editor of the *Quarterly Journal of Pure and Applied Mathematics,* President of the Cambridge Philosophical Society, and the most active promoter in Cambridge of research in pure mathematics. (He had founded the *Messenger of Mathematics* as a journal for the publication of research by undergraduates.) Glaisher was the teacher who was afire with the beauty of mathematics itself—hang the competition of the Tripos. Thomson tells how fascinating his lectures were, and how they made him for a time quite enthusiastic about elliptic functions.[2]

Outside the classroom Glaisher was a good listener and a good talker, with a humor that Thomson calls kind but discriminating. His published papers deal with elliptic functions and a variety of other topics in pure mathematics and theoretical astronomy. When Glaisher died in 1928 G. H. Hardy, whose standards were very high, concluded that he was "underestimated as a mathematician. He wrote a great deal of very uneven quality, and he was old-fashioned, but the best of his work is really good."*

Hardy (in *A Mathematician's Apology* [Cambridge: Cambridge University Press, 1940]); and W. H. Young (see I. Grattan-Guinness, "University Mathematics at the Turn of the Century: Unpublished Recollections of W. H. Young," *Annals of Science* 18 [1972]: 369–75). Young was Fourth Wrangler in 1884. I agree with these authors about the Tripos, but would warn against jumping to the conclusion that no budding mathematician could gain anything from studying mathematics at Cambridge.

*J. W. L. Glaisher was born to research. His father was James Glaisher, F.R.S., a

H. M. Taylor, Whitehead's tutor, was, like him, the son of a clergyman-schoolmaster. Nineteen years older, he was an earnest, modest, conventional Victorian. He appears in Gwen Raverat's *Period Piece* as the "Mr. T." who in 1883 wooed her mother, a beauty from Philadelphia; he had money, virtue, and good looks (with the flowing beard that was then admired); but she found him too short, and dull company. Being rejected in favor of George Darwin, he never married, but lived alone in the large house he had rashly bought. In his fifties Taylor became blind. Relieved of college responsibilities, he did his best mathematical research and was elected to the Royal Society; he also developed and extended Braille for use in mathematical textbooks; and he exercised his administrative abilities as Mayor of Cambridge. An admirable man.

Taylor was the College lecturer on geometry, which became Whitehead's favorite among the branches of classical mathematics. I cannot say that Taylor was responsible for this; some experts believe that the reason why some mathematicians lean toward geometry and others do not is to be bound in brain structure. However, accounts of Taylor mention his strict standard of verbal and logical accuracy, and it is likely that Whitehead's insistence on this in his mathematical writings owes something to him. Also, Taylor is known to have imparted enthusiasm for hydrodynamics,* the only branch of applied mathematics on which Whitehead published papers.†

W. D. Niven, who would be elected F.R.S. in 1882, was the same age as Taylor. He gave Whitehead the most valuable part of his education in mathematical physics. Except for Cayley, the Cambridge of

---

distinguished meteorologist who was famous by virtue of his balloon ascents to make observations. He named his son James Whitbread Lee Glaisher in honor of his colleagues in the founding of the British Meteorological Society, S. C. Whitbread and John Lee. The son, like Whitehead later, became President of the Association for the Advancement of Science's section on Mathematics and Physics. And, like Whitehead, he had a long career at Cambridge but never became a professor. Thomson laments his decision, at the height of his powers, to accept a ten-year position as one of the four tutors at Trinity, which made his research subject to continual interruption. When Cayley died in 1895 the Chair of Pure Mathematics went instead to Whitehead's friend Forsyth. (To be sure, Glaisher was not in Cayley's class; neither was Forsyth, or anyone else in Britain.) Glaisher then made himself an authority on faïence and pottery; he bequeathed his collection—the finest anyone had assembled—to the Fitzwilliam Museum.

*Horace Lamb caught it and went on to write the authoritative treatise on the subject, which he dedicated to Taylor.

†His papers are described in Chapter XII, section ii.

1880 was far behind Europe in pure mathematics—but not in mathematical physics. Stokes, Adams, and Rayleigh lent distinction. And Cambridge was ahead of Europe in the area that had become crucial. Clerk Maxwell had arrived in 1871, and published his great *Treatise on Electricity and Magnetism* two years later. Though the work sold widely in Cambridge, Maxwell's theory was slow to gain acceptance even there. His professorial lectures had been of little help toward understanding his formidable book, and he was such a poor lecturer that very few had gone to hear him—two in 1878–79. After his death late in 1879, when he was only forty-eight, there was naturally a marked increase of interest in his magnum opus. This was maintained by a course of lectures, mainly on it, that Niven gave for two terms each year. Niven had been Maxwell's pupil and close friend; he later edited both the second edition of the *Electricity and Magnetism* and Maxwell's *Scientific Papers*. His lectures were public and attracted practically all the mathematicians of the university.

Besides dons, advanced undergraduates came to hear Niven; attention was called to his lectures, alone among nonprofessorial ones, in the 1880 edition of the *Student's Guide*. Thomson's memory of them is worth quoting.

> Niven was not a fluent lecturer nor was his meaning always clear, but he was profoundly convinced of the importance of Maxwell's views and enthusiastic about them; he managed to impart his enthusiasm to the class, and if we could not quite understand what he said about certain points, we were sure that these were important and that we must in some way or other get to understand them. This set us thinking about them and reading and re-reading Maxwell's book, which itself was not always clear. . . . The best teacher is not always the clearest lecturer but the one who is most successful in making his pupils think for themselves, and this Niven . . . certainly did.[3]

Whitehead's thinking for himself about Maxwell's *Treatise* led him to choose it as the subject of the research dissertation he wrote for his Trinity College fellowship in 1884. Niven was then in Greenwich, having been appointed Director of Studies at the Royal Naval College in 1882. The rest of his career was spent there, but he kept up his friendships with his Cambridge pupils.[4] Whitehead was fortunate to have been one—probably in 1881–82.* This intimate acquaintance

*His command of mathematics was then quite adequate for the purpose. It is possible that Whitehead also heard Niven on Maxwell in 1882–83. Probably Niven commuted

with Maxwell's theory of electromagnetic phenomena, such as Whitehead could have got from no one else, was important for all his future work on the philosophy of physics. Maxwell had tried to explain all the known facts on mechanical principles, and emerged with his beautiful set of equations of the electromagnetic field. Early in the first course of lectures that Whitehead gave to students of philosophy—at Harvard in 1924—he essayed an untechnical exposition of Maxwell's equations.[5]

None of the other four members of the Trinity College mathematical staff in 1880–83 are credited with any mathematical achievements. One, Glaisher's contemporary W. W. Rouse Ball, wrote primarily on the history of mathematics,[6] which English scholars had neglected except for Newton's great role; he also produced a fine book of mathematical recreations and puzzles, which became deservedly popular.[7] Whitehead became much interested in the history, less in the recreations. He found Cambridge—Trinity especially—intensely aware of its glorious past. Rouse Ball was the prime authority on Newton, though not the only one; H. M. Taylor wrote the article on Newton for the great eleventh edition of the *Encyclopaedia Britannica*. Ball's best book is perhaps *A History of the Study of Mathematics at Cambridge*.[8]

Nothing need be said about the three members of the Trinity College mathematical staff not yet named: Joseph Prior, Richard Appleton, and Thomas Dale.* Among the Cambridge professors, Stokes and Cayley come into the Whitehead story.

From 1849 until his death in 1903, G. G. Stokes of Pembroke College held the chair that once was Newton's; he had been Kelvin's teacher. College tutors, who practically forbade their mathematical honors students to attend professors' lectures, urged them to go to Stokes's, which in the early 1880s were given on either light or hydrodynamics, in the Easter term. J. J. Thomson's recollection of the lectures on light shows why they were recommended—and how very different things were a hundred years ago.

---

to Cambridge to give his lectures then. They were announced for the last time, in both the *Cambridge University Reporter* and *Trinity College Notices*. No other Trinity Lecturer covered electricity and magnetism, which had been listed among the subjects to be included in the most advanced Part of the Tripos Whitehead would take.

*"Little Joe" Prior was most successful as a tutor. Appleton moved from a mathematical to a theological lectureship. Dale did not live long.

> For clearness of exposition, beauty and aptness of the experiments, I have never heard their equal. He had only the simplest apparatus at his command, no light but that of the sun, no assistant to help him. He prepared the experiments himself before the lecture, performed them himself in the lecture, and they always came off.[9]

I have found no record of Whitehead ever participating in an experiment, and his name is not on any list of those who went to the Cavendish Laboratory. His acquaintance with physics as an experimental science was acquired in Stokes's classroom.

In the 1880s Arthur Cayley was without doubt the greatest mathematician alive in England. He had been turning out papers for forty years, even when he was practicing law; thirteen large volumes were needed to publish them after his death. More important, he was the chief creator of several new branches of pure mathematics, such as matrix algebra, the theory of invariants, and hypergeometry. Clerk Maxwell called him the man

> Whose soul, too large for ordinary space,
> In *n* dimensions flourished unrestricted.

Whitehead, in writing about mathematics, always emphasized its pursuit of the utmost in generality; when he came to write about the proper method of metaphysics, he incorporated this characteristic.

In 1941 I asked Whitehead whether Cayley was one of his teachers at Cambridge. He answered that he was not, but that this meant nothing, for he "knew him, used to have lunch with him, etcetera." I did not then know that Cayley lectured but one term out of three, and only on the problems that currently interested him. Nor did I know how Cayley lectured—not at a blackboard, but seated at one end of a table, with a quill pen and some large sheets of paper, a few young dons beside him. Small wonder that the undergraduate Whitehead did not attend. Cayley was a Fellow of Trinity, and Whitehead's important friendship with him belongs mainly to the years when Whitehead was a young Fellow. But Cayley was an important part of the Cambridge in which Whitehead studied.

Reading descriptions of Cayley the man reminds me of Whitehead: constant work, great modesty, perfect courtesy, and dislike of controversy. A considerable facial resemblance has been pointed out to me. You need only give Cayley's nose a little aquilinity when comparing photographs.

We come now to the man from whom Whitehead got most of his mathematical training—his coach. In all probability he was coached by Dr. E. J. Routh, F.R.S.—the greatest in the series of great coaches that stretched back a century and a quarter. In the eighteen years preceding Whitehead's arrival in Cambridge, he had trained most of the top Wranglers and *all* the Senior Wranglers. He had himself been Senior Wrangler in the year when Clerk Maxwell was Second. Almost all of the abler students now went to him as a matter of course. Routh was the chief teacher of a whole generation of Cambridge mathematicians.

Coaching arrangements were strictly private, and no list of Routh's pupils from 1880 to 1883 seems to exist.* Three items of indirect evidence make it reasonable to believe that Whitehead was one of them. The Preface, dated November 4, 1882, to a new edition of one of Routh's mathematical works[10] thanks Whitehead and E. G. Gallop (another Trinity honors student who was beginning his third year) for their help in correcting many of the proof sheets; this suggests that Whitehead was one of his best advanced pupils. When Routh died in 1907, Whitehead was one of thirteen who represented the Royal Society at the funeral service in the University Church, Great St. Mary's.[11] Many years later, explaining his reservations about Einstein's theory of relativity, Whitehead employed "a metaphor used by Dr. Routh in lecturing on metacentres."[12]

Coaching with Routh was quite different from anything done today, or done at that time outside Cambridge. It was very far from cramming. He divided his crowd of pupils into classes of eight to ten according to their abilities, and three times a week gave each class lectures that by common consent were the best in Cambridge on Tripos subjects. J. J. Thomson, who could ill afford Routh's fee of £36 a year, wrote, "Though no doubt most of us went to Routh because we thought that we should get a higher place in the tripos by doing so, the teaching itself was well worth the extra expense."[13] The lectures were methodically planned to impart to a would-be Wrangler, in the course of his years with Routh, all the mathematics that could be useful in the examination. That, however, was only part of Routh's system. Between meetings the men had to solve a group of hard problems. At the beginning of each lecture Routh dealt rapidly with the work handed in, noting its imperfections from the standpoint of the Tripos. The better students were rewarded by being required to write a weekly paper consisting of a dozen more problems. The

*The time when money was slyly deposited in an inconspicuous place in the coach's room was past, but the time when Trinity would collect from its students on behalf of their coaches had not yet arrived.

Cambridge name for a coach's pupils—"beasts"—was supremely apt for Routh's.

Both Thomson and Karl Pearson published several pages in which Routh is remembered with great appreciation.[14] He was a modest, friendly man; he did not limit his mathematical talk with Pearson to Tripos requirements. Part of Thomson's description of him may remind us of Immanuel Kant:

> The regularity of Routh's life was almost incredible; his occupation during term time could be expressed as a mathematical function of the time which had only one solution. . . . The fact that year after year he gave the same lectures at the same time did not make him stale as it would most people. . . . His way of taking exercise was as regular as his lectures: every fine afternoon he started at the same time for a walk along Trumpington Road; went the same distance out, turned and came back. His regularity was not, as might perhaps have been expected, accompanied by formal and stereotyped manners; these were very simple and kindly and we were all very fond of him.[15]

In term time, Routh taught up to six classes daily. It was a marvel of physical endurance. The added marvel was that in the vacations he produced some first-rate contributions to applied mathematics: a fundamental one in dynamics was published three years before Whitehead came to Cambridge.[16] The book on which Whitehead helped with proofreading, originally a textbook, was a definitive cyclopedia of the dynamical part of theoretical physics.

Glaisher and Taylor made real contributions to Whitehead's mathematical education, but Routh and Niven gave him the most.*

---

*In the Whitehead corpus there is just one recollection of a teacher, and this was written not by Whitehead but by Lucien Price, in *Dialogues of Alfred North Whitehead*. Price reports Whitehead as saying to him on June 19, 1943 (dialogue XXX): "I went up to Cambridge in 1880. For a boy of nineteen I was a good mathematician. My teacher was a pupil of Clerk Maxwell, who had died only a year or so before, and he, too, was eminent." From the first two sentences one would suppose that Whitehead was about to refer to the excellent mathematics master he had in his last year at Sherborne School; but that master was neither a pupil of Maxwell's nor eminent in the required sense. "My teacher" must mean someone at Cambridge. Anyone who studied there in those times would naturally mean by his teacher, his coach. In Price's manuscript (now in the Columbia University Library) "J. J. Thomson" is written after "my teacher," but is crossed out. Thomson did some coaching in the early 1880s, and in *R&R* (p. 79) he names two of the men he trained; if Whitehead had been one, he would surely have been named. Also telling is the fact that Thomson was not a pupil of Maxwell's. Whitehead followed the remark Price quoted with a reference to Maxwell's work, and then to what was being done in electrical theory at Cambridge a decade later; it is in this connection

ii

Whitehead gave Trinity College good evidence of his progress toward wranglership when he took the annual college examination for honors students in the spring of 1882. He was again a Prizeman, again rewarded with books worth £35; but this time he was not in the middle of the first class, he was almost at the top. The five-day examination was set up, like the Tripos, as a writing race. There were so many problems to be worked that no mere human could come anywhere near making a perfect score. That would have been 1,400 points; no one but Whitehead (with 735) and Gallop (with 774) topped 600.

In Whitehead's third undergraduate year, the one important event was the Tripos Examination at its end. The preceding term was always devoted to reviewing everything that had been learned, and to endless practice in solving problems. It was especially in this term that the *Cambridge Review* expected to receive letters demanding a stop to the Monday evening hour of ringing the big bells of the University church. How could the pleasure of a few bell-ringers outweigh the havoc done to the mental concentration of many? (Every place in Cambridge is hard by every other place.) Tripos men ordered tea and wet towels from their gyps, while others looked forward to boat races between the colleges and the social activities of May Week. The examinees, however, were not the only men keenly interested in the outcome of the mathematical tripos. The waiters at Trinity College ran sweepstakes on it.

The race began at 9:00 A.M. on Monday, May 28; after three hours, a break for lunch, then two and one-half hours more. This was repeated on Tuesday and Wednesday, when Part I of the Tripos concluded. A week later, the class list was published; everyone who had passed qualified for graduation. But the real contenders had run only the first mile of a two-mile race. Part II began on Thursday morning, June 7, and ended on Saturday afternoon.

The subjects covered in Part I, "the first three days," were the same as they had been for generations: Euclid and the elementary parts of

---

that he would have mentioned Thomson. A third interpretation is more promising. When remembering Maxwell's work, Whitehead would think of Niven, who was his teacher on it, had been Maxwell's pupil, and was himself eminent. Finally, and most probably, Whitehead may have had his coach in mind. Routh was eminent and was associated in the minds of Cambridge men with his exact contemporary, Maxwell; by some slip their relation could have been transformed in Price's notes to a teacher-pupil relation.

algebra, plane trigonometry, conic sections, dynamics, hydrostatics, optics, and astronomy. And, of course, certain basic sections of Newton's *Principia*. The traditionally required part of Scripture was "Newton I, II, III," that is, the first three sections of Newton's Book I. The well-worn volume that every student kept on his shelf was not Motte's translation of the *Principia* but an edition of "Newton I, II, III" put out with notes, illustrations, and a collection of problems by Percival Frost of St. John's College a few years earlier.* The candidate had to be able to state and prove any proposition or lemma whose number was given, then quickly solve a problem involving it—the "rider." He was limited to the purely geometrical methods of the *Principia;* throughout Part I of the Tripos, use of the calculus was prohibited. This was Cambridge's way of insisting on mastery of elementary mathematics.

In Part II of the Tripos, Whitehead had to show his mastery of the differential and integral calculus, spherical trigonometry, analytic geometry of three dimensions, the easier parts of differential equations, and advanced parts of the subjects of Part I. To us that does not sound difficult; what we should call advanced mathematics was reserved for Part III, which those wranglers who hoped to be professional mathematicians would take in January. It was the insistence on speed that made every Part of the Tripos an ordeal. The man who had to stop and think about the book work would not get far; his fingers ought to be dispatching it while he was thinking about the rider. Cambridge, in all departments, traditionally emphasized the kind of knowledge that can be precisely tested, and exact knowledge was her ideal. In mathematics she assumed that what you really *know,* you can apply instantly and surely. Not the understanding that underlies research, but *training,* was what the mathematical student was given, and had now to demonstrate. It was not part of any coach's business to arouse enthusiasm for mathematical research. To a pupil who had a bit of that, Routh gave the solid and substantial knowledge to serve as its basis, and practice in tackling problems with precision and finding the shortest proofs. If any teacher had started to spout about creative thinking, in the fashion of the twentieth century, he would have been laughed at. The Almighty in His wisdom—and mercy—could no doubt have put into the minds of academic leaders ways of upholding the ideal of

*Euclid's *Elements* also were studied, in a recent Cambridge text, the excellent one issued, with notes helpful to students, by Isaac Todhunter of St. John's twenty years before.

exact knowledge other than the Tripos way, with its glorification of competitive speed. But competition had always been the rule at Cambridge. The *Ordo Senioritatis* was one of the most ancient institutions in academia. The chief factor in ranking the graduates used to be prowess in disputations (carried on in Latin). In an age of mathematical science, this gave way to speed in figuring out and writing down solutions of mathematical problems. Leslie Stephen—a Cambridge man through and through—wrote in 1903, just before the end of a life devoted to literature, history, and biography, that to him "Senior Wrangler" was "still a title exciting an almost superstitious veneration."[17] The winner of that title was chaired around his college, to which he had brought the highest of honors.

By 1890 the Tripos system had become an object of deserved scorn from many mathematicians, especially pure mathematicians. It was pointed out that young mathematicians in German universities were not asked exactly how they had placed in an examination, but what they had done or written. True, when Karl Pearson, a working applied mathematician, went to Germany, he was glad to have had Cambridge's training in solving problems.[18] A few others may have felt as he did; but too many allowed themselves to be led by their training for the Tripos to see little in mathematics save a surprisingly effective set of tricks.* And there can be no doubt that an excessive portion of the mathematical talent of Cambridge dons was diverted to the ingenious devising of Tripos problems, many of them artificial in the extreme. When Whitehead, thirty years later, in pleading for the reform of mathematical teaching protested the setting of silly problems to schoolboys, he was protesting something from which he had suffered much at the university level. In his years as a Cambridge don he was active (with Forsyth) in the movement to bring sanity into the study of mathematics by abolishing the order of merit.

Back to June 1883. After Whitehead handed in his last paper in Part II of the Tripos, he would have to wait ten days to learn exactly how he had made out. He could watch the boat races on the last three days of the week devoted to them; he could—and very likely did—hear the President of the Royal Society, Professor Huxley, deliver the Rede Lecture for 1883, "The Origin of the Existing Forms of Animal Life: Construction or Evolution?"; he could do reading he had postponed; he could walk and talk as much as he liked. But Whitehead was by

*Bertrand Russell is the extreme example; see Chapter XI, section i, below. Among Cambridge critics who were more typical, working mathematicians rather than theorists of mathematics, Glaisher and Forsyth may be mentioned.

nature an anxious person, and though he had no dreams of being the Senior Wrangler, he very much wanted to be well up on the list. The ten days and nights of suspense were quite hard on him. There is a story in the family that his friends were so worried that they took turns staying with him every night. This could have happened during the final period of waiting, or in the week between Part I and Part II of the Tripos, or when he took, or waited for the results of, Part III the following January. The story is completely vague, and I cannot recover it as it was first told, no doubt by his wife. I can only say that I suspect it was exaggerated.

At noon on Tuesday, June 19, the contest came to its ceremonial close in a crowded Senate House. The Senior Wrangler was announced, came forward amid deafening applause, and knelt before the Vice-Chancellor to be admitted to the baccalaureate degree. The winner was G. B. Mathews of St. John's College. This was something of a surprise, in that Mathews was the first Senior Wrangler in twenty-three years to be trained by a coach other than Routh. Then the Second Wrangler was named and came forward; he was Gallop. Robert Lachlan of Trinity was third. Whitehead and John Chevallier, also of Trinity, were bracketed Fourth Wrangler. There were thirty-four Wranglers in all. The Second Class honors students ("Senior Optimes") came up next in their order of rank, thirty of them. The next procession, of Third Class men ("Junior Optimes"), also thirty in number, ended with a bit of ritual horseplay: the one ranked lowest, traditionally called the "wooden spoon," on receiving his degree clutched a large wooden shovel-shaped object handed down from the undergraduates' gallery and carried it off in triumph. Finally came the names and places, relative to the men, of the five women who had taken the Tripos. There were special cheers when it was announced that Miss Perrin, of Girton College, had made a score between those of the twentieth and twenty-first Wranglers. Of course, no woman was admitted to the B.A. degree in those days. Even the announcement in 1890 that Miss Fawcett, of Newnham, had outscored the Senior Wrangler failed to change that situation.

Whitehead made some kind of bet with one of his friends about how well he would do in the Tripos. When he won it by coming out Fourth Wrangler, he promptly said that it had been made only in jest; but the loser insisted on presenting him with a set of Sir Walter Scott's works.*

---

*This story came from Jessie Whitehead, to whom her father's Scott eventually passed.

Whitehead was probably pleased to place no lower than Fourth Wrangler. If we look over the honors lists for the half-century preceding the disappearance of the order of merit, we find that G. H. Hardy and several other Cantabrigians who became well known as mathematicians were Fourth Wranglers. Among the Wranglers of 1883, Mathews was the only one who achieved stature as a mathematician comparable to that achieved by Whitehead.

iii

Not much of the summer of 1883 was spent on holiday. Whitehead had finished off his college mathematics; he now had until the second week of January to prepare for the advanced part of the Tripos, Part III. In it order of merit would be eliminated, and performance would be judged simply as First Class, Second Class, or Third Class. One who earned a First had good reason to hope for a career in a university, and would soon head toward the competition for a fellowship in his college. Those who failed to earn a First Class usually settled for a schoolmaster's life, or took their leave of mathematics. The subjects of the examination were arranged in four groups—A, B, C, and D; to earn a First Class, the candidate had to do passably in all four and show "eminent proficiency" in one. This type of examination had been initiated two years earlier, as a result of pleas by reformers who saw the absurdity of examining all honors students in all established branches of advanced as well as ordinary mathematics. In effect, the reform separated what we call "graduate work" from examination for the B. A. In the first year under the new regulations about half of the Wranglers attempted Part III, and less than half of these were awarded a First Class. The University's Special Board for Mathematics had to make it known that the Wranglers were too slow to appreciate Cambridge's intention of setting a high standard for advanced mathematics, and of encouraging those who hoped to be mathematicians to start specializing in the fields that most interested them. The company that took Part III with Whitehead consisted of only about one-third of the June Wranglers.*

For Cambridge, with its Newtonian heritage, mathematical physics was most of mathematics. In Part III of the Tripos, only Group A

*The name "Part III" was dropped a few years later. Thereafter, Parts I and II of the Tripos Whitehead took were called Part I, and the professional examination was called Part II; the two were taken in June, usually one year apart.

consisted of topics in pure mathematics. There is no record to show which group Whitehead chose to concentrate on. We know only that the next term he attended J. J. Thomson's lectures on electromagnetism, and that in the summer he wrote an excellent fellowship dissertation for Trinity College on Maxwell's theory. It is reasonable to suppose that when preparing for his final Tripos examination he paid most attention to Group D, the one that included electricity and magnetism.

The examination, which began January 7, 1884, and ended on the ninth, consisted of two three-hour sessions each day. One of the six sessions was devoted to essays: from a list of twelve topics (three from each group), two were to be chosen. One of the Group D topics was "the electromagnetic theory of light"; another was "magnetic induction in soft iron, and its hydrokinetic analogies"; both topics could be handled well by anyone who knew Maxwell on electricity and magnetism, as I think Whitehead did.

The other five sessions presented the usual host of questions with riders, but also the opportunity to avoid those (about two thirds) that the candidate thought would give him the most trouble. This was a much better type of examination than Parts I and II had been. The coaches did not undertake to train candidates for it. The students were on their own, with such help as they could get from professors' or other open lectures.

Of the twelve candidates who wrote this examination, Whitehead was one of seven who made First Class. His friend Frank Morley had entered upon the Tripos in June with high hopes, and so was disappointed by being named Eighth Wrangler. This was quite a respectable rank, but when Morley failed to get a First Class in Part III he had a breakdown, then took a school job. He escaped from it after three years by emigrating to the United States, where he became the professor of mathematics at Haverford College and later at the Johns Hopkins University. His sons—Christopher, Felix, and Frank—went to Oxford, not Cambridge. (They went as Rhodes scholars.)

iv

The culminating competition for Whitehead was the one Trinity College held early in October 1884. This was the time—just before instruction began in the autumn term—when the College annually elected several recent graduates to six-year "prize" fellowships. Fellows were paid a dividend that was determined annually in the light of

Trinity's investments; in 1884 it was £225. Until Whitehead became a Master of Arts three years later—not a degree to be earned—he would receive only two-thirds of this sum, but that itself would be enough to live on. The winner of a fellowship might do anything he pleased; he could go to London to prepare for the Bar. If he chose to stay in Cambridge, he was given rooms in College and dinners in Hall. When the six years were up, he remained a Fellow if he had become a Lecturer in the College, as Whitehead did.

To Whitehead, the fellowship would not only put him in the company of the sixty Fellows who (with the Master) *were* Trinity College; it would provide a glorious opportunity to do mathematical research. In fact, he would have to show research promise to win it. In the old days prize fellowships had been awarded chiefly on the basis of performance in the Tripos. Trinity was the first college to outgrow that habit and substitute its own written examinations. Trinity now also required—placing the greatest weight on it—a research dissertation, on whatever subject the candidate pleased; the electors could get reports from suitable referees. Eligibility for the prize expired three years after one became a Bachelor of Arts.* A few men succeeded on their first try, more on their second or third; many never succeeded. On the off chance that he might be one of the lucky few, Whitehead entered the competition in 1884. He never liked to postpone facing a challenge.

For the subject of his fellowship dissertation Whitehead, as I said in section i, chose Clerk Maxwell's *Treatise on Electricity and Magnetism*. In view of the permanent significance of that treatise for the philosophy of physics, we might hope to make a good beginning of Whitehead's intellectual biography with what he wrote about it in 1884. Alas, we cannot know what he wrote. Trinity College did not begin to preserve fellowship dissertations until 1896, and Whitehead did not keep anything for long. He does not refer to the dissertation in any of his writings, though he often refers to Maxwell's work. If in the fame of his old age some young scholar asked him about this, his answer would surely have been that Maxwell's work was of the highest importance, and his own dissertation of no importance. He had no autobiographical inclinations, and was the sort of man who has no place in his mind for work that he did when young and passed beyond.

*That is, three years after being inaugurated a Bachelor of Arts. On Commencement Day in 1883 Whitehead became a Bachelor designate; the inauguration ceremony was held the following April. Thus Whitehead could try for a fellowship in October of 1884, 1885, and 1886.

We cannot be as sure as we should like to be about Whitehead's subject. Trinity did not keep memoranda from the fellowship electors of 1884, or other documents about Whitehead's dissertation. The reason for believing that he chose to write about Maxwell's *Electricity and Magnetism* is that Bertrand Russell said so in print. Russell, remembering his early concern (in 1896–97) with the question of matter's relation to space, referred to "Clerk Maxwell's great book on electricity and magnetism" and said, "This book had been the subject of Whitehead's Fellowship dissertation, and Whitehead urged me to prefer its views to those of Boscovich."[19] As this is not the sort of thing on which Russell's emotions might have affected his memory, I think we can trust his statement about Whitehead's dissertation subject, especially since he added in a footnote a general recollection from his own experience: "On this ground, Whitehead was always regarded at Cambridge as an applied, rather than a pure, mathematician."

Unless a source of further information unexpectedly turns up, we shall have to remain ignorant of what we most want to know, namely, what Whitehead took as his main task in writing about Maxwell's *Electricity and Magnetism.* That is a big, two-volume work, and it is unlikely that he dealt with all parts of it. Although it is by no means limited to the theory of electromagnetism, the most plausible conjecture would be that this was Whitehead's main subject. And for all we know, much of the dissertation may have dealt with theoretical developments fathered by Maxwell and published after the *Treatise* appeared. There had been a fair number of them. One of the most important was J. H. Poynting's general theorem, published early in 1884. Poynting showed how to express the flux of energy at any place in the electromagnetic field by a simple formula in terms of the electric and magnetic forces at the place. This discovery was especially significant to Whitehead, and remained so. His 1931 presidential address to the American Philosophical Association[20] includes a section on the flux of energy in physical nature, in which he said:

> The discussion of the Poynting Flux of Energy is one of the most fascinating chapters of Electrodynamics. Forty-seven years ago, when a young graduate student, I first heard of it in a lecture delivered by Sir J. J. Thomson.

I believe that this is the only specific experience concerning science, occurring in Whitehead's student days, that he explicitly recalled in his published writings.

One thing that can safely be said about Whitehead's turning to

Maxwell's theory of electromagnetism for his dissertation subject is that he was plunging into what Cambridge, thanks to Niven, looked upon as the mainstream of mathematical physics. Fifteen more years would pass before the paramount importance of Maxwell's theory was acknowledged elsewhere.* But as J. J. Thomson wrote in *A History of the Cavendish Laboratory, 1871–1910,*

> To the younger school of mathematical physicists in Cambridge who had been brought up on Maxwell's *Electricity and Magnetism,* the generality and, in essentials, the simplicity of Maxwell's theory had for some time carried conviction.[21]

At the time of Whitehead's dissertation, both theoretical work, like Playfair's, and experimental work—most notably by Thomson—had recently been done in the light of Maxwell's theory. I hazard the conjecture that the generality and simplicity of the theory were what made it attractive to Whitehead the mathematician.

On Monday morning, September 29, 1884, Trinity's fellowship candidates, having submitted their dissertations, sat down at the tables in Hall to write their examination papers. The examination ran for six consecutive days. Unlike the University, with its Tripos, the College neither published nor preserved the texts of its fellowship examinations. The one Whitehead took was mathematical, but probably included a paper on some nonmathematical subject of his own choos-

---

*In Britain, phenomena that indicated support for the theory had been observed and reported even in Maxwell's lifetime. (See T. K. Simpson, "Maxwell and the Direct Experimental Test of His Electromagnetic Theory," *Isis* 57 [1966]: 423ff.) But the test that would be decisive was extremely hard to design and carry out. Electromagnetic waves had to be electrically produced in such a way that their velocity could be measured. Maxwell's theory required the velocity to be the same as the known velocity of light. In Germany, Helmholtz suggested this experimental problem to his pupil, Heinrich Hertz, who published his results in 1887. They convinced the scientific world of the truth of Maxwell's theory.

In 1911 the physicist Arthur Schuster published a historical survey that is still fairly authoritative. In *The Progress of Physics in Thirty-three Years, 1875–1908* (Cambridge, 1911) he easily defended the reluctance of the scientific world outside Cambridge to accept Maxwell's theory before Hertz's work (p. 45). Thomson later recalled Cambridge's reception of that work: "By us Hertz's discovery was received with the welcome which might be expected from those who found their most cherished convictions confirmed by a series of beautiful investigations. . . . On the Continent, where Maxwell's theory had, for the most part, met with but little support, Hertz's discovery created a revolution in the study of Electromagnetism" (*A History of the Cavendish Laboratory, 1871–1910* [Cambridge, 1910], p. 87).

ing.* There is no way of knowing what subject he chose, or if, in fact, such a paper was required.

The two Fellows who served as both Examiners and Electors were Thomson and Forsyth. It is likely that they asked Niven to give them his opinion on the Maxwell dissertation Whitehead had submitted. There could be no more competent judges of his merits.

On October 9, five men were elected Fellows: Whitehead,† another mathematician, and three classical scholars. Trinity College admitted the new Fellows in a simple ceremony on Saturday, October 10. It is not certain that Whitehead managed to be present. He had so little expectation of being elected on his first try for a fellowship that when the examinations were over he left Cambridge on holiday without giving his address.[22]

Before the month was up he found himself a Teaching, as well as a Prize, Fellow of Trinity. In short, 1884 was the lucky year for young Whitehead as a mathematician and an academic man. It also brought him the most valuable element in his education as a philosopher—his membership in the private discussion club known to outsiders as "the Apostles." That calls for a chapter of its own.

---

*This was the case when Thomson took the fellowship examination four years earlier.

†In his "Autobiographical Notes" Whitehead, writing from memory at eighty, wrongly gave the year as 1885.

CHAPTER VII

The Cambridge Apostles

i/Election to the Cambridge Conversazione Society ("the Apostles").

ii/History of the Society: George Tomlinson, F. D. Maurice, Tennyson. The mid century. Henry Sidgwick. Verrall, Maitland, Ward.

iii/H. H. Turner, Walter Raleigh, J. K. Stephen, and others who were members when Whitehead was elected.

iv/The Apostles and Whitehead's character.

v/Goldsworthy Lowes Dickinson. H. B. Smith. Jack McTaggart. Roger Fry. Other Apostles.

vi/The records kept by the Apostles. Whitehead's attitudes toward change, adventure, the multifariousness of things and qualities, and uniformity.

vii/Whitehead's religious convictions.

viii/Whitehead's views on telepathy, progress, and other questions discussed by the Apostles. The Apostles after Whitehead.

On May 17, 1884, Alfred North Whitehead was elected a member of the most elite discussion club in the English-speaking university world—the Cambridge Conversazione Society, commonly known as "the Apostles." Its Saturday night meetings were devoted to free and perfectly frank discussion of ideas. As we shall see, it was not a debating club. Use of that easy label is one of several mistakes that mar most published accounts of the Apostles. The elite character of this club has also given accounts of its late-nineteenth-century history a jaundiced press today. It is essential to maintain a sense of history, and also to refrain from blaming the Apostles for being a mere discussion club.

Whitehead found that "the Society"—as it was called by those who belonged—had six active members. (The notion that there were always exactly twelve is another error; the original founders numbered twelve, but that had long been merely the outside limit, seldom touched.) Of the six, three were undergraduates; the others (and Whitehead himself) were recent Bachelors of Arts, one already a Fellow of Trinity. It is quite inaccurate to call the Apostles an undergraduate club, though the great majority of its members were undergraduates when they were first elected to it.

Election was for life. After anywhere between one and a dozen years—most often after three, four, or five—an Apostle gave up his active membership for honorary membership. This was called "taking wings"; he was thereafter an "angel," and would join the discussion, perhaps even read a paper, at any meeting he chose to attend, but was not obliged to come to any. One or more angels were present at most meetings.

For sixty years before Whitehead's election, the Society had been distinguished by the intellectual stature of its members. But it was not its intention to pick out, each year, the two or three men who gave the strongest signs of becoming eminent. Nor was it concerned to spot those who appeared likely to be Senior Wranglers or Senior Classics. Intellectual originality and honesty, and having an open mind, were the traits most sought for, the chief ingredients in "the Apostolic character," as it was called. All the active members took many months in getting to know an interesting undergraduate very well before proposing him for election.

Membership was secret, and the meetings were secret. Nothing that was said behind the locked doors was to be repeated to nonmembers as expressing so-and-so's real views. Like other secret societies, the Apostles had some ritual and a private language. In autobiographies and biographies of its distinguished members so much has been said about this private discussion club that Paul Levy calls it "the most important and best publicised secret society since the Freemasons."[1] Books have been written about it—by outsiders, of course. The present author, also an outsider, is devoting this chapter to it because the Apostles were indubitably, as Whitehead said, "a wonderful influence" on him.[2] My sources include various memoirs and biographies, and the best account of the Society yet published, that in Levy's book, *Moore: G. E. Moore and the Cambridge Apostles*.[3] But I should not have had this chapter if an Apostle of the generation after Whitehead, Professor R. B. Braithwaite, had not (many years ago) looked at the Society's records with me and permitted me to take notes on Whitehead's part in its meetings. Professor Braithwaite also disabused me of the notion that the Apostles were a debating club, and otherwise helped me form an accurate view of their proceedings.

The conduct of their meetings was governed by a firm tradition. No differences more serious than that between tea and coffee—and in the time of serving either one with anchovies ("whales") on toast—are noticeable in the memories concerning them that have been published over the decades. Every Saturday night during term, the Apostles gathered, a few minutes before the ten o'clock closing of the College gates, in the rooms of the active member who at the previous meeting had agreed to write a paper on a question chosen at that time, such as "Is it now as it was in the beginning?" or "What about telepathy?" or "Is self-love the immediate motive of all action?" or "Is one life enough?" All kinds of questions were discussed—the more philosophical, the better; current political issues were seldom chosen. The host was the Moderator. His paper usually set out his convictions and the reasons for them; at a minimum, it laid out ideas for discussion. After reading his paper, which usually took about twenty minutes, the host called on each of the others, in an order settled by drawing lots, to stand on the hearth rug, facing the circle of his "brothers," and say exactly what he thought, holding nothing back.

And no speaker was constrained—as he would be in a debating club—by any rule of relevance. After the host responded to what had been said, a vote was taken. This was sometimes on the question addressed by his paper, but more often it was on a quite different one, which, arising from the discussion, most intrigued the members at that time. The division was not based on a motion on which action might be taken. It merely recorded the convictions of each man present. Finally, the one who would be Moderator at the next meeting was required to submit four questions (one, at least, in humorous terms), and the active members present decided on which of the four he should write his paper. If he failed to produce one, he might give a supper for the brothers in his rooms; or the Secretary might read out for fresh discussion one of the few papers the Society had kept from an earlier period.

A basic assumption among the Apostles was that when a few close friends were perfectly frank with each other, the clash of their opinions would really bring about that learning from others which is supposed to occur in the discussion of ideas. A later Apostle, E. M. Forster, said in his biography of another, Whitehead's contemporary Lowes Dickinson, "The young men seek truth rather than victory." That was a simplification, but not entirely wrong. Insofar as anything can lead a man to put out of his mind the impression he is making when he talks, the Apostles' meetings did so. The give-and-take is described in various reminiscences as at once earnest, humorous, and joyous. The Apostles had fun. Some memoirists smile at the assurance that intellectual integrity was in their keeping, and at the self-importance in the youthful discussions of deep questions. These were outweighed by the presence of more curiosity, detachment, honest tolerance, and honest expression of feeling than would be possible in later life.*

The Apostle Henry Sidgwick, just before he died in 1900, included in an autobiographical fragment a short account of his beloved Society which was true in its time (and Whitehead's) and has become classic.

> Absolute candour was the only duty that the tradition of the Society enforced. No consistency was demanded with opinions previously held—truth as we saw it then and there was what we had to

*Consider the fate of the Metaphysical Society, founded in London in 1869 to discuss the conflict between religion and science. Its forty members—including several former Apostles—were too famous, and too middle-aged or old, to learn much more from each other than how completely they disagreed. The miracle is that the Metaphysical Society lasted eleven years.

> embrace and maintain, and there were no propositions so well established that an Apostle had not the right to deny or question, if he did so sincerely and not from mere love of paradox.[4]

Did not a few years in the Apostles provide an excellent internship for future philosophers? Since Sidgwick's period, freedom to deny or question any proposition has become well established among professional philosophers. But they, like other scholars, show much concern about consistency with at least their own previous opinions. Whitehead did not. He wrote to formulate the truth as he saw it then and there, on the particular subject of his inquiry.

ii

Whitehead looked at everything historically, but had a layman's carelessness about some historical details. I shall try to give a correct account of the history of the Apostles.

The Conversazione Society was founded in 1820 by an undergraduate at St. John's College who never made a name for himself, George Tomlinson.* He and some friends wanted to explore philosophical questions. The University was doing nothing—indeed, until Whewell's time did nothing—to promote the study of such questions; it merely examined all students on Paley's *A View of the Evidences of Christianity*. Tomlinson and eleven friends took the path of self-education through discussion. No more than eight men were elected in any subsequent year, and by 1840 the Society had become still more selective.

In 1824 its center moved to Trinity College. A large majority of its members thereafter were men of Trinity; in Whitehead's time, of Trinity or King's. The move became in retrospect a second founding, because F. D. Maurice of Trinity was then in his first year as a member and became the acknowledged leader of the Society. Maurice invigorated it, and established that devotion to sincerity and rationality which its members have tried to maintain ever since. The spirit of the Society in Whitehead's time was Maurice's accomplishment. It is sometimes said that he and his great friend John Sterling together founded the Apostles.† In fact, Maurice was the thirtieth

---

*He became Bishop of Gibraltar.

†This mistake occurs in the articles on Maurice in the *Encyclopaedia Britannica* (11th ed.) and in the *Dictionary of National Biography;* the articles were written by Leslie Stephen. Although Stephen was not an Apostle, he should have known better.

Apostle; Sterling, the forty-fourth, was not elected until 1825, two years after Maurice. Sterling was a man of great promise, but is now known chiefly as the subject of Carlyle's biography. The brief mention of Maurice's Christianity in the second chapter of the present volume (in connection with his admiring follower, Whitehead's uncle Henry) suggests how far ahead of his time the man was. He eventually became the Knightbridge Professor of Moral Philosophy at Cambridge, as well as an outstanding Anglican theologian. In his student days he could not sincerely subscribe to the Thirty-nine Articles, and so left Cambridge without a degree.*

When young Maurice came to Trinity College he had already read Coleridge's *Biographia Literaria.* In the Apostles he ardently defended Coleridge's metaphysics and Wordsworth's poetry against the utilitarianism of Bentham, which some favored. He succeeded, and anti-Utilitarianism was for a time the dominant strain in the Society. The men they venerated, besides Coleridge and Wordsworth, were Plato and Kant, Shakespeare and Goethe. The members were pleased when jeering outsiders started calling them "the Apostles." Might they not, on going out into the world, be apostles of transcendental wisdom to the Philistines, whom they privately called "Stumpfs"?[5] In 1828 Maurice and some friends bought a relatively new journal; in the year and a half that they owned the *Athenaeum* it was written almost entirely by Apostles under Maurice's editorship. By Whitehead's time, this period in which the Apostolic spirit was that of a troubled, questing, almost masonic society that would reform the world was long past. But not forgotten. Not until the Edwardian age would many of the Apostles look upon it as merely amusing. And Whitehead would always hold a low opinion of intellectual elitists who, like some in Bloomsbury, were unconcerned with the social system.

It is surprising that Whitehead does not mention Maurice anywhere in his writings. Probably he had come to set him aside as a confused thinker, which he was. When Whitehead at eighty wrote a few sentences about the Apostles, he made the mistake of saying, "The club was started in the late 1820's by Tennyson and his friends."[6] Actually, the Society elected Tennyson, as the seventieth Apostle, in 1829. The brothers were much taken by the poems he had been writing, often recited them, and urged their publication; perhaps this was the chief importance of the Society to him. His active membership was short; early in 1830 he formally resigned, and became "an honorary member

*This requirement was not abolished until 1871.

extraordinary." Arthur Hallam, the sixty-eighth Apostle, was then the heart and soul of the Society. The publication of *In Memoriam,* in which five stanzas describe what we can recognize as a memory of Hallam in an Apostles' meeting, probably made it inevitable that "Tennyson and Hallam" should become the ones most often mentioned as forebears in the biographies and memoirs of later Apostles. Whitehead knew *In Memoriam* almost by heart.

For a dozen or more years after Maurice's election, the chief interests of the Apostles were religion, poetry, and political radicalism. The mere fact that this band of young men was said to engage in unrestricted discussions of religion was enough to cause it to be considered subversive of religion and social order at Cambridge.* By the middle of the century things were quite different. In the mid forties, the Apostles had elected first Henry Maine, then two sons of the aristocracy who became statesmen holding high positions in Disraeli's and Gladstone's cabinets—the Honorable William Harcourt and Edward Stanley, fifteenth Earl of Derby. No Apostle was ever Prime Minister, but Harcourt almost was. In the mid fifties Maine, Harcourt, and the future jurist Fitzjames Stephen (older brother of Leslie Stephen) enjoyed a substantial influence on general public opinion as writers for a new weekly, the excellent *Saturday Review of Politics, Literature, Science, and Art.* Within the Society, Maine's membership was naturally followed by that of two other great historians of law, Frederick Pollock and F. W. Maitland.

From the 1840s through Whitehead's time, the Society was not strongly marked for long by any one philosophical perspective. In the early fifties, Clerk Maxwell's interest in speculation found expression in the meetings, and he was the leading member. With the long, active membership of Henry Sidgwick—1856 to 1865—the Society reached a peak. Sidgwick was the perfect incarnation of the spirit of the wholly dispassionate, uncommitted, but hopeful pursuit of philosophic truth. He was the greatest Apostle in the nineteenth century—unless you place a higher value on the ardent leadership of Maurice, from whom he was different in every way. I don't; but perhaps each was the right leader for his time. Before he took wings, Sidgwick read the paper of the evening on thirty-six occasions. In the autobiographical fragment from which I have already quoted, he said of his election to the Apostles that it "had more effect on my intellectual development than any

*Their activity was viewed with alarm in the parliamentary debates of 1834 on the proposed admission of non-Anglicans to Cambridge degrees.

one thing that happened to me afterwards."[7] The chief, though not the only, product of his thought was his great treatise, *The Methods of Ethics*. Whitehead was bored by the book but admired the man. Sidgwick was a power in the University, and did more than anyone at Cambridge for the education of women.

In Sidgwick's time the Apostles elected two other men who were influential within the Society and afterward did much for the University: Oscar Browning and Henry Jackson.* In the late sixties, the young mathematician W. K. Clifford—philosophically an empiricist and agnostic—was an important member. Clifford was a genius who died at thirty-three. The most active Apostle contemporary with Clifford was an applied mathematician, James Stuart. He was dedicated to introducing two novelties at Cambridge, and succeeded in establishing both of them. The first, University Extension—the teaching of workmen and other nonacademic persons by dons in lectures outside the University precincts—was on his mind when he was an Apostle. He easily aroused the interests of the brothers, and this was not temporary; in subsequent decades many who became dons gladly gave such lectures. Stuart's second achievement, against great opposition, was the addition of Engineering to the subjects of study at Cambridge.

In this rapid survey of the Apostles' history before Whitehead's time, three men must be noticed among those elected in the seventies: Arthur Verrall, Frederick William Maitland, and James Ward. In his life at Cambridge Whitehead saw a good deal of them. Verrall attended meetings of the Society for many years, and called his election "the best thing that ever happened to me in my life." He was the liveliest of classical scholars, and is given the chief credit for starting the now customary habit of approaching Greek dramas as works of art, not as mere texts for scholars or grammatical exercises for students. Verrall was brilliant, unorthodox, and lovable. Maitland was witty, solid, and lovable. His pioneering was in medieval English law, which he made essential for the understanding of England's history. He had studied the moral sciences under Sidgwick, whom he revered, then studied law. Although he had a great influence on legal studies, Maitland was primarily a historian—the ablest one in England, Acton said. History was not, as it was for Pollock, who collaborated with Maitland on one work, the handmaid of the law. Whitehead's philoso-

*See Chapter V, section vii.

phy of history owes something to his early contact with Maitland's genius.

James Ward, older than Verrall or Maitland, was thirty-three when he was elected. He came from a Calvinistic home in Liverpool, and had trained for the Congregational ministry before escaping to Berlin, where he studied philosophy and lost his orthodox faith. Then, ten years older than most undergraduates, he entered Trinity College to read for the Moral Sciences Tripos. In 1875 the single Trinity fellowship in that field went to him, Maitland being the loser. Ward made his reputation first in psychology; in it he gained a position comparable to that of his contemporary William James; both had studied physiology, and together they killed associationism. He then developed a system of pluralistic, idealistic metaphysics, capped by a cautious theism. A melancholy puritan by nature, Ward had a judicious intellect. Some passages in a paper he read to the Apostles in 1879 remind us of Whitehead the philosopher.[8] Whitehead could not have protested better than Ward against dogmatic finality—"that finality which regards our most general conceptions as if they were handcuffs on a once vagrant body of facts now at length secure and well-known to the savant."

Verrall, Maitland, and Ward, like Sidgwick and Henry Jackson, spent their mature lives at Cambridge, and all five gained professorships (Verrall, in English literature).

When Whitehead became an Apostle, the Society had nurtured distinguished philosophers, mathematicians, classical scholars, educators, statesmen, historians, an outstanding theologian, and a great poet—not to mention others who would have a place in a complete history. One gap is noticeable: natural science outside mathematical physics—perhaps because it was only beginning to have a place in the Cambridge curriculum. Charles Darwin was not an Apostle. Frank Balfour, whose great promise was noticed in Chapter V, was briefly an active member in the mid seventies. I find no other biologist of importance among the Apostles of the nineteenth century.

In the nature of the case, a discussion club like this draws more on young men in the humanities than on mathematical scientists. But Levy's complete list of the Apostles from the beginning to 1914 shows very much more than a sprinkling of Wranglers. Though none of these were as important in the Society as Maurice, Hallam, or Sidgwick, several, as we have seen, were leading members in their time. One would never guess this from the narratives that have appeared in print (including the one in Levy's *Moore*). Their authors are literary men outside Cambridge, for whom Cambridge mathematicians other than Newton are almost nonexistent.*

Of the Apostles' meetings, Whitehead in his "Autobiographical Notes" wrote: "Older members who had 'taken wings' often attended"—Maitland, Verrall, Henry Jackson, and Sidgwick—as did "casual judges, or scientists, or members of Parliament who had come up to Cambridge for the weekend."[9] Henry Sidgwick, in Whitehead's day still the Apostle *par excellence,* said in his late autobiographical fragment: "The tie of attachment to the society is much the strongest corporate bond which I have known in life."[10]

Sidgwick was not implying that he had given and received a great deal of help on the ladder of success. It is a mistake to assume that the Apostles, being members of a secret society, must have been under special obligations to help each other advance their careers. Because the conduct of their meetings provided unusual opportunities to perceive intelligence and honesty, an Apostle had a better basis—other things being equal—for recommending a brother than for recommending a non-Apostolic friend for a job. But we must not forget that judgment of men on their comparative merits was one of the strictest duties in the Victorian morality of these men, especially in the academic world. I have not come across a case of Whitehead receiving preferment because he was an Apostle, and I do not believe he would even think of recommending one man over another because the first man was an Apostle, the second not.

The bond of the Society was strong in many cases besides Sidgwick's. E. M. Forster dedicated his biography of Lowes Dickinson to "Fratrum Societati." Apostles who had been active members for several years generally remained throughout their lives well aware of their fraternal relation to the others. An example, not atypical, occurred in a 1916 letter from one angel, Whitehead's contemporary H. H. Turner, to another, Bertrand Russell. Turner could not agree with Russell's position on conscientious objection to the war. "But the Society must not shirk any of the consequences of the absolute freedom of thought which it prizes and inculcates: so that we must agree to differ. Yours fraternally."[11]

---

*Levy's neglect of the mathematicians is understandable, since the center of his interest, G. E. Moore, remained all his life wholly uneducated in mathematics. (Russell urged him to take private instruction from Whitehead, but he did not.)

iii

There is, we shall see, no way of learning about the impact of the Apostles on Whitehead except from the descriptions of the men who participated in meetings with him. I begin with Turner, for it was he who proposed Whitehead's election.

The lives—and the characters—of these two make a sharp contrast. Turner was a brainy Yorkshireman with great energy, even bounce. Half a year younger than Whitehead, he was already well ahead of him in his career. He had finished public school—at Sherborne's rival, Clifton—a year before Whitehead did, and when he took Trinity's entrance scholarship examination with him in 1879, turned in a nearly miraculous performance that won him an immediate Scholarship on the Foundation. Before the end of his freshman year he was elected to the Apostles. In the Mathematical Tripos of 1882 he was Second Wrangler, like Maxwell and Clifford before him. He was now working in astronomy and giving practical instruction in it on a University appointment. Within ten years he would be the Savilian Professor of Astronomy at Oxford.

Whitehead took twenty years longer than Turner to become a professor. He was a loner among mathematicians, cultivating a large field—abstract algebra—on his own. Turner was a superb late-Victorian example of the kind of scientist we now know so well, who causes a science to advance rapidly and systematically by leading its cohorts in universities and international societies.* He also put his heart into scientific education for workingmen. He had keen interests outside science—in music and art (he was the son of an artist), but *not* in religion. Whitehead naturally took to religion, but not, on his own, to music and art. Both were men of strong feelings and convictions, but Turner's self-confidence was ample, Whitehead's not. Turner, at this time President of the Magpie and Stump Debating Club, was an easy and forceful speaker, Whitehead was not. Turner let it be known that if, after his death, colleagues chose to drink a glass to his memory, let them drink what they pleased, but if any wished to know his preference, "let it be strong ale." One cannot imagine Whitehead saying this.

---

*Turner, while still a young man, was one of those chiefly responsible for the International Photographic Map of the sky. He later added the collection and analysis of earthquake records to his activities. He died in harness, at the 1930 meeting of the International Union of Geodesy and Geophysics.

In 1884 the active Apostle most admired, the delight of the Society, was Walter Raleigh. He was a very tall, thin man, an exuberant genius with extraordinary subtlety and spontaneous humor. Until Miss Evelyn Wade came into Whitehead's life, Walter Raleigh was the most vivid person of his own generation that he knew. An almost exact contemporary of Turner, he was equally precocious—not in mathematics but in English literature, which had not yet a place in the Cambridge curriculum; he read for the Historical Tripos. When he matriculated at King's in the autumn of 1881, he already had a B.A. from University College, London. At the beginning of his second year the Apostles tapped him. His humor and gaiety led their meetings away from the Victorianism of Sidgwick and his period. Even Whitehead, basically a Victorian, escaped to some extent; to that extent, Raleigh was a permanent influence on him.

Fortunately, a paper Raleigh read to the Society in 1882 was published after his death by his son Hilary. The title piece of a selection of his lighter essays, *Laughter from a Cloud,* it addressed the question, "Is sense of humor or personal integrity more potent for pleasure to its owner?" Raleigh took his stand with the humorist, as against the man of virtue. The latter, he said, is at bottom an optimist, convinced that human nature is high and holy. Raleigh, descended from Scottish Convenanters and the son of a well-known Congregationalist preacher, remarked that Heaven and Hell are now "no more than stage properties." In this play we call human life, humor yields a true aesthetic pleasure. At twenty-one, Raleigh showed in this paper some of the qualities that made the brilliance of his lectures, when he became the first professor of English literature at Oxford, legendary.[12]

Raleigh had no taste for theory or system; his interest as a literary critic was not in movements but, he said, in "live men." He wrote live books on half a dozen of the great English poets, and was himself a good poet on occasion. He was also a historian, and thought his best book was *English Voyages of the Sixteenth Century*. He loved England, hated everything German except German music, and was writing a heroic, official history of the Royal Air Force when he died in 1922.

Something of the gaiety and forthrightness of Raleigh's talk can be seen in his letters, two volumes of which were published by Lady Raleigh. (He had been knighted, and greatly enjoyed being Sir Walter Raleigh.) The fun it was to be with him is illustrated in his son's Introduction to *Laughter from a Cloud;* he remembers that at a family sing-song one summer evening his father was moved to set English classic poetry to the tunes of nursery rhymes, and that Gray's "Elegy" was sung to the air of "Tom, Tom, the Piper's Son."

Arthur Hamilton Smith, a year older than Whitehead, had been an Apostle for two years when Whitehead was elected. He devoted his quiet life to Greek antiquities at the British Museum, and to promoting Hellenic studies. As Keeper of Greek and Roman Antiquities, he was responsible for the successful protection of the entire collection, including the Parthenon Marbles, during the First World War. Smith's publications were mainly catalogs, the most important being, perhaps, his complete edition of the Parthenon Marbles. His achievements and Turner's include good examples of work that is neither political nor theoretical, yet essential to the advance of civilization: the design of instruments. Turner introduced the heliostat, a device for making the sun stand still during photography at eclipses. Smith invented the cyclograph, a special form of camera for making a panoramic view of the entire periphery of an object such as a vase or a statue. Although these inventions are far removed from the Apostles' discussions, noticing them reminds us that the technological mind was one of the variety of types of mind that existed in the Society.* Published accounts of the Apostles overemphasize the literary type.

In the 1880's expansion of Britain's control of lands overseas was a fact of life. Protectorates were multiplying, Egypt was captured, General Gordon was besieged and died in Khartoum, the Colonial Office would need good Oxford and Cambridge graduates for decades. It would be strange if no Apostle devoted his life to the service of the Empire. The one who did so, among Whitehead's fellow Apostles, was H. F. (Harry) Wilson. He had come to Trinity College two years before Whitehead, from Rugby, and was a top football player. He read for the Classical Tripos, had some scholarly interests, and became a Fellow of Trinity at the same time as Whitehead; but he had already begun the study of law. It was as legal adviser, and later as administrator, that he served the Colonial Office, especially in South African affairs. During and after the First World War, the Royal Colonial Institute claimed him, as Secretary and Joint Editor of its journal, *United Empire*. Twice knighted, he seems to have been guided by the sentiment he artlessly expressed in the conclusion of a youthful poem:

---

*I don't think Whitehead ever invented anything, but his son North wrote *The Design and Use of Instruments and Accurate Mechanism: Underlying Principles* (New York: Macmillan, 1934).

". . . there is work to do! / To England, to yourselves, to God—be true!" Wilson was also devoted to Cambridge; his public service in his old age was as a member of the Cambridge University Statutory Commission.

We can be more brief about the other Apostles who were active members when Whitehead was admitted. James Duff Duff, an unaggressive scholar, took his baccalaureate with Wilson, two years ahead of Whitehead. His election to the Society came late—after he had been a Fellow of Trinity for a term, and only three months before the Society elected Whitehead. For about twenty-five years he was on the College staff as a Lecturer in Classics. Lytton Strachey, whose tutor he was, nicknamed him "Plum Duff," because he "cooes like a dove."[13] Duff was a Latinist of some distinction, and widely read in modern literature; he translated some Russian authors.

Almost every Apostle can be credited with achievements beyond his regular professional work. For example, Arthur Hugh Clough, son of the short-lived poet and an Apostle from 1883 to 1886, spent his life as an examiner in the Government's Department of Education, but also designed Walter Raleigh's house outside Oxford.

William Wyse, another classicist and gentlemanly scholar, was Duff's exact contemporary, and won his Trinity fellowship at the same time. He must have been the more interesting of the two undergraduates, for the Apostles elected him in his second year. (He took wings shortly before Whitehead's election.) Less fortunate than Duff in his career, he spent a dozen years as private tutor in a coaching establishment* before he was called to Trinity as a Lecturer, and ill health ended that after six years. Wyse was a fine Greek scholar, known particularly for his work on the text of Aristotle's newly discovered "Constitution of Athens."

The angel who most often took part in meetings with Whitehead was J. K. Stephen, son of Sir Fitzjames Stephen, and so a first cousin—a much older cousin—of Virginia Woolf. He came from Eton to King's in 1878. Jim Stephen—"Jem" in the family—had a bright, affectionate nature, and everybody loved him. He was easygoing, and did not exert himself much in academic competition, though he made a First Class in the Historical Tripos of 1882; that was a switch from his real love, Classics. One of his few publications is *Living Languages*, a defense of compulsory Greek at the universities. He was now study-

*For two of these academic years, 1892–94, Wyse was also Professor of Greek at University College, London.

ing law, and would be admitted to the Bar. His legal career was short, for he began to lose his mind—the result, it was said, of brain injuries from being struck in the head by a windmill sail—when he was scarcely thirty; he died soon after. Stephen's portrait, which hangs in King's College, shows a massive athletic frame and a striking face; he was, among other things, a fine actor. The saddest memory of him at Cambridge is of his magnificent performance (1882) in the title role of Sophocles' *Ajax,* under Verrall's direction. "Brilliant" is the inevitable adjective for Jim Stephen, as for Walter Raleigh. He published many delightful poems about Cambridge life.*

One of Stephen's closest friends was Harry Goodhart, another active angel who died in his thirties. In 1883, a Fellow of Trinity, he was a top football player, and also tutor to the Duke of Clarence, who was best man at his wedding. Goodhart was a Greek scholar; Trinity appointed him a Lecturer at the same time as it appointed Whitehead in mathematics. After six years he became a professor at Edinburgh. I know nothing more about him except that he was obviously "in" as a young man at Cambridge.

Equally "in" was William Herrick Macaulay, a good oar, all-around athlete, son of a cousin of Thomas Babington Macaulay and future uncle of Rose Macaulay. He was several years older than the other Apostles, for he had taken a degree at (and become a Fellow of) the University of Durham before he enrolled at King's. He spent the rest of his life there, as Fellow, Lecturer in mathematics, and administrator. Macaulay was a competent applied mathematician, though not quite of Royal Society caliber. Three times he refused to sit for a college portrait because he did not deserve the honor; finally he let Roger Fry paint it. It shows his magnificent presence and his reserved

---

*The outside world identifies J. K. Stephen as the wit who in 1890 dashed off twelve lines of verse that made the *Cambridge Review* famous overnight:

Will there never come a season,
That will rid us of the curse,
Of a prose that knows no reason
And an unmelodious verse;
When the world shall be delivered
From the clash of Magazines,
When the inkstand shall be shivered
Into countless smithereens;
When there stands a muzzled stripling,
Mute, beside a muzzled bore;
When the Rudyards cease from Kipling,
And the Haggards ride no more?

manner. I should say that this austere man must have been one of the least spontaneous participants in the Apostles' discussions.

In fact, Whitehead came to the Apostles at a relatively quiet time in their history; Raleigh was the only active member who positively sparkled. In addition to the ones already noted, there was Harry Cust, elected a year earlier. He would have sparkled at the meetings—he was a clever, worldly aristocrat who would become an M.P. and a journalist—if he had come to them. And there was Theodore Beck, a delightful friend of Whitehead's who had been elected in his second undergraduate year, and taken the Mathematical Tripos with him (but failed to be a Wrangler). Beck could not come to any of the meetings in Whitehead's time because he had gone to India to be the Principal of a college in Aligarh. In May 1884 the Apostles' membership was quite low. Since October, Goodhart, Wyse, Turner, and Macaulay had taken wings. The brothers now tapped Alfred North Whitehead.

iv

Except when Macaulay or Turner was present, Whitehead was the only mathematician in the discussions in which he took part. It happened—and was quite unusual in the history of the Society—that no mathematical students were elected for several years to come; none until 1892, when C. P. Sanger and Bertrand Russell were chosen. (Whitehead was the one who urged the brothers to get to know Russell; he had seen his brilliance on reading his entrance scholarship examination.) Thus Whitehead's membership in the Apostles was the climax of his Cambridge education in the humanities. And these years were of some importance in the education of his character. E. M. Forster said something about that kind of education when he described the type of discussion society to which Goldsworthy Lowes Dickinson was elected in 1885 (he maintained the rule of secrecy by not naming the Society):

> Certainly these societies represent the very antithesis of the rotarian spirit. No one who has once felt their power will ever become a good mixer or a yes-man. Their influence, when it goes wrong, leads to self-consciousness and superciliousness; when it goes right, the mind is sharpened, the judgment is strengthened, and the heart becomes less selfish.[14]

Whitehead's character was very far from rotarian, and I cannot imagine him as a yes-man. Substantial credit for his sharpness of mind and soundness of judgment should be given to the Apostles' meetings; because of lack of material about his earlier life, we cannot say how great the credit should be. As for his unselfishness, I incline to think that his was fairly well established before he became an Apostle, and then became irreversible.

In one respect Whitehead was quite different from his fellow Apostles. He was a loner. This is the essence of his character. A man who forms really close friendships will figure in the memoirs and biographies of his friends. Only in Russell's autobiography does Whitehead play an important role; this will be discussed in Chapter XI. Among his contemporaries—we must never forget that Russell was eleven years his junior—Whitehead had a great many friends, but no close ones. Paul Levy's scrutiny of memoirs and biographies of the Apostles who were contemporary with Whitehead led him to the same conclusion; he found that Whitehead does not figure in any of them. Levy calls this "somewhat curious."[15] I should say, singular. His friends had good discussions, discussions of almost any subject, with him, but he did not give himself without reserve to anyone.

Thus no source of substantial information about Whitehead's personality emerges from his membership in the Apostles. What does emerge is evidence of his philosophical views, from records of the meetings in which he took part.

V

Before presenting that evidence, let me complete the description of his fellow discussants by attending to the ones who were elected in the course of his active membership. I do not wish to burden the reader with more potted biographies, as they will doubtless be called, than necessary. I want only to make real the men, of so many different kinds, with whom Whitehead in his youth discussed philosophical themes. The academic society in which we live is so different from Whitehead's Cambridge that some fullness is needed. If Whitehead when young had kept a journal in which he expressed his feelings about his Apostolic friends, we should have something much more valuable than anything I can offer.

After Whitehead's election, the Apostles chose no one for nine months. Then Dickinson was proposed by Raleigh and elected. "Goldie" had been gripped by Shelley and Plato, and had written a number of respectable poems; eventually Shelley and Plato turned his main interest into political philosophy. But at this time he was floundering. He had been an undergraduate at King's and had taken his baccalaureate the year before, in Classics. Knowing that he did not want to devote his life to classical scholarship, he was looking for ways in which a quiet, academically minded man could do good in the world. He had been brought up in the tradition of F. D. Maurice's Christian socialism; his father, a portrait painter, was one of the founders of the Working Men's College. It was some time before Goldie came regularly to the Apostles' meetings, for he was working on a cooperative farm with a few other young Cambridge men. A failure at that, he studied medicine for two years, but sensibly gave up the idea of practicing it when the dissertation he had written on Plotinus was thought worthy of a prize fellowship at King's. Gradually he found that he could write, not for scholars but for the educated man, on modern history and on political ideas, always hoping to turn his readers toward understanding opposed ideals and—especially after the shock of 1914–18—ameliorating conflict between nations. He thought, with Plato, that a thinker's place is in the world, but as his biographer, Forster, says, he "could never slap it on the back or stand it a drink."[16] In his Fellow's rooms at King's College he wrote articles and books incessantly. He is often called the spiritual father of the League of Nations, which of course deeply disappointed him. He died in 1932.

Goldie was a philosopher, but not a professional philosopher. He was a seeker for the ideal that should guide human life. (He early became disenchanted with established Christianity.) He read Hegel with his fellow Apostle McTaggart, and became for a time an idealist in the metaphysical sense also: one who finds mind, or the spiritual and ideal, of central importance in the nature of things. Led by Hegel to believe that the history of mankind is understandable, he went on to spend his mature life using his understanding of the past to influence the future, in the way I have described. He also traveled a good deal—he was a bachelor by nature, in fact, homosexual—to see other civilizations, and to see his countrymen as others see them. When he was awarded a Travelling Fellowship, the purpose of which was to widen the Fellow's mind in the interest of international peace (this was before 1914), Rupert Brooke, who admired him, said, "If they widen Goldie's mind any more, it'll break."

There are three things about Goldie that make me like him: he had an abiding sense of the beauty and wonder of the world, an undogmatic mind, and a strong aversion to the murderous conflicts that divide men. In these respects, and in his courtesy and gentleness,

Whitehead was very like him. Whitehead himself, always fond of him, told Harvard students that although Dickinson was not a profound thinker, wherever he was there would be a discussion going: he would start and sustain it, without going center stage.[17] It is reported that his pale, studious face would break into a delightful sudden smile, revealing an intensely human being.[18] Instead of disagreeing with you, he would say, "I think I see your point of view." For several of his books Dickinson chose the form of a dialogue or a symposium. In his autobiography he said that the object of the most famous of these, *A Modern Symposium* (published in 1905), was "to raise the mind above the fighting attitude"—"Plato's object long ago." Although *A Modern Symposium* was republished in 1967, its tone is too Edwardian, its prose too well modulated, for our times and our taste.

After Dickinson, the Apostles next elected a third-year undergraduate, Henry Babington Smith, younger brother of Arthur Hamilton Smith. The Smiths were a Cambridge family closely related to the Babingtons and Macaulays. Indeed, the families that composed what Noel Annan has called "the intellectual aristocracy" of Victorian England were well represented in the Apostles. H. B. Smith was one of a type of man that Cambridge produced in considerable numbers, though the Apostles boasted a smaller proportion of them: a degree with honors in classics, a clerkship in a government department, advancement to a policy-making position, some achievements in it, and knighthood. Phrases like "quiet efficiency" and "sober judgment" appear in articles on them in the *Dictionary of National Biography*. Smith was a schoolmaster for a few years before he became a clerk in the Treasury. A post in India took up parts of the lives of many good university graduates in those days; in his thirties H.B. was Private Secretary to the Viceroy, Lord Elgin, whose daughter he married. Later, he was Secretary to the Post Office. But most of his life's work was concerned with international finance—with matters like the Ottoman Public Debt at the turn of the century, and the Financial Mission that Britain sent to the United States in 1915. As Finance Minister of India he became famous for introducing a tariff on English cotton. He was a director of the Bank of England when he died in 1923.

In May 1886 Whitehead nominated for membership a Trinity freshman who was reading for the Moral Sciences Tripos and had a consuming interest in philosophy, John McTaggart Ellis McTaggart. There is reason to think that his part in the Society's discussions was more significant for the germination of Whitehead's later philosophy than that of any other member.

Jack McTaggart came to philosophy very early. His father, a county judge in East Anglia, died when he was four; his mother, who brought him up and to whom he was always devoted, was a clergyman's daughter who had become a professed agnostic. The boy was expelled from his first preparatory school for arguing against the Apostles' Creed and announcing that he did not believe in God. (He remained an atheist all his life.) At thirteen he was reading Kant and Herbert Spencer.

When McTaggart joined the Apostles he was an outspoken materialist, empiricist of John Stuart Mill's school, and political radical. A few months later he discovered Hegel, and soon made a one-hundred-and-eighty-degree turn. He became, and remained, an unorthodox Hegelian, a rigorous rationalist, an idealist who saw ultimate reality as a community of immortal souls, and an extreme Tory. Meanwhile, he had become the leader in the Apostles' discussions, and remained so for five years. He would have been their leader whatever his views were, for he was the ablest reasoner, the quickest wit, and prone to remarks that were delightful and exasperating in equal measure. Not since Henry Sidgwick's active membership had there been a period which historians of the Society could name after one man. McTaggart was their "Pope," as Sidgwick had been. Not that they were at all alike, except in lucidity. Sidgwick never stopped hopefully searching for the truth in all quarters; McTaggart soon reached a point at which he had the truth, and knew that he had it.

In 1886 he was tall and thin, with large expressive gray eyes in a large head, shy, and uncouth. He seems to have been agoraphobic, and perhaps had a slight curvature of the spine; but his friends were unable to resist the temptation to attribute his strange, crablike walk, in which he kept his backside to the wall, mainly to his having been repeatedly kicked at school. Completely unathletic, he had absolutely refused to play football. He won out, and at his public school, Clifton, was excused from this requirement. A pariah for a time, he gained admiration in the Clifton Debating Society, and made friends. On becoming a Hegelian he found it easy to cherish what he had triumphed over—the system, institutions and their rituals—as composing the most valuable feature of civilization. Ever after, a heretical opinion gave him as much pain as a bad argument. The English public school, Trinity College, the House of Lords, the national Church, the Empire, were above criticism. McTaggart's virulent patriotism in the First World War caused an estrangement from his dear friend Lowes

Dickinson, and he broke off relations with Bertrand Russell. (Whitehead remained a good friend of both McTaggart and Russell.)

Friendships were very important to McTaggart. He enjoyed them with persons of all classes and interests; he was good-natured and liked people. He was himself a man of no type, a pure individual: sentimental, profound, witty, absurd, combining strong feeling with great logical acuity. He should have been a biographer's delight, but the only biography—a short one written by Dickinson half a dozen years after McTaggart's death—is quite unsatisfactory. Forster, in his biography of Dickinson, tells why: "McTaggart was a remarkable figure, possibly a great man, certainly a very strange one, and, biographically speaking, such a man needs rather ruthless handling if he is to come alive. Dickinson only brought sensitiveness and piety."[19]

The influence of the young philosopher on Whitehead was not limited to the time when both were active members of the Apostles,* for McTaggart's career was entirely at Trinity College. As a candidate for a Prize Fellowship (in 1891, the year in which he took wings) he submitted a dissertation on Hegel's dialectic. Whitehead, lecturing at Harvard forty years later, was fond of quoting the remark made by Henry Sidgwick when he was reading that dissertation: "I can see that this is nonsense, but what I want to know is whether it is the right kind of nonsense." Apparently Sidgwick concluded that McTaggart on Hegel was the right kind of nonsense, for McTaggart won the fellowship. He continued to work on the elucidation of Hegel until 1910, which happens to be the year in which Whitehead left Cambridge. When Whitehead became famous as a philosopher, he acknowledged that he had been influenced by Hegel without having read him, and explained, "I was an intimate friend of McTaggart almost from the very first day he came to the University, and saw him for a few minutes almost daily, and I had many a chat with Lord Haldane about his Hegelian point of view, and I have read books about Hegel."[20] The chats with Lord Haldane did not occur until well after 1910. The leading books on Hegel in English were three by McTaggart. He must be accounted the prime source of Hegel's influence on Whitehead.

As Trinity College's chief lecturer on philosophy for many years, McTaggart was remarkably successful. There is much testimony that he conveyed the importance of gaining a worldview, and of clarity in

---

*See Chapter X, section viii.

philosophic argument. Because of him, Bertrand Russell as a young philosopher was an idealist for two years. McTaggart, not Sorley, was the idealist philosopher Cambridge took seriously, the counterpart of F. H. Bradley at Oxford.

After two decades devoted mainly to elucidating Hegel, McTaggart set out to demonstrate, without any reliance on Hegel, the truth of the metaphysical convictions that his logical analysis of Hegel's principles had suggested to him and that he had held for many years. Whitehead read in manuscript several chapters of the first volume of "dear old Jack's" *The Nature of Existence* (published in 1921). The second volume was finished, but not revised, when McTaggart died suddenly at fifty-eight; it appeared posthumously, completing the most tightly reasoned system in the whole history of British metaphysics. Few have deemed it a credible system; only the arguments for the unreality of time kept the attention of philosophers for very long. Unduly neglected by them, McTaggart's metaphysics has drawn attention in other quarters; thus it has been argued that Virginia Woolf's novels of the 1920s should be understood in its light.[21] The present chapter, however, is not the place to attempt a sketch of McTaggart's metaphysics. Let me say only that it plainly had its origin in his emotions, proceeded by careful, detailed argument from allegedly self-evident principles, and in its conclusion embodied a mystical vision of eternal life. His metaphysical work, as well as his role in making Whitehead aware of Hegel, deserves notice from commentators on Whitehead's metaphysics.

At Clifton one of McTaggart's friends had sought him out just because others laughed at his walk and his refusal to play football. This was Roger Fry, a tall boy with a shock of wild hair who loathed public schools and their standards. In 1885 they went to Cambridge together and shared lodgings, though Fry enrolled at King's. They remained intimate friends, and Goldie became a third. In May 1887 Roger Eliot Fry was elected to the Apostles.

His family belonged to the Quaker strand in the Victorian "intellectual aristocracy." The father, Sir Edward Fry, was a distinguished judge and a botanist who had been elected to the Royal Society. The religious tests had deprived him of an Oxford or Cambridge education; he hoped that his son would become a scientist too, and a fellow of a Cambridge college. So Roger studied natural science, which partially satisfied his itch to be doing something with his hands, while the beauty of "the backs" delighted his eyes. Not until he had gained a First Class in both Parts of the Natural Science Tripos (with concentration in botany) did he take training in drawing and painting. But he had the habit of watching the changing light on the fields, and he had already done a good deal of sketching, even tried painting in oils; "his first picture," says Virginia Woolf in her biography, "was it seems a portrait of Lowes Dickinson."[22] Roger Fry was developing the faith in the light of which he would live, the faith, in his own words, "that the aesthetic pursuit is as important in the long run for mankind as is the search for truth."[23] This nascent faith made him different from the other Apostles.

The air that he had breathed at home was as severely Victorian, and as pious, as Whitehead's had been. First McTaggart, then Dickinson and the Apostles, wakened his mind. He prized the Society for the complete freedom in its meetings to question anything and everything. He loved speculation and argument, but he got the greatest possible pleasure from his eyes and ears. After the first meeting he attended, he persuaded Goldie to spend the rest of the night with him out in "the backs" of King's College. As he wrote to a friend, "We lay out under the willows till four o'clock watching the moon set behind gorgeous clouds and the sun rise and hearing the birds wake up one after another."[24]

Fry's tremendous vitality brought adventure wherever he was. For example, in 1888 he helped start, design, and edit a new paper, the *Cambridge Fortnightly*. In its short life, Goldie contributed several poems, Whitehead an interesting fantasy.*

Two failures to win a fellowship at King's—he did not try very hard—suggested that a don's life was not for Fry. By the time he took wings he was ready to go to Italy to see pictures, and to Paris to learn how to paint better. In later life he considered himself to be first a painter, then an art critic. Others—and posterity—reversed the order. Fry was not the sort of painter who paints instead of verbalizing; all his life he was talking or writing—most often to get people to see what he saw in an Old Master, or to convey his enthusiasm for new aesthetic achievements singled out by his extraordinary sensitivity to visual forms. All his life he was a fighting constructive rebel, always starting something new, usually to force the English to take visual art seriously, in their homes as well as in museums. The story of his successful efforts is well known, though it is easy to forget that in 1910, when Fry organized what he called a "Post-Impressionist" exhibition, scoffing laughter was the standard British response to

*It is described in Chapter VIII, section vi.

Cezanne. Kenneth Clark opined that so far as one man can change taste, Fry did.[25]

I resist the temptation to say that the priority which Whitehead in his philosophy gives to beauty over goodness and truth owes much to the fact that Roger Fry was a "brother." At least three quarters of the meetings that Whitehead attended took place before Fry was elected. Also, literature was the only art that was much discussed by the Apostles. Virginia Woolf shrewdly guessed that for them "literature was half prophecy. Shelley and Walt Whitman were to be read for their message rather than for their music."[26] Whitehead's emphasis on the aesthetic in his mature philosophy has several sources, of which the chief is his wife, Evelyn. That will be discussed later, but it is fair to say at this point that his contact with Roger Fry did something to prepare him for this influence of hers.

I pass over as not pertinent here Fry's association with Bloomsbury, which belongs to the last twenty-four years of his life. He died in 1934.

For completeness' sake three other Apostles should be mentioned. Between H. B. Smith's election and McTaggart's that of H. F. W. Tatham occurred; he came to very few meetings. Between McTaggart and Fry, Arthur Beresford Cane was chosen. He was a little more active; a scholar at Trinity, he read history; then he studied law and was called to the Bar. A more important man, so far as the history of the Apostles is concerned, was the one who followed Fry, Nathaniel Wedd of King's; he became a friend of Whitehead's and a great friend of Dickinson, McTaggart, and Fry, but was not elected until February 1888 and so scarcely enters into the subject of this chapter.

vi

Each year the Apostles chose one of the angels to be their President. His only function was to preside and speak at the gala dinner for angels and brothers that was held every June, at a place he chose in or near London. For that occasion the most recently elected member was given the title of Vice-President. The Society's only other officer was the Secretary, who kept the minutes of the Saturday night meetings in the "Book of the Society." These were not minutes in the customary sense of the word. To write down the course of an evening's discussion as it went on would have been wholly incompatible with the purpose of the meetings. The Secretary recorded only (in addition to any election or resignation that occurred) the date, the Moderator's name, the subject of his paper, and the question on which those present (including angels) divided when the give-and-take ended. Everyone then signed his name in the appropriate column, or signed as not voting, and if he chose, wrote a single remark or a phrase of comment after his name. The Secretaries also preserved a few of the most interesting papers that had been read to the Society in its long history. Whitehead was Secretary—succeeding Harry Wilson in that office—from February 21, 1885, to February 12, 1887, when he was succeeded by H. B. Smith. In the middle of this period Oscar Browning gave the Society a cedarwood chest—the "Ark"—in which the Book of the Society and whatever else was being kept were housed thereafter.

Professor Braithwaite found no papers by Whitehead in the Ark. It would have been surprising if he had. As I said in an earlier chapter, Whitehead was a late maturer. The brilliant addresses for which we remember him, such as those he published on English education, came from a mind that had mulled over the topic for decades. Whitehead was not always brilliant, as Raleigh and McTaggart were. A paper for Saturday night discussion with friends who were his equals or superiors in knowledge would have been kept only if it was striking and had not been dashed off. (Whitehead once gave a supper for the brothers in his rooms because he had not finished the promised paper.)

No Whitehead papers in the Ark, then; and of course *he* did not keep any. But his votes and comments, as recorded in the minutes, tell us something about his philosophical convictions when he was young. That is well, for searching has turned up no other source. An intellectual biographer is in his case unusually dependent on private documents, because Whitehead did not write out the philosophical ideas for which he is now best known—his ideas on becoming, being, perishing, and God—until he came to Harvard as Professor of Philosophy in 1924, aged sixty-three. He had indeed been writing a philosophy for ten years before, but only as a foundation for physical science; this limitation was deliberate. Long before that, he had written on logic, which is traditionally a branch of philosophy; Whitehead, however, wrote only on *symbolic* logic, treating it as an algebra among other algebras. Some Whiteheadians have combed his English essays on education, which began in 1911, for his metaphysical ideas; however, they are at best partly and implicitly present in those essays. There are no manuscripts or journals to help us find the beginnings. Letters, then? The only letters Whitehead wrote were those he felt obliged to write. He never wrote letters to discuss philosophical ideas, still less to express his personality.

I do not doubt the value, for a full understanding of Whitehead's philosophy, of studying his pre-philosophical publications, and have myself done some of that.[27] What we lack, and what only the Apostle's minutes offer a bit of, is direct expression of philosophical opinions in his early years as a mathematician. I do not doubt that a man who composed a philosophical system in the seventh decade of his life entertained some ideas about it, embryonic at least, in earlier decades. The third decade is almost always the one that most repays study. For this reason Whitehead's votes and comments in these minutes are not only indispensable; they are of the most useful vintage. I must again express my gratitude to Professor Braithwaite for letting me see them.

Whitehead took wings when he resigned as Secretary in February 1887, but he kept on coming to meetings until May 1888, after which he rarely came. Since his membership in the Apostles began in May 1884, there is a period of four years in which the Book of the Society gives us, for almost every meeting, a brief indication of Whitehead's views. In that time, no question was voted on more than once. That fact, and the brevity of each entry, rule out any possibility of finding a change of Whitehead's mind on any topic over these four years. I shall therefore group the convictions he expressed by topics rather than set them out in chronological order, and I shall do no more than roughly indicate dates except when there is some reason for being specific.

One of Whitehead's convictions came out strongly at the end of the meeting that began on February 14, 1885, and ended in the early hours of his twenty-fourth birthday. The question put for the division was, "Does the devil exist, or is he merely loathsome?" No doubt a variety of talk led up to this, not simply the wish to force a choice between literal and liberal ways of reading passages about the devil in the Bible and other religious texts. Whitehead, at any rate, did not respond in that way. He was one of the two members who voted Yes, the devil exists. What he then wrote in the Society's Book shows that he was upholding a philosophical view, not a religious belief of fundamentalists. "He is the Homogeneous." Most simply put: there really is a lot of sameness in the world, and its complete dominance would be the devil's victory over life. A state of affairs without concrete differences to cherish is abhorrent; drifts toward such a state—always going on—should be opposed as one would oppose the devil.

Two weeks later, Whitehead read the paper of the evening. Of the four questions he had proposed, the brothers' choice fell on "Is it now as it was in the beginning?" The wording first suggests that Whitehead dealt with the opposition between the doctrine of evolution and the Book of Genesis, but I doubt that he did this, for it is unlikely that in 1885 any of the brothers rejected evolution. The only clue to what was debated is that the idea of homogeneity came again to the fore in the final question, "Is a homogeneous universe a failure?" Whitehead and two others voted Yes. Four voted No. There might have been argument about Herbert Spencer's conception of evolution as progress from an indefinite, incoherent homogeneity to a definite, coherent heterogeneity. But the discussion—especially Whitehead's discussion—is more likely to have been about the speculative application of the second law of thermodynamics to the physical universe as a whole, with its prospect of the "heat-death" of the universe, of energy differentials flattening out until an inert state of perfect equilibrium is reached.

The talk at the next meeting also was about the process of the universe. The final question was, "Democritus or Heraclitus?" That is, which of these two philosophers do we think came closest to the truth about the nature of things? Whitehead was one of the majority who gave the nod to Heraclitus. He thus rejected atomic materialism—unfortunately (for us) without setting down a word of comment. His vote, however, is evidence of an early preference for a rich, "thick" metaphysics over a spare reductionist one. For Democritus change was universal, but it was only motion. Only atoms and the void were real; the whole qualitative side of our experience was illusory; becoming and perishing (which is what Whitehead's mature metaphysics is all about) were but the joining and separating of the eternal atoms. But for Heraclitus change was not mere change of place, it was incessant transformation; the world was to be pictured as an ever-living fire, destructive and creative, with perpetual tension between opposites that arise and perish. To know how much of a Heraclitean Whitehead was, we should need to find some expression of comparison with Eleatic or Platonic metaphysics, but the record of the Society's meeting during his active membership includes none.

The records of these three meetings strongly suggest—I do not say demonstrate—this about the general outlook of the young mathematician: he had a positive attitude toward change, tension, the multifariousness of things and qualities; his hackles rose over the notion of unrelieved uniformity; and he rejected the "spatialization" of change. I use Bergson's word because I want to make a point. When students (and some scholars) find these features in the philosophy of science which Whitehead began to write thirty years later, they most often assume that their source was nothing but Whitehead's reading of

Bergson. My point is that in 1885 Bergson was just beginning to find his philosophy, and none of it had been published.

On various issues in human life, the stands that Whitehead took were similar in kind. In the late spring of 1885 he answered No to the question, "Shall we truckle to our environment?" That was the most natural answer, but the question was not an idle one, for the conservative, easygoing Jim Stephen voted Yes. Three years later a meeting ended with division on a question loaded the other way: "Shall we run our heads against brick walls?" Yes, said Whitehead. Throughout his Apostolic years he took a position on the side of diversity and adventure whenever the question concerned their contrast with uniformity and the status quo. He voted that the world ought *not* to agree about tastes; on another occasion, that we should not "be normal." Nor should we "accept Morrison's Pill for all stomachs." To "What shall we do with social experiments, encourage, make, discourage?" Whitehead answered, "Encourage." This was in June 1884; he was not, at least not then, enough of an activist to answer, "Make them."

It is easier to hold a philosophy of change if you believe there is a criterion of progress. Whitehead did not accept that comfort. When Dickinson read a paper in June 1886 and the closing question was, "Have we a criterion of progress?" Whitehead, dissenting from his view, voted No.

The contrast between thought and feeling came up several times. "Shall we cut off the tail?" was a way of asking whether we should try to be purely rational beings. Whitehead voted against cutting off the tail, and wrote, "The operation is too dangerous." (Readers of *Adventures of Ideas* know how hard-headed rationalism evoked his skepticism and alarm, and anyone who looks into Whitehead's view of Bertrand Russell's social philosophy will find that this reaction dominated it.) In the autumn of 1886, when Whitehead read a paper entitled "Is the playwright the best historian?" the closing question was, "Facts or feelings?" "Feelings," he said. The idea that feeling is central, and can itself be cognitive, is a major theme in his mature philosophy. Back in March 1886, at a meeting in which Arthur Hugh Clough was Moderator, the question for the division was, "Is a poet a prophet or a piano?" Whitehead, dissenting from Clough, declared that a poet is a prophet. His lifelong attitude toward poetry—toward great poetry, at least—was serious, in one way too serious: he heard it not as music but as an expression of deeper meanings than ordinary language can convey.

The only classic archrationalist whose name turns up in records of these meetings is Hegel. In October 1887 a paper by Dickinson gave rise to a vote on the question, "Shall we read Hegel?" At this time Dickinson was reading Hegel's *Logic* with McTaggart, and was worried about what to make of it. He now voted No, with the comment, "but *I* shall." Whitehead, a late arrival that night, wrote in the record book, "urge others to do so and to read other metaphysicians as well." Note that he urged *others* to read Hegel; if Whitehead himself had already tried, he had looked only at Hegel's remarks on mathematics and been turned off.[28] Two years later, at the only meeting Whitehead attended in 1889, a discussion of the question, "Shall we live to learn or learn to live?" issued in a vote on "Shall we turn Tom Jones into Hegel?" In those two years the Apostles had heard a good deal about Hegel from McTaggart, but Whitehead had not gained firsthand knowledge of the Master; he never did.[29] Naturally, he voted against turning Tom Jones into Hegel.

We may wonder just what the brothers meant by turning Tom Jones into Hegel. This illustrates the risk in trying to use Whitehead's votes and occasional comments to find his philosophical views when he was an Apostle. When these votes and comments deal with the same general topic on several occasions, and show a convergence of the bits of evidence, general conclusions can reasonably be drawn, and I have done so. Coming now to the general topic of religion, this is again possible.

vii

In their religious views, the Apostles of 1884 to 1888 were a mixture of believers and nonbelievers, with the former in the majority, and with Whitehead emphatically one of the majority. "Do we believe in God?" was the closing question on May 2, 1885. Walter Raleigh had read the paper, and twice the number of brothers usually in attendance had come to hear him. Maitland, Jim Stephen, and three others declared that they did not believe in God. (McTaggart, whose atheism has been noted, was still at Clifton.) But there were nine affirmative votes, Whitehead's, Raleigh's and Dickinson's among them. Two years later, a paper McTaggart read to the Apostles was followed by division on whether a personal God is a satisfactory explanation of the universe. McTaggart and two others said it was not; Whitehead and one other votes Yes. Young Whitehead believed in God the Creator.

I turn back to 1885 to note that some weeks after the vote on belief in God, Whitehead, answering Yes to the question, "Shall we transcend our limitations?" wrote in the Society's Book, "I want to see God." The beatific vision was important to him. The following October he gave a paper on the question, "Is one life enough?" The question voted on was "Do we desire immortality?" Whitehead accompanied his affirmation with the note, "I want a sort of a something." He wanted immortality of some kind, not necessarily the specific kind in which his father and grandfather believed. In the Lent term of 1887, McTaggart led a discussion of the idea of Nirvana. The final question then took the form, "Should we object to be annihilated?" Yes, said Whitehead, dissenting from McTaggart.

In the two meetings next to be noticed, I am not sure how to interpret Whitehead's vote. When he read a paper entitled "Shall we wash each other's feet?" the voting question was, "Shall we be humble?" To his Yes, Whitehead added, "thinking of our ideal." I suppose that in this paper Whitehead included some discussion, however brief, of Christian humility; but his discussion, and that of his brothers, may have centered on a different kind of humility, the intellectual humility which the Apostles tried to maintain toward each other; that may have been the meaning of "our ideal." In another meeting, Whitehead voted Yes on the question, "Is the Bible a more than ordinarily moral book?" I am uncertain about the point of the question. I suppose he was saying that the Bible contains extraordinary moral insights; but it may be that the question concerned the difference between theistic books and merely moral books, and was badly phrased.

Early in 1888 there were three meetings in which the final question was clearly religious. To "Is the ground on which I stand holy ground?" Whitehead gave the orthodox answer, "It is not." To "Shall we worship the Father, or the Son, or Barabbas?" Whitehead replied, "In order of merit." As he had the alternative of not voting, I suppose he was saying that the question was foolish. On "May we whitewash the Creed?" he voted No. This was the last vote on a question having to do with religion that Whitehead would cast in the Society for many years to come.

In an earlier chapter we saw that Whitehead's paternal grandfather and uncles had given him excellent Victorian models of useful lives conducted in the light of Christian faith. In the incessant conversation that was the Cambridge mode of education, he heard arguments for atheism from friends whose faith had loosened. The arguments did not make him a skeptic. But although he consistently affirmed his religious belief to fellow Apostles, acceptance of the Anglican Church, in which he had been raised, ceased to be automatic—for other reasons. At Cambridge as at Oxford, though less so, religious young men talked about the relative merits of Roman Catholicism. Perhaps the Anglican Church had everything except religion, with regard to which it offered only a socially useful simulacrum. After Whitehead read a paper on May 15, 1886, the question in the closing hour was, "Should Churchmen go to Rome?" "Yes," he declared, "or in the other direction." Note that the question was not whether *we* should go to Rome, but whether Churchmen should. "Churchmen" could have meant either members of the Established Church or its clergy. Whitehead himself, a member in 1890 and for some years thereafter, seriously considered abandoning it for the Roman Catholic Church; this will be taken up in Chapter IX. There can be no certainty about what he meant by "in the other direction." Perhaps he meant toward Nonconformism, or at least the broadest of the Broad Church positions in the Anglican Communion. It seems more likely, since a strong minority of the Apostles were unbelievers, that what Whitehead meant by going "in the other direction" was the dropping of religious beliefs, until one came to rest in the agnosticism of Leslie Stephen or W. K. Clifford—the two are practically identical. (This was what Whitehead himself did, ten years later, at the conclusion of his long, serious flirtation with Rome.) Unlike Clifford, Leslie Stephen, to his regret, was never elected by the Apostles, though his brother and nephew were. But his position was as well known to the Apostles of 1886 as that of any intellectual outside the Society could be. He had made current the word that Huxley coined. As a fellow of a Cambridge college in the 1860s, he was in holy orders; the *Essays in Free Thinking and Plain Speaking* and *An Agnostic's Apology,* which he wrote after relinquishing orders, were very well known.

Whether or not we interpret "the other direction" as referring to agnosticism, what is excluded by Whitehead's comment is the idea that an Anglican, whether communicant or priest, should find his right place, and relief from any religious uncertainties he might feel, in Anglo-Catholicism. That was what Whitehead's brother Henry—seven years older and ordained in 1879—had done; he was called a Tractarian. The conjecture is probable that in saying—perhaps impulsively—that Churchmen should go either to Rome or in the other direction, Whitehead had not only his brother but also Cardinal Newman in mind. Henry and Alfred felt close to each other; it is a pity that no nineteenth-century letters between them are extant. In the early 1880s Henry was a Fellow and lecturer in classics at Trinity College, Oxford; Newman was an honorary Fellow and was annually invited there. Alfred had never met him, but it would be surprising if he had not by 1886 read his *Apologia,* and perhaps the earlier *Essay on the Development of Christian Doctrine* and the later classic on the logic of religious belief, *A Grammar of Assent.* We have no independent evidence on *when* Whitehead read Newman, only evidence that he was impressed. When the question of his own possible conversion to Roman Catholicism came to a head, early in 1890, he went to the Oratory at Edgbaston to see him.

In a meeting of the Apostles three weeks after this one, a vote was taken on the question, "Can lying be dispensed with?" The Moderator, Jim Stephen (a freethinker) said that it could. Whitehead, voting No, wrote, "It is the nearest approach to truth—sometimes." Newman's words, given great prominence by Kingsley in their controversy twenty years earlier, were, "It is no more than a hyperbole to say, that, in certain cases, a lie is the nearest approach to truth."* It does seem that Newman was on Whitehead's mind in 1886.

viii

What did Whitehead, when an Apostle, think about the age in which he lived? At a meeting of the Society held late in May 1887—a few weeks before the celebration of Queen Victoria's Golden Jubilee—the Secretary, H. B. Smith, read out a paper that F. D. Maurice had presented some sixty years before, when George IV was on the throne. The brothers then divided on the question, "Has the period in which we live any claim to the character of a pre-eminently enlightened age?" Presumably "the period in which we live" meant the present generation, in contrast to that earlier time and to everything preceding it. One would expect a consensus in the affirmative, with some expression of reluctance: Dickinson voted, "Yes, alas." But Whitehead, and only Whitehead, voted No. He wrote nothing in the Society's Book, and I know of nothing outside it, to explain this. What was behind it? After all, it was extreme to assert that the present age has no claim to the character of a preeminently enlightened age. What of the claims that could be made with respect to criminal law, medicine, mathematics, physical science? Did Whitehead think these of slight importance compared to something else (religion?), of su-

*Kingsley had taken this sentence from Newman's *Sermons on the Theory of Religious Belief* and set it down for attack on the title page of the piece that provoked Newman's *Apologia.*

preme value? Did he rate some earlier period higher in the general enlightenment of its best-educated people? Was he too cautious, or—even at twenty-six—too wise, to subscribe freely to the common opinion? It is maddening that we have no basis for offering anything other than these guesses.

This was not the only time that Whitehead took a position in lone dissent from his fellows. In November 1886 the Apostles discussed telepathy; of the six brothers present, he was the only one who thought it improbable. Jim Stephen had read the night's paper, "What about telepathy?" and this was also the question on which the vote was taken, in these terms: was telepathy proved, plausible, improbable, or disproved? No one was rash enough to claim that it had been disproved, but three thought it plausible, while Stephen and McTaggart signed their names in the "proved" column. It must be remembered that psychical research was almost a fad in Cambridge and London. Many of the best people took it up. Henry Sidgwick had been President of the Society for Psychical Research for three years after its founding in 1882. Remembering that Sidgwick wanted to believe in Christianity but found its truth unproved, and noticing that neither Stephen nor McTaggart believed in God, I am inclined to agree with the opinion, which others have advanced, that Cambridge men who believed in telepathy had found themselves a substitute for religion. But Whitehead had religion, and probably felt that eagerness to find evidence for telepathy was a vulgarization of the spiritual. It has also been suggested that by taking up the cause of parapsychology, an educated man attacked the supremacy of scientific materialism. At this time the Newtonian world machine was believed to be in full control of Nature. So far as I know, there is no evidence that Whitehead was as yet the least skeptical about it as physics, or critical of its reign in human life.

We must notice that Whitehead responsed Yes to the question, "Shall we have a theory of morals?" The emphasis, I take it, was on "theory"; a negative answer would mean that one preferred to rely on hunches. Whitehead's desire for a theory of morals will surprise those who know only the opinion he held in his Harvard years, namely, that every occasion is unique, and so transcends the general rules which it is the business of a moral theory to lay down. I believe that he came to hold this opinion sometime after he married Evelyn Wade, late in 1890; she delighted in, and emphasized, the uniqueness of every human occasion.

Being British university men, the Apostles were up on the political questions of the day. In their meetings they did not thrash out specific issues in the way a debating society would, but dealt more with different general attitudes. No doubt talk about these was triggered by specific issues, but the records do not show what these were. The positions Whitehead took in the divisions are marked by both patriotism and common sense. In May 1885 he read a paper on the question, "Shall we sing 'Rule Britannia'?" In the vote, which was on the same question, he and all present said Yes. Not surprising. In April 1887 H. B. Smith discussed the critical question, "Are we more virtuous than our neighbors?" The division was on "Are Britons the salt of the earth?" Whitehead voted Yes, with the notation, "wishing to qualify this statement." Abolition of the monarchy, discussed in December 1885, was a practical question (the Prince of Wales had brought the monarchy into bad repute) with philosophical aspects. Whitehead read the paper, "Shall we abolish monarchy?" When the division was taken—on the same question—he voted Yes.

There remain to be considered Whitehead's positions on several interesting questions. I shall group them together as miscellaneous questions.

The notion of progress was much in the air, and Dickinson read a paper on it in 1886. To a philosophic mind, the crucial question is whether we have a criterion of progress. That in fact was the question on which the Apostles divided. Dickinson said *he* had a criterion; Whitehead, that we have none. Another paper by Dickinson, early in 1887, seems to have dealt with the relation between the world of scholarship and the world outside; Goldie, as we have seen, inhabited both worlds. The division was on the question, "Study or marketplace?" Whitehead's vote was sensible, and the future practice of his life would be in perfect accord with it. "Study," he wrote, "with windows"; and he drew this little sketch:

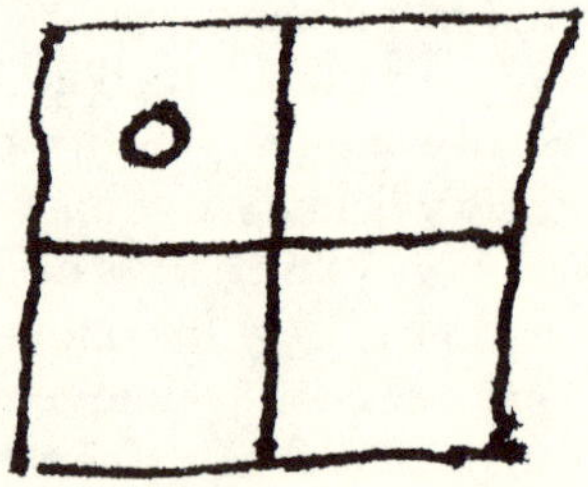

Two weeks later the topic of primitivism (in the philosophical, not the aesthetic, sense) came to the fore in the question, "Did the savage enjoy himself?"* Whitehead, whose mind was alive but realistic

*It is likely that the background of the discussion included, besides ideas of Rousseau's and the Enlightenment, the new work that J. G. Frazer had undertaken. Frazer was a well-known Fellow of Trinity; his first classic book, *Totemism,* was published in 1886.

and balanced, never extreme or effervescent, answered No. This cast of his mind is perfectly illustrated by his answer (also in 1887) to the question, "Is enthusiasm the chief good?" "Yes," he wrote, "suppressed and tamed." A perfect, if trivial, example of the sensibleness of Whitehead's answer occurred when a vote was taken on the question, "May we read 'Dolores'?"—the controversial poem by Swinburne, who was much admired by McTaggart. Dickinson voted No, everyone else Yes. Whitehead wrote after his Yes, "Shall know when I have read it."

The meetings of the Apostles in Whitehead's active years include five in which the division concerned women, sex, and marriage. The most logical place to report Whitehead's votes on these questions is the chapter (IX) dealing with his marriage.

The Apostles have gone on, generation after generation. The place that the metaphysician McTaggart held as their leading spirit was next filled by his philosophical opposite, G. E. Moore. After him, leadership was shared by Lytton Strachey and John Maynard Keynes. Under Strachey's influence, atheism was so dominant that the Society would think twice before admitting a known Christian. In the early 1930s, the suspicion that they were subversive, which had been leveled at the Apostles in the 1830s, would have a basis; Anthony Blunt got Guy Burgess elected, and would have liked to turn the active membership into a Communist cell.

The Cambridge Conversazione Society still meets. Many angels hold important positions in public life as well as in the British universities. As always, many outsiders, from jealousy or other causes, look upon the Apostles with some disdain. A change that promises to be permanent occurred in 1970 when the first woman was elected to membership.

CHAPTER VIII

The Young Mathematician

i/Teaching Fellow at Trinity.

ii/A. R. Forsyth. Whitehead's visit to Germany.

iii/Young Whitehead's lack of publications. Grassmann's *Ausdehnungslehre*. Whitehead on the *Ausdehnungslehre* and other new algebras.

iv/First publication: "On the Motion of Viscous Incompressible Fluids." Second paper on this subject. "Whitehead's paradox."

v/Teaching at Girton. Whitehead's support of Foreign Missions. His support of Toynbee Hall.

vi/Three fantasies written by Whitehead and published in Cambridge.

Whitehead's many years as a teacher of mathematics at Trinity College began in 1884. In his "Autobiographical Notes" he says only (misremembering the year), "In the autumn of 1885, the fellowship at Trinity was acquired, and with additional luck a teaching job was added."[1] He told Lucien Price that but for the vacant lectureship he "would probably have taken a teaching position in a Public School and never have gone much farther."[2] I don't believe he would never have gone much farther; his older brother Charles, who didn't, lacked his gifts. But there *was* luck. We saw at the end of Chapter VI how surprised Whitehead was when he was awarded the fellowship on his first attempt, which happened to few candidates. In all probability his dissertation had unusual merit. In any case, Whitehead benefited from two special circumstances.

The first, concerning teaching, actually preceded the fellowship election. In Chapter V it was noted that although students paid tuition to their colleges, they also—as they had for a century—hired private tutors ("coaches") to give them two or three years' intensive training for the Mathematical Tripos. Since the 1870s there had been much talk among dons—especially at Trinity—about this scandalous situation, committees had been appointed to plan an attack on it, and in the autumn of 1883 Trinity took action. The College began an "experiment," as the *Cambridge Review* called it, and in the Lent term of 1884 this was "in full working order."[3] The experiment consisted in setting up, in addition to the established curriculum, several classes of about eight honors students each, the same from year to year, in order that these students should find it unnecessary to hire coaches; each was entitled to as much individual attention as he needed. The experiment was implemented by asking some members of the staff to do extra work (J. J. Thomson was one of these) and by hiring—at so much per pupil—two promising men not on the staff. Trinity hired Frank Carey, who had been Third Wrangler in 1882, to take one class, and in March 1884 hired Whitehead to take another. Whitehead's first teaching was thus an effort to do for undergraduates what had been done for them by Routh and other established coaches with long experience. The experiment did not make a dent in the coaching system, and after three years it was stopped. A radical change in the examination system—still more than twenty years in the future—was needed to make recourse to mathematical coaches a merely occasional practice.

The second piece of luck in Whitehead's becoming a Teaching Fellow of Trinity was the almost simultaneous occurrence of several vacancies in the College's mathematical staff due to deaths and resignations. The staff had been fairly stable, and after 1884 there were no vacancies for several years. But in 1884 Thomas Dale died, and W. D. Niven, who had gone to Greenwich, began giving all his time to the Royal Naval College. R. C. Rowe, a Fellow of Trinity who had become a professor at University College, London, was lured to return, but he died in September 1884. At the end of the year the staff lost Dale's replacement, J. J. Thomson, owing to his appointment to the Cavendish professorship in the University.

One vacancy was filled by appointing A. R. Forsyth to a lectureship that began in January 1884; we shall have more to say about him. In October of that year Whitehead was asked to continue taking a class in Trinity's scheme to make coaches unnecessary. This was merely for one term; he would have to prove his worth. But when Thomson was lost the College Council took this action: "Mr. Whitehead to continue as Assistant Lecturer and to deliver certain lectures in place of Professor Thomson."[4] There are no records to show what lectures these were, but there can be no doubt that this was a bit of an honor, and a fortunate one for Whitehead. (Perhaps this action of the Council is what Whitehead was thinking of in "Autobiographical Notes.") He must have done a good job; for the Easter term his appointment was again renewed, and he was asked for the first time to teach two of the standard courses in Trinity's curriculum, Advanced Hydrostatics and Newton I, II, III (i.e., the first three Sections of Book I of the *Principia*). For 1885–86 he was given a full year's appointment, and so was fairly started in a Trinity teaching career that would last without interruption for twenty-five years thereafter.

ii

As one year of lecturing followed another, Whitehead would be asked to teach most of the branches of applied mathematics. But he was not a hired slave; it was now customary at Trinity to give young teachers some opportunity to follow their own interests. In the

Cambridge University Reporter of June 10, 1887, it was announced that Mr. Whitehead, in the Lent term of 1888, would give a course of lectures, open to students from all colleges, on "Grassmann's *Ausdehnungslehre,* with special reference to its applications." Grassmann's great work had not previously been the subject of a lecture course at Cambridge. That it should be so now was a sign of the new climate in which mathematics was being pursued there.

That climate was largely due to the presence of Forsyth. He became Whitehead's closest friend among the mathematicians, and was godfather to his first child. Later, he played a part in important changes in Whitehead's life—in 1910 and 1914. What kind of man was Andrew Russell Forsyth?

He was as unusual among mathematicians as McTaggart among philosophers, and although he became Cayley's successor in the Chair of Pure Mathematics, his name is now permanently in eclipse. He had a strange mind and a character definitely not Apostolic, for he was prim, pompous, and without humor or spontaneity. The son of a ship's engineer, he was of pure Scottish blood; short, with prominent eyes that were as dark and piercing as eyes can be, in a round face. Forsyth was under constant nervous tension, and appeared to be fierce but was only precise, and in fact helpful to everyone. He was the soul of correctness, neatness, and order. Add that although he could be interested in many things, he could not be interested in philosophy or theology, and we may wonder whether he meant much to Whitehead. I do not think he meant *very* much to him, but he meant much to Whitehead's career in England.

Forsyth had won a major scholarship at Trinity when he took the entrance examination, two years before Whitehead. He had the audacity to go to Cayley's lectures, as Whitehead did not; he was the only undergraduate to do so. A rapid rise to fame began with the Tripos of 1882; he was the Senior Wrangler. This was at once followed by a Trinity fellowship, for which he wrote a dissertation (published as a book, *Theory of Functions*) that showed promise of originality. His other gifts prevented its fulfillment. Few creative mathematicians have read exhaustively in their fields before making their discoveries. Forsyth was a voracious reader in English, French, and German, and he had a phenomenal memory. A prodigious worker, he spent his life organizing the great variety of his mathematical reading, and using his skill in algebraic manipulation to solve equations. Thus he carried the theory of differential equations to completeness in six volumes, without making any fundamentally new discoveries.

Forsyth had a strong beneficial influence at Cambridge for about twenty years after he joined Trinity's staff. The great recent advances in mathematics, Cayley's (and Sylvester's) expected, had been made in Germany and France, but Cambridge was ignorant of them until Forsyth urged his advanced students, and young colleagues like Whitehead, to read the men he had read—Weierstrass, Felix Klein, Poincaré, and others. He revealed a whole new world to the young mathematicians of Cambridge.

Whitehead not only started reading the Germans; he visited Germany to hear mathematical lectures. At Cambridge, the "Long Vacation" ran from mid June to the first week of October. Jessie Whitehead was told that her father spent a Long Vacation in Germany. North Whitehead believed that he was there for a German academic year or half-year. As the record at Trinity shows that he taught every term from 1884 to 1910, the visit must have been made in a Long Vacation. The German university year was very different from the English; it consisted of a winter term from mid October to mid March, and a summer term from mid April to mid August, when the German vacation began.* Whitehead must have left Cambridge as early as possible in June, got permission from German professors to start hearing their lectures when they were one-third over (which should not have been difficult), and spent the rest of his Long Vacation here and there, probably in Weimar and other German cities.

The most likely year of Whitehead's foreign study is 1885, though 1886 is almost as likely; 1887 and 1888 are possibilities, and so is 1884—if he got his fellowship dissertation finished and submitted early. We do not know to which German university he went. Berlin attracted the most visitors; in the summer term of 1885 Kirchhoff lectured there in mathematical physics, Weierstrass and Kronecker gave lectures and a seminar, and Helmholtz lectured on electricity, magnetism, and light from an experimental point of view. Berlin's great rival in mathematics, Göttingen, could not match that, but Voigt and Schering gave summer-term lectures; in 1886 Klein (often a recipient of Forsyth's hospitality at Trinity College) moved from Leipzig to Göttingen. Whitehead might have gone to Leipzig in 1885 to hear the author of the Erlangen program for unifying geometries, or to Göttingen later. Grassmann was dead, but Klein and other Germans were now promoting the study of his work. At Heidelberg there

---

*To study in Germany during the German winter term, Whitehead would have had to miss the Michaelmas term and most of the Lent term at Cambridge.

were lectures in 1885 by the world's leading authority on the history of mathematics, Marcus Cantor. Georg Cantor was at Halle, lecturing then on higher geometry, and higher arithmetic as an introduction to the theory of types of order. The summer term offerings in German universities were so rich that it would have been strange if an unmarried young Cambridge mathematician had not used at least one of his Long Vacations to study there. It should be added that Frege was at Jena; he had published his *Begriffsschrift* and *Grundlagen der Arithmetik,* but we lack a reason to think that Whitehead knew of them.

Inquiries I have sent to all these universities about young Whitehead have yielded only negative replies. This proves only that after a century in which many records were destroyed, I have failed to get any official confirmation. There is no doubt that Whitehead heard mathematical lectures in a German university. Students who saw him and his wife on Sunday evenings at Harvard did not hear him talk about this, for Mrs. Whitehead was Germanophobic.

iii

Whitehead's mathematical career shows a feature which is quite unusual. We should expect a first-rate mathematician, as soon as he got his fellowship, to seize upon one mathematical problem after another, figure out solutions or contributions toward solutions, and publish them—some papers every year—in the professional journals. But Whitehead's mathematical bibliography shows a blank for almost the first five years of his fellowship. There was no shortage of problems in the area of his dissertation.* I doubt that there was a shortage of publishing space.† In January 1887 Whitehead was elected a Fellow of the Philosophical Society, which was the great scientific organization at Cambridge; he never read a paper to it. In April he became a member of the London Mathematical Society; he presented one paper there—ten years later.

---

*Whitehead *may* have tackled one or more of these problems in 1885 or 1886, but he published nothing. At Trinity College he had ample opportunity for talks with the leading physicist who was doing work in the light of Maxwell's electromagnetic theory, J. J. Thomson (already a Fellow of the Royal Society); I have, however, no evidence that either man contributed ideas to the other, and I doubt that this occurred.

†Papers by Whitehead's contemporaries were being published in the *Messenger of Mathematics, Quarterly Journal of Pure and Applied Mathematics, Philosophical Magazine,* the Cambridge Philosophical Society, and the London Mathematical Society.

If he was a first-rate mathematician—by which I do not mean a mathematical genius, for his work does not quite permit us to call him that—he was a strange one. In the last chapter I indicated that he was a loner. I wonder how much he talked mathematics in casual conversation with his friends; my guess is that he talked most readily about the things the Apostles discussed. I picture him as not by nature addicted to attacking specific problems of the sort that his mathematical colleagues had on their minds, and as more interested in the new branches of mathematics, created in the preceding half-century by Galois, Grassmann, Cayley, Hamilton, Riemann, and Boole, which had enormously enlarged the scope of the science. (Whitehead was one of the first men at Cambridge to give a course of lectures on non-Euclidean geometry.) His uncle Henry Whitehead is said to have had "the keen enthusiasm of the hunter in out-of-the-way fields of knowledge," and to have inherited this from his father, schoolmaster Thomas Whitehead.[5] Alfred North Whitehead was more a hunter and explorer than a problem-solver. With these interests, I suspect, went some degree of hero worship: of Maxwell, Hamilton, Grassmann. It is likely that soon after he saw (probably in 1886) that Grassmann's algebra was the most general algebra ever produced by the mind of man, he was bitten by the idea of doing justice to it. If that should take years, let it. Other young mathematicians could carry on the workaday pursuit of research and publish stacks of papers; Whitehead was a proper Victorian.

Hermann Günther Grassmann was born in 1809 and died in 1877. The *Ausdehnungslehre,* which he published in 1844,[6] was written just before Hamilton made his famous discovery of quaternions. That discovery is contained as a special case in Grassmann's broader conception. No fame among mathematicians came to him in his lifetime. His publisher could not sell the book, and Gauss made only a lukewarm response to the copy sent him. A second exposition of the *Ausdehnungslehre,* which he published in 1862, fared no better. (Whitehead preferred the earlier one.)[7] Grassmann never won a university post, and had to content himself with teaching elementary mathematics in his native Stettin (now Szczecin). There is no more striking case of genius-neglect in the history of mathematics.*

It must be admitted that Grassmann's expositions were hard to follow. And when he explained that he was offering not a "real"

*In his old age Grassmann turned to Sanskrit philology, in which his work received immediate appreciation.

science—that is, one which is to be tested by external reality—but a "formal" one, for which consistency alone is required, this encouraged the dismissal of his book as "too philosophical." The idea that a consistent theory of *n*-dimensional space, like Grassmann's, has uses beyond those of three-dimensional geometry, had not arrived. Whitehead's lectures in 1887 and 1890 on the *Ausdehnungslehre* paid special attention to its applications.

How did Whitehead come to Grassmann? His own reading of recent German mathematics might have done this. He may first have become curious when he was writing his dissertation on Clerk Maxwell's *Electricity and Magnetism.* Maxwell almost certainly never read the *Ausdehnungslehre,* but he breifly discussed[8] an important paper, "Neue Theorie der Elektrodynamik,"[9] in which Grassmann incidentally called attention to it. Cayley knew about Grassmann's masterpiece, and conversation with Forsyth stimulated Whitehead's interest. But the strongest local stimulus, in all probability, came from another young mathematician at Trinity, Homersham Cox. In 1883 Cox had published a long, suggestive paper (which Whitehead later praised highly[10]), "On the Application of Quaternions and Grassmann's *Ausdegnungslehre* to Different Kinds of Uniform Space"—that is, to elliptic and hyperbolic as well as Euclidean space.[11] As a dissertation, it had won him a Trinity fellowship in 1881. Cox did not become a member of the mathematical staff, and although Whitehead must have known him, we know nothing about the content of their discussions. In 1891 Cox moved to India as professor at Allahabad.*

Whitehead also knew papers dealing with Grassmann's calculus of extension (the usual English name for the *Ausdehnungslehre*) that had been published by W. K. Clifford of University College, London, and Arthur Buchheim of Oxford. In his "Applications of Grassmann's Extensive Algebra" Clifford expressed his "profound admiration of that extraordinary work."[12] He was Grassmann's first English admirer, a Fellow of Trinity and an Apostle, but he died the year before Whitehead came to Cambridge. Buchheim's papers were published in 1886–88. The most promising of the young Oxford mathematicians, he died in 1888, not yet thirty. Buchheim had been active in the

*Cox was a Fellow of Trinity until 1886. He began the study of medicine, but gave it up. J. J. Thomson remembered him as "one of the clearest-headed men I ever met but remarkably absent-minded" (*R&R,* p. 41). So was Whitehead. Cox wrote a few more mathematical papers before he went to Allahabad. His Indian students loved him. He died in India in 1918.

London Mathematical Society; Whitehead had chances to talk with him there.

In England, it was up to Whitehead to carry out the exploration of Grassmann's ideas.

In doing this he would be working toward the satisfaction of one of his lifelong interests: discovery of the fundamental nature of mathematics. When did this interest arise? When he first studied higher mathematics; he had the meditative type of mind that could not be content with the formulae and their applications. The traditional definition of mathematics ran: the science of number and quantity, or as it was sometimes put, of discrete and continuous magnitude. By the time Whitehead became a Fellow of Trinity the inadequacy of this view of mathematics had been repeatedly shown. Thus, the field of number was enlarged when—a precise, systematic way of interpreting and using expressions involving the square root of -1 having been found—"complex quantities" were accepted. In 1833 a Cambridge mathematician, George Peacock, introduced the term "symbolical algebra" as a name for algebra which removes the restriction of its elements to numbers. Three such algebras appeared in the 1840s, each a landmark in the history of mathematics. Besides Hamilton's theory of quaternions and Grassmann's calculus of extension, there was the algebra of logic, created by George Boole in *The Mathematical Analysis of Logic* (1847); in it the logical relations between classes and between propositions were represented by equations. "Quantity" had come to mean *whatever* may be operated upon according to fixed, mutually consistent laws. Whitehead took the inadequacy of the traditional definition of mathematics for granted. He wondered what should replace it.

In the Lent term of 1890 Whitehead again offered his course of lectures on the *Ausdehnungslehre*. At this time he envisaged a large scheme of work, different parts of which occupied him throughout his years as a mathematician in England. The evidence of this is in a letter that I have published, with analysis and commentary.[13] Whitehead wrote it to the Provost of University College, London, on March 16, 1912, to state his qualifications for the Chair of Applied Mathematics. He said:

> During the last twenty-two years I have been engaged in a large scheme of work, involving the logical scrutiny of mathematical symbolism and mathematical ideas. This work had its origin in the study of the mathematical theory of Electromagnetism, and has

> always had as its ultimate aim the general scrutiny of the relations of matter and space.

Maxwell had concluded his *Electricity and Magnetism* with an argument for the necessity of a medium of propagation; his treatise was devoted to the details of its action. We may speculate that it was either while Whitehead was writing his fellowship dissertation on Maxwell's treatise or soon thereafter that there grew in his mind a strong desire to work out eventually a theory of the relation between matter (conceived as electrical and perhaps involving an ether) and space—in other words, between electromagnetic physics and geometry.

But the first task in Whitehead's large scheme of work was to make a detailed investigation of systems of abstract mathematical ideas that were being symbolized in various ways and applied to physics. Maxwell liked quaternions but was troubled by quaternion "methods."[14] Now, vector analysis, in one form or another, had appeared as one alternative; Grassmann's calculus was another. In recent decades the field of mathematical physics had seen a proliferation of symbols, formulae, and rival methods that were more or less clear. A detailed comparative study of the chief algebraic systems, in which the utility of each would be explored, was needed. And in the end we might see the general nature of symbolic reasoning, and so of mathematics.

Whitehead proposed to carry out the task just described in his first book, which he eventually named *A Treatise on Universal Algebra.*[15] Most of the first of the two volumes he intended would be taken up with the exploration and extension of Grassmann's ideas. In the Preface he wrote, "The book is the product of a long preparatory period of thought and miscellaneous reading." This reflective reading, which led him to entertain his "large scheme of work," was his main occupation in the late 1880s.

iv

In the dozen years from Whitehead's winning of his Trinity fellowship until he sent the first volume of *Universal Algebra* to the Cambridge University Press, he published only two papers. Both appeared early in 1889, in the *Quarterly Journal of Pure and Applied Mathematics.*[16] The papers were two parts of a single investigation that had nothing to do with Grassmann or with Maxwell. It concerned the general equations of motion for viscous fluids. Until 1918, Whitehead published no other paper in applied mathematics.

I suppose that his mathematical interest in the laws of fluid motion was first aroused by H. M. Taylor and by Stokes's lectures.* In 1885 Whitehead wrote an essay—never published, and so lost—on the general equations of hydrodynamics. It was his entry in the university competition for Smith's Prizes, awarded annually to Bachelors of Arts for the two best essays on any subject in mathematics or natural philosophy (i.e., physics).† Whitehead's essay was not thought one of the two best, but the judges took the unusual step of declaring that it deserved honorable mention; that is how we know of his special interest at this time in the general equations of hydrodynamics.

Viscous fluids (e.g., oils), because of their resistance to flow and their adherence to surfaces of vessels or obstacles, require separate treatment. Stokes had worked out the differential equations almost forty years earlier.‡ Complete solution of them is extremely difficult; but if the squares and products of the velocity components are discarded, as Stokes himself did, solutions applicable to slow motions are readily obtained; there was not much interest in rapid motions. This was the situation in 1889. Whitehead thought it worthwhile, taking Stokes's equations as a first approximation, to show a procedure for taking the second-degree terms into account—a procedure that could be applied repeatedly, so as to find a solution correct as far as terms involving the $(n + 1)^{\text{th}}$ powers of the velocities, when a solution involving their n^{th} powers is known. He gave his first paper, "On the Motion of Viscous Incompressible Fluids," the subtitle "A Method of Approximation." The method might be useful, he said, by making it possible to find the true limits of applicability of the general equations through comparison with the results of experiments. Readers of his philosophical magnum opus, *Process and Reality,* published forty years later, will remember his insistence that a metaphysical theory be stated "with the utmost precision and definiteness," so as to enable one to deduce its consequences and compare them with observations.[17] In all his philosophical writings he emphasized the seeking of ever-closer

---

*See Chapter VI, section i.

†These prizes originated in the bequest of Dr. Robert Smith, Professor of Astronomy and Master of Trinity in the middle of the eighteenth century. They had until recently been awarded on the basis of a special four-day examination; among the mathematicians, winning the First Smith's Prize was considered as great an honor as being the Senior Wrangler.

‡Because Claude Navier had derived them in 1821, they are sometimes called the Navier-Stokes equations.

approximations toward the precise, complete truth, which human beings can never reach.

In this paper Whitehead did not go beyond a compact presentation of the mathematical details of his method of approximation and two first-step applications of it—to the flow in a pipe, and to the motion of a viscous fluid in which a sphere is rotating. Do not jump to the conclusion that since he was writing two years after the Michelson-Morley experiment his real subject must have been the earth's motion through the luminiferous ether! He does not mention the ether, nor light, and had no hidden subject.

Whitehead's second paper, "Second Approximations to Viscous Fluid Motion," dealt entirely with the application of the method in the very important case, first investigated by Stokes, of the steady motion of such a fluid past a fixed spherical obstacle.* His attempt to improve upon Stokes's formula gave him a solution which, he pointed out, could not stand. One of the boundary conditions of the problem was that at an infinite distance from the sphere the motion of the fluid should reduce to uniform streaming, that is, be unaffected by the obstacle; in Whitehead's solution the component of the velocity that is due to his retention of second-degree terms did not meet this condition. He had not made a mathematical mistake; he had really shown that a second approximation to Stokes's solution is impossible.† The cause of Whitehead's paradoxical result was not discovered for many years.

The subsequent history of his paper, as I have learned from a scientific colleague at Johns Hopkins University, Professor Bruce Marsh, is fascinating. I could find but one reference to it in British scientific literature of the next decade. Finally, in 1910, a Swedish mathematical physicist, C. W. Oseen, showed its significance. Oseen proved that Stokes's equations did not have uniform validity as a first approximation; they broke down for motion at a considerable distance from the spherical obstacle; and he worked out equations that did not. In later decades the rise of aerodynamics stimulated research on the motion of viscous fluids past objects. The unacceptable conclusion of White-

*As Whitehead remarked, another statement of the same problem is to find the motion of a sphere moving steadily in a straight line through a viscous fluid that is otherwise at rest. Thus Millikan, in his famous experiment of 1911 to determine the electronic charge on an ion, used Stokes's formula to calculate the drag on an oil droplet falling through ionized air.

†Whitehead did not say this, but concluded, wrongly, that his result indicated that discontinuities must arise in the flow field.

head's paper was christened "Whitehead's paradox."[18] In a classic paper of 1957—ten years after the end of Whitehead's long life—two mathematicians at Trinity College observed that there is little reason to doubt that his iterative procedure, if applied to Oseen's equations rather than Stokes's, would eliminate the paradox.[19] They then presented the method that is now customary; it is a matching technique for combining the results of two series of approximations, one repeatedly correcting Stokes's equations applied to the fluid near the obstacle, the other repeatedly correcting Oseen's equations applied to the fluid at greater distances.* In this way, rather than by Whitehead's straightforward mathematics, successive approximations are reached to cope with phenomena that are more complex than he imagined.

I have no reason to believe that in 1910 Whitehead was still interested in hydrodynamics; he was beginning a long process of wrestling with what was meant to be the fourth volume of *Principia Mathematica,* with Whitehead solely responsible for it. Everyone knows of Bertrand Russell's discovery, early in their collaboration, of a contradiction (usually called "Russell's paradox") in set theory which affected the logical foundations of arithmetic and had to be resolved if the *Principia* was to be written. It took Russell six years to devise a theory of logical types, in a form that avoided the paradox and satisfied him, though the theory did not end discussion of the paradox. The contrast with Whitehead is amusing. I wonder whether Whitehead ever knew that a paradox in the theory of a branch of applied mathematics was named after him. He discovered it—without calling it a paradox, or realizing how pervasive it was—before Russell came to Cambridge as a freshman. Then he got busy on his *Universal Algebra.* Escape from "Whitehead's paradox" waited upon Oseen's work, twenty-one years after Whitehead's paper. Comparison of the results of experiments with deductions from Oseen's equations[20]—the sort of eventuality he hoped for when he wrote the paper—appeared only in 1919, forty years after the paper.

V

In May 1888 Trinity College raised Whitehead from his humble position as an Assistant Lecturer; this made him one of the six Lec-

*Technically, this is a power series expansion based on the "Reynolds number" (the product of velocity, density, and sphere diameter, divided by the coefficient of viscosity) for the flow.

turers on its mathematical staff. He was quite willing, in fact preferred, to lecture in the first college hours of the morning. That autumn, he took on an afternoon teaching job at Girton College, two miles northwest of Cambridge.

The purpose of Girton's founder and ruler, the formidable Emily Davies, was to educate women to the point where they could take Cambridge examinations, and to show that they could perform as well as men. Preparation for the Mathematical Tripos was provided by a resident instructor and several Cambridge dons, who at this time usually lectured gratis. Girton's financial records do not show which ones did so. As I have no reason to think that Whitehead needed a small addition to his income, I suppose that he was not paid, but simply did his part to extend a Cambridge education to women, now that he was a full-fledged Lecturer at Trinity.

The Cambridge dons at Girton were called tutors; what they did was essentially what Whitehead had done when he took a class in Trinity's scheme to abolish coaching. He was quite good at this, for he always took a personal interest in the strengths and weaknesses of each pupil, male or female. Bertrand Russell, remembering Whitehead as a teacher of himself and his friends, wrote, "He would elicit from a pupil the best of which a pupil was capable."[21] Girton students who had Whitehead as their tutor considered themselves fortunate.[22] Presumably he felt some nonmathematical interest in some of them, but I doubt that he did anything about it. The regulations that Miss Davies imposed on student life would in any case have made that extremely difficult. Girton was then an enclosed female institution. No men were allowed within its walls except the dons, and they only during their hours of teaching.

The unfortunate fact is that knowledge is lacking of Whitehead's personal life as an unmarried don. His habit of not writing letters has covered up his tracks so far as friendships with women—or men—are concerned. When he wanted to communicate with someone, he went to see that person. He was secretive about his private life—insofar as he had a private life.

The young mathematician had the faith of his Anglican forebears. His concern with religion was expressed in talk, and by contributions to the Church Missionary Society. Sir Edmund Whittaker wrote in the Royal Society's obituary of Whitehead: "As an undergraduate, he talked openly and often to his friends about religion, and especially about his interest in Foreign Missions. There was (he said) some difficulty in comprehending and following some of the divine commands, but the injunction to go out into all lands and spread the gospel was as plain as words could be."[23] I do not know how much support Whitehead gave to Foreign Missions when he was an undergraduate, but when he began to earn his own keep, in 1885, he started to give a guinea a year, which would have been considered a substantial contribution in those times. The Church Missionary Society was run by the Evangelical party of the Church of England; that was his father's party. The Society for the Propagation of the Gospel, which was run by the High Church party, received nothing from Alfred; his brother Henry, who was seven years older, would have preferred support of it. When Alfred was taking the Mathematical Tripos, Henry was leaving a responsible academic post at Oxford to devote the rest of his long life to the Anglican Church, in India; at one time he was head of the Oxford Mission in Calcutta. But there is no record of Alfred engaging in any activity on behalf of Foreign Missions other than making financial contributions.

In the mid 1880s a different kind of missionary activity had got under way in the slums of East London, and Whitehead became involved in it. An account of its origin and character will round out our view of young Whitehead's life. In his old age, talking with his wife to Lucien Price, he recalled that although the conditions of life in the years from 1880 to 1910 were ideal for moderately comfortable university people, there were "a great many things which needed changing." His point was, "We intended to change them and had set about doing so."[24] One of the facts that stood in the way of change was the educated middle class's gross ignorance of the life of the poor in the great cities. Whitehead's uncle Henry had known their life when he was a priest in Stepney and Limehouse. In the early eighties Samuel Barnett knew it well, from many years as a Vicar in Whitechapel. A few books had been published. Then, in the autumn of 1883, a great journalist, W. T. Stead, launched a successful campaign in the *Pall Mall Gazette* to arouse the public about the misery in London. Many young men who were living civilized lives in the universities felt the pricks of a social conscience. In 1884 Barnett founded Toynbee Hall, the mother of all settlement houses.* There, students or recent graduates of Oxford and Cambridge, especially those who did not yet know what to do with their lives, lost their ignorance of the poor by living

---

*Visits to Toynbee Hall inspired Dr. Stanton Coit to found the first American one, the Neighborhood Guilds Settlement in New York City, and Jane Addams to found Hull House in Chicago.

in their midst and doing a great deal of educational and social work. Whitehead did not take up residence in Toynbee Hall, and I cannot suppose that he considered doing so. He liked the mathematical and educational work he had begun, and I assume he felt that this was the most valuable thing he could do. I also assume that when he said to Harvard students in the 1930s that for the great majority of men it was important to find, grow in, and keep at a job year after year after year, Whitehead was not only being sensible, he was recommending what he had done in his own life. He was simply not the sort of man who abandons his chosen daily work to give himself utterly to something quite different that has captured his heart. What Whitehead did for Toynbee Hall was to serve for more than two years as Secretary of the Cambridge General Committee of the Toynbee Association. He was good at bringing people together for a worthwhile purpose.

Barnett made his first appeal to university men at Oxford, in February 1884, with great success. Social idealism was strong there, thanks to T. H. Green, Ruskin, and others. They had inspired a religious dedication to social reform in the breast of Arnold Toynbee, an economic historian and tutor at Balliol College who had died young in 1883.* Three months after Barnett's success at Oxford, but long before the opening of the Hall named in memory of Toynbee, there was an enthusiastic meeting at the Guildhall in Cambridge. Whitehead's friend D'Arcy Thompson and Professor James Stuart (an Apostle) were among the speakers. Soon Trinity had its own Toynbee Association and guiding committee, of which Whitehead was a member in 1886. It was after a meeting of it in his rooms, early in 1889, that he became Secretary of Cambridge's General Committee.

In the middle and late eighties, support for settlement houses was almost fashionable; and it had a patriotic color, for the mutual isolation of the uneducated city poor—it was only in 1870 that the three R's became compulsory—and the educated middle class was an obvious weakness in the British nation. (So one of the speakers at the initial Cambridge meeting was the Prince of Wales' oldest son, then a Trinity undergraduate.) A symptom of the movement's popularity was the incredible sale of Mrs. Humphry Ward's early novel, *Robert Elsmere;* it came out in February 1888, and soon sold more than a

*Toynbee's *Lectures on the Industrial Revolution of the Eighteenth Century in England* (originally given to workingmen, posthumously published in 1884) established the popular use of the phrase "the industrial revolution." The historian Arnold Joseph Toynbee was his nephew.

million copies. It was the story of a young priest who left the Church of England to found a Brotherhood of those devoted to social Christianity without church dogma.

Groups of alumni of public schools and women's colleges, as well as university men, took to founding settlement houses. Most of these were to some degree religious, but Toynbee Hall was not. Barnett, a Low Churchman like Whitehead's father, insisted that a settlement was not a mission. It carried on many activities, with no set program other than emphasis on cultural and civic rather than vocational education. Whitehead may have given a few University Extension lectures on mathematics, though that is doubtful.

After 1891 his name disappears from the Annual Reports of Toynbee Hall. He had married, started to raise a family, begun work on a big mathematical treatise—and so stopped drumming up support for an institution whose success, by that time, was assured. He did not cease to approve of Toynbee Hall, nor did his wife; their daughter remembers meeting Barnett.

The *Cambridge Review,* in its news of extracurricular activities from 1885 to 1890, mentions Whitehead *only* in connection with Toynbee Hall. In choosing to work for it during his bachelor years, he was simply doing the sort of thing suggested by the model of a useful life that had been set before him by his father and paternal uncles and grandfather. Later, the Bloomsbury set's lack of interest in "outcast London" was one reason why Whitehead did not approve of them.

vi

Scanning the nontechnical English journals in which a young Cambridge don might publish a short piece, we find, in this period, three written by Whitehead. Two appeared in the *Cambridge Review* in 1886 (on February 10 and May 12); they are signed "W," and were identified as his by Whitehead's daughter, Jessie, when I was compiling a bibliography in 1941. The third, signed "A. N. W.," appeared in the *Cambridge Fortnightly* in 1888. All three are written in the first person as explicit fantasies. If he ever tried his hand at fiction again (which I doubt), he did not publish the results.

The earliest piece is "A Celebrity at Home: The Clerk of the Weather." ("Celebrities at Home" was a popular feature in Edmund Yates's "Society" weekly, *The World.*) It tells of conversations with a sad old gentleman who "has been thrown out of employment by the progress of science." The Clerk of the Weather had been under contract to deliver down on the earth all the water drawn up by the sun; apart from that, he managed the weather as he pleased. Sometimes he was able to send the rain away so a child could play out of doors. Alas, most children forgot him when they went to school. He tells Whitehead, "It is a horrible thing to exist only in other people's thoughts. You are absolutely at their mercy; . . . I go wandering about from mind to mind, like the soul of a departed Positivist." To make things worse, "Now I am no longer the Clerk of the Weather, for those Kew people have made me resign, so what am I?" (Charles Chree of King's, who had placed immediately below Whitehead in their Tripos, had taken up meteorology, and would soon become Superintendent of Kew Observatory.) The Clerk denounces the new scientific weather predictions as so much "tomfoolery." Whitehead then says, "It seems to me that there was a good deal of nonsense in your time," whereupon the old man stamps off, deeply offended. The sketch ends, "I have not seen him since, and I am afraid he is not getting on very well, for he has but few friends, and is rarely spoken of. So I thought it well to set down on paper what I can remember of him, as it would be a pity if the old gentleman were quite forgotten."

Whitehead was all for the improvements in human life that come from the application of scientific discoveries, but he liked to think of the old ways too, and to hope that they would not be quite forgotten. The young mathematician was fond of history. His fantasy is more charming than it appears in my account, but it is not the work of a man who writes because he has to. Just as Whitehead made only one speech in all his years as a member of the Cambridge Union, so his writing outside his profession was very occasional. No English editor kept in touch with him. This first piece was printed in the small type which the *Cambridge Review* used for space fillers and book reviews. It deserved better. His second sketch, "Davy Jones," was printed in type of proper size. It too presents an old-time character, is equally charming, and perhaps a bit more accomplished. The text shows unmistakably who wrote it, for it begins, "One fine summer's afternoon I was walking on the shore near the North Foreland," which was in his father's parish on the Isle of Thanet; and when the narrator tells Davy about "the fish-torpedo" (the self-propelled device invented by Robert Whitehead), Davy Jones congratulates him warmly "on my relationship, as he supposed, to the inventor, assuring me with great politeness that his interest in the next sea-fight would be vastly heightened by the effective use of the weapon." He had not really much interest in naval battles, saying, "It is a miserable way of obtaining men—dead before they reach me; as far as I am concerned they might just as well have died on shore. Two men on a raft give me more innocent amusement than ten battles." Davy had had great disappointments: the survival of Noah's Ark, of Ulysses ("a mean-spirited, tricky fellow, always baulking just expectations"), of Aeneas, and, most grievous, of Caesar and all his fortunes. But he had praise for modern commerce. "Such dash, such enterprise; none of that petty calculation which, though it may save a life or two, is the sure sign of a small mind." This fantasy, like the other one, ends with its subject expressing resentment and walking away. The narrator had asked him what he thought of Plimsoll (the member of Parliament known as "the sailor's friend") and Chamberlain. " 'They are more offensive to me,' he answered warmly, 'even than Jonah's whale.' "

Whitehead grew up beside treacherous waters. In 1927 he wrote, "I always suspect that Julius Caesar and his Roman successors had colossal luck in getting across and in getting back."[25] This fantasy is not nostalgic; Whitehead was taking pleasure in contemplating the role of chance in human events.

The *Cambridge Fortnightly,* edited by his friends and fellow Apostles Roger Fry and Nathaniel Wedd, lasted for five issues. Whitehead's fantasy, "A Visitation," was published in the fourth, dated March 6, 1888. He begins mildly enough, by remarking how pleasant it is to watch family parties as they stroll through a Cambridge college in the May term. Then Whitehead passes to the pleasure of rambling alone through the quads and communing with the ancient buildings on a summer afternoon while the students are away boating or playing cricket. "But sometimes I have visitors of my own to attend to." He imagines conducting the united guilds of Corpus Christi and of the Blessed Virgin Mary through Corpus Christi College, which they founded in 1352.

His visitors are startled by the dress of the college porter, but understand that such things must change in the course of five hundred years, and when a Fellow of the college turns up in flannels and a boating blazer, carrying a tennis racket, they think him the Master of the college, "or at least a priest fresh from some festal service in honour of their peculiar saint." Whitehead introduces the company to the Fellow, who takes over the conversation with them. He says, "You will find that in some points we have deviated from the old paths. Religion is now based more directly upon study of the Holy Scriptures than it was perhaps in your times." The master of the guild observes that this is a very great change, but that with the approval of

His Holiness the Pope even the Scriptures, he supposes, might be sparingly read. He is next informed, "'We have discovered that the authority of the Pope himself is not based on any sound scriptural warrant.' 'You do not obey the Pope?' gasped out the master. 'But the patron saint of our college—tell me—you worship her?'" The negative answer is not finished when, with a faint "Oh!" the master of the guild vanishes, and his companions with him.

The story includes a reference to the abolition of the ecclesiastical character of Cambridge colleges, which was effected by statute in 1871 and 1882. The Fellow of Corpus Christi is represented as having opposed this change, protesting against disregard of the pious wishes of founders; so he felt he deserved well of the visitors. When they vanish he turns on his heel and departs, not wanting to talk about *their* pious intentions. Whitehead is left alone, to reflect that Cambridge is haunted by ghosts from its pre-Reformation past.

Is this fantasy nostalgic? I find it hard to judge with any confidence. The question is prejudiced for me by my knowledge that for several years Whitehead carried on a private debate with himself about changing his religious commitment from Anglican to Roman. That will be described in the next chapter. I should guess that "A Visitation" was published almost two years before the private debate got fully under way. But as I noted earlier,* there is reason to suspect that by 1886 he was sometimes wondering whether the religion of the Church of England was the real thing. Whitehead never published another ghost story. Why did he choose the fourteenth century for this one? Perhaps because he grew up taking pleasure in the contemplation of beautiful old buildings and what went on in them when they were new. This in itself is not nostalgia so far as religion is concerned.

All through the 1880s, Whitehead lived pleasantly in Trinity College. After his first two years as an undergraduate he was moved from the noisy rooms in Whewell's Court on Trinity Street to the quiet of Nevile's Court, where he occupied C.4, in the middle of the south side. From 1886 through 1890 he enjoyed a large, high-ceilinged set of rooms (C.2) there. His position as a Teaching Fellow of Trinity was secure. But as he approached his thirtieth year, his mind was troubled.

---

*See Chapter VII, section vii.

CHAPTER IX

Whitehead's Thirtieth Year

i/The question of religious commitment. Whitehead's visit to Cardinal Newman. Possibility of conversion to Roman Catholicism.

ii/Whitehead's attitude toward women and sex.

iii/Evelyn Wade.

iv/The Wade family.

v/Evelyn's independent spirit.

vi/Whitehead and Evelyn Wade: Contrasts and complements.

vii/Proposal and marriage.

viii/Whitehead's increased self-confidence. His *Treatise on Universal Algebra*.

The thirtieth year of Whitehead's life began on February 15, 1890. We know nothing about that day, but we can say that doubts about himself had been growing in his mind, and that he wondered what shape his life should have. A year later, only one question still hung fire: what religious commitment he ought to make. In every other respect, his path was clear and he was on his way when his thirtieth birthday arrived.

His daily life for the last five years had been taken up by the two things he most enjoyed—teaching, and mathematical research. But he was worried by what he took to be his lack of success in both. He found himself horribly inferior to his friend Forsyth, who was only two years older. The comparison was inevitable, but unfair to himself, for Forsyth was phenomenal: his lectures were letter-perfect, and his rise as a research mathematician was meteoric—indeed, too bright to last. Still, Whitehead could not blink the fact that in these years he had not accomplished much. He was content to have published only two papers while he was absorbing and mulling over Grassmann's work, but he seems to have lacked the self-confidence to start writing the full-scale treatise on it that he planned. He was happy with Cambridge and Trinity College. Except for the relatively few professorships and college headships in the English academic world, there was no higher honor anyone could hold in it than that of being a Fellow of Trinity. Whitehead felt that he had not proved himself worthy, and from time to time he talked to older Fellows about resigning his fellowship. Fortunately this was not encouraged.

He was still a bachelor. This did not worry him. Nor was it unusual for a don of twenty-nine. Men of the middle class married later in those days than they do now, and did not propose marriage until they could support a wife and raise a family. Also, Whitehead was a romantic who would marry only for love. If he had been seriously in love with someone for a time, we have no record of it. When it would happen in the future, God only knew.

If marriage for love was not his destiny, perhaps the life that God intended him to lead was that of a monk, pursuing his beloved mathematics in a monastery rather than in Trinity College.* He was now thinking of leaving the Church of England for the Roman Church. This became public knowledge only after his death, when Bertrand Russell, who was his most intimate friend through the two middle decades of Whitehead's life, said in a BBC talk about him: "He was at all times deeply aware of the importance of religion. As a young man, he was all but converted to Roman Catholicism by the influence of Cardinal Newman."[1] We do not know when this influence began, but, as was noted earlier,† Whitehead may have had Newman's *Apologia* on his mind in 1886. And the *Grammar of Assent* offered a reasoned answer to the final question: What, of highest positive value, lies outside the competence of mathematical demonstration and inductive inference? Newman was not only a keen thinker; he was a historian, and a saintly man who drew one's sympathy.

I am not aware of any admirers of Newman among Whitehead's Cambridge friends. His admirers elsewhere were inclined to say that no Cambridge man could appreciate him. Cambridge men (except Whitehead) said, Yes, common sense and the dry light of reason are stronger here.

Whitehead, however, must have had earnest discussions with his Anglo-Catholic brother, Henry. Harry, as the family called him, visited his parents in Broadstairs in the summers; he had gone to Calcutta in 1883 to be the Principal of Bishop's College, an institution founded sixty years before by the Society for the Propagation of the Gospel to train native preachers, catechists, and school teachers. In 1889 he took on the added responsibility of heading the Oxford Mission to Calcutta, and extended its work beyond the city. He had given up his fellowship and position as Bursar at Trinity College, Oxford, to spend his life in a way that Alfred admired. Harry had seen Newman.‡ Would that there were some extant record of the talks that Alfred, on

---

*I am saying *only* that this thought sometimes occurred to him. My source is Jessie Whitehead, who had it from her mother. Mrs. Whitehead may have been tempted to magnify the difference she made in his life.

†See Chapter VII, section vii.

‡Newman, having been elected an Honorary Fellow of his old Oxford college, Trinity, visited it in February 1878. Although Henry Whitehead was then on leave of absence, the College annually invited Newman to the Gaudy on Trinity Monday. There is no reason to doubt that Henry Whitehead saw him in 1880, and perhaps again in the next two or three years. See President of Trinity College, "Whitehead and Newman," *Oxford Magazine,* May 8, 1949, p. 290.

summer visits to Broadstairs, had with his favorite brother! All we can be sure of is that Harry, however sympathetic to Newman, maintained that his reasons for leaving the Church of England were insufficient. Finally, Alfred decided to go to the Oratory at Edgbaston, near Birmingham, in the hope of talking with the Cardinal.

Whitehead's son North told me he had it from his father that he made the trip and saw Newman. There is no written record of the visit.[2] The Oratory did not keep a visitor's book until after the turn of the century, when Newman's room had become something of a pilgrimage point. If Whitehead, no diarist, wrote down anything about the occasion, it is not extant. And at no time in his life was he given to talking much about anything he considered private; his religious quest was decidedly private. But something came out one Sunday evening in the academic year 1932–33. It was near the close of an "at home" for his Harvard students; the circle that was talking to him and the circle talking to his wife had become one. She remarked on the unfairness of Lytton Strachey's brilliant *Eminent Victorians,* and followed this by saying that when "Altie" was young he had an interview with one of them, Cardinal Newman. Whitehead sat quietly, and said nothing. The strange silence was broken by a student asking him what Newman was like. Whitehead was still silent, then suddenly said, "He was wonderful." "But," he added, "he was very old then."[3] The same impression, of an event so significant and personal that Whitehead did not want to talk about it, comes from Lucien Price's report of a conversation in September 1945. Sir Richard Livingstone and Whitehead were discussing university education. When Livingstone brought up Newman, Whitehead said that he had once seen and spoken with him; but when Livingstone asked if he could remember what Newman said, Whitehead answered, "It is too long ago," looked preoccupied for a moment, then abruptly changed the subject.[4]

I cannot say when his religious impasse reached the point at which he took the train to Birmingham. Mrs. Whitehead, in telling their children about the visit, connected it with his proposal of marriage to her. Their daughter, Jessie, told me she was given to understand that her father at the Oratory talked with another priest, who, seeing that something other than Rome-or-Canterbury was on his mind, asked him, "Why don't you first marry your girl?" and responded to "I don't think she'd have me" with "Have you asked her?" If the story is true, the only likely time for Whitehead's trip to Edgbaston was early in the summer of 1890. (Newman died on August 11.) But I would not rule out the possibility that this dialogue was supplied by Mrs.

Whitehead to flesh out the account her husband gave her. If Whitehead saw Newman before he met her, his visit could have occurred in the first half of 1890, or even in 1889.*

Newman had been very frail, quite deaf, and almost blind since 1886: not up to an interview with a stranger. After a very few minutes Newman's secretary would have intervened and led Whitehead away, to talk with someone else about joining the Catholic Church. The result, to Whitehead's mind, was inconclusive. But he never forgot his few minutes with Newman. And the man remained high in his esteem. In philosophical lectures at Harvard he recommended reading him, and sometimes went so far in praise of Newman's merits as a thinker as to call him "the most profound mind of the nineteenth century."

ii

Before describing the woman to whom Whitehead successfully proposed marriage (whether weeks or months after his visit to Newman), let me set down my only available information about young Mr. Whitehead's attitude toward women and sex. These subjects were discussed in six meetings of the Apostles not mentioned in Chapter VII. This was in the years 1884–88, after which he went to almost no meetings. At one of the six, a paper was read on the question, "Shall we alter the marriage service?" At the end there was a division on the question, "Do we agree with Peter, himself a married man?" Whitehead voted No. He would not want to think of his wife as "the weaker vessel," subject in all respects to his kindly authority. Rather, each should gain from the other. On another occasion there was a vote on whether marriage should be terminable. No, he said. I take this as less a sign of unbending dogmatism than a sign that he entertained an idealistic conception of marriage. Incidentally, the moral philosopher

*We may set aside, as either wishful thinking or misinterpretation of what he was told, Bertrand Russell's notion that the visit never occurred. When I wrote Russell in 1959, asking whether the influence of Cardinal Newman on Whitehead which he had mentioned in *Portraits from Memory* involved personal contact, he replied (September 26, 1959) that Whitehead "was about to go to Birmingham to consider conversion when he fell in love with Mrs. Whitehead and forgot the Cardinal." This is at variance with the statement in Volume I of Russell's *Autobiography* (p. 150) in which he illustrates Whitehead's liability to "impulses which were scarcely sane" thus: "Before he met Mrs. Whitehead he had made up his mind to join the Catholic Church, and was only turned aside at the last minute by falling in love with her."

Henry Sidgwick, who happened to be present, disagreed with him; but Sidgwick was a much older man who knew of divorces that had been made in heaven and marriages that could not be legally terminated when a spouse became hopelessly insane. One day in 1885 there was a division on "Is purity binding to the same degree on men and on women?" Whitehead voted No, with the comment, "always admirable." My guess is that when he married he had no prior experience of sexual intercourse; in those days, that would not have made him unique. When a paper was read on the question, "Would one sex be better than two?" the vote was on the question, "Shall we abandon the system of cooperative reproduction?"—in other words, would it be better, if it were possible, to perpetuate the race without sexual relations? Whitehead voted Yes. It is not far-fetched to see this as an indication that he was afraid of sex. Certainly he felt that a man's life should be devoted to other things; he voted No on the question, "Is our chief function to produce spermatozoa?" which closed a meeting in 1888 when Roger Fry read the paper of the evening. Finally, it is amusing, in retrospect, to note that the question voted on at the very first meeting of the Apostles he attended, in May 1884, was, "May we offer you a cigarette, Madam?" He was not a smoker and never became one. Neither was his bride, but she emphatically became one.

iii

Her name was Evelyn Wade. It appears on their certificates of marriage as Evelyn Ada Maud Rice Willoughby-Wade. Whitehead met her while visiting his parents at Broadstairs soon after the end of the academic year in June 1890. His aunt Susan was living in Bleak House—yes, Dickens's Bleak House, then called the Fort House—with her daughters, May and Annette, who had invited their friend Miss Wade to visit them.

Evelyn was almost five years younger than Alfred. In contrast to his bright blue eyes, hers were black. She was half an inch taller. I know of no photographs that were taken in 1890; I must rely on a series of family photographs beginning in 1893, and their daughter's memory that Mummy's somewhat curly hair was a striking auburn. Every feature of the longish face shows strength; here was a woman who would never bow to circumstances. You would not call her pretty, but handsome. She was high-spirited; Whitehead liked that. Stunning and not unseductive in looks, she cared a great deal about appearances, had an instinctive tendency to dramatize, and under other circumstances could have been a successful actress. Her face could be demure on occasion, but normally showed an aristocratic pride. "Gallant" was her highest word of praise, and her values were close to those of the Three Musketeers.

Evelyn Wade cared a good deal about family, not about her father and mother personally—he was dead, and she didn't get along at all with her mother—but about family status. That was why Alfred thought she would not have him: she would be marrying into the middle class.

iv

Evelyn's mother was a Harley: Lucy Maria Rice Harley. Her family, according to Evelyn, was that of the earls of Oxford, the first of whom was the intriguing statesman Robert Harley, briefly popular in Queen Anne's reign. Unfortunately the genealogical reference books do not mention Lucy Maria Rice Harley. I conclude that she was either an illegitimate and unacknowledged daughter of the sixth and last earl, Edward Harley, who is recorded to have died without issue, or an illegitimate granddaughter of the fifth earl, Edward Harley, whose beautiful wife had so indifferent a reputation that her children were called "the Harleian Miscellany" (her name was freely coupled with Byron's and the Duke of Cumberland's).* I do not doubt that Evelyn Wade's mother was a Harley, but her descent from the earls of Oxford does not seem to have been recognized by law.

I know nothing about the character of Mrs. Wade except that Evelyn could not stand her, so the two did not visit. There is a story in the Whitehead family that she once horsewhipped a man.

The Wades were so different from the Whiteheads that it will be worthwhile, in view of the influence Evelyn would have on Alfred's perceptions of human life, to see who they were and what they did.

They had been landed gentry in County Meath, Ireland, since the reign of Charles II. Some were High Sheriffs, but most had Army careers. Perhaps Field Marshal George Wade (1673–1748) was one of them, as Jessie Whitehead asserted. From Burke, one learns only that he died unmarried but acknowledged two natural sons. Since he fought in Spain when young, this would explain Evelyn's

*According to Cockeyne. "The Harleian Miscellany" is also the name given to the collection made by the bibliophilic first and second earls of Oxford and purchased by the Government as an original part of the British Library.

occasional remark that there was Spanish blood in her veins. Anyhow, he was a brave soldier, and should be praised for the judiciousness and slowness with which he disarmed Highland clans (while building military roads). In the mid nineteenth century, there was a man, the gallant Right Honorable Arthur MacMurrough Kavanagh, who was related to the Wades. Jessie urged me to learn about him if I would understand her mother's values. Born with only rudiments of arms and legs, he made himself an expert horseman, writer, painter, and sailor; he traveled widely, and was in Parliament fourteen years. Also, he helped tenants, by rebuilding houses and introducing lace-making.

Evelyn's paternal grandfather was a Colonel who served and married in Ireland. The oldest of his three sons, Sir Thomas Wade, was now a professor at Cambridge. He had not gone up the academic ladder to get there! After one year as an undergraduate at Trinity he had entered the Army, and after a few years got himself into a regiment that was under orders for active service in China. He loved to study languages, and made learning Chinese his business during the six months' voyage. Soon he was interpreter to the garrison at Hong Kong. Giving up the Army for the Civil Service, he spent many years in Peking as a successful negotiator with the Chinese. Between crises he wrote books on the language. When he retired he settled in Cambridge, gave his library to the University, and was appointed the first Professor of Chinese.

Evelyn's other Wade uncle also was successful, in a less exciting way; he became a banker. Of his many children, none entered the academic profession. Evelyn's father, Arthur Robert Willoughby Wade (no hyphen), was the youngest son and the failure in the family. The only life he knew was that of the military, in which his career was undistinguished and rather short. When he was nineteen his father bought him an ensign's commission in the Army. At twenty-one, he got by purchase a lieutenancy in the Duke of Albany's Own Highlanders, later called the Seaforth Highlanders; that was how officers' ranks were filled in those days. But it was seven years before he had enough money to buy a captaincy, in 1853. He was then on "home service." Wade enjoyed his captaincy one year only. The regimental records show that he was not cashiered, but it looks as if he was asked to sell his commission and leave the regiment, probably for being too familiar with another officer's wife. Since Evelyn had several siblings, only one of whom was certainly legitimate (as she was), it appears that her mother had been his mistress for several years before they married.

They were living in County Westmeath, on very little money, in the autumn of 1865. Anti-English feeling was running high, and after shots were fired into his house ex-Captain Wade took his family to France. On the boat his pregnant wife went into labor. The boat docked at Honfleur, across the Seine from Le Havre, and there Evelyn was born on December 18.

The family stayed in France, living at Pont-Aven, in Normandy. There was nothing her father could do but offer his military services to foreign governments. He is believed to have died in Spain in the course of the Carlist wars of the 1870s. Evelyn always referred to him as Captain Wade of the Seaforth Highlanders. But she did not talk about him, or her mother or siblings. She made her own life.

V

At Pont-Aven the servants were French, and Evelyn was schooled in a convent at Angers, so French was her native tongue. Though the Wades were Anglicans, the convent school was natural enough in the circumstances. The confessional exercised her creative imagination; she had to confess *something*. She was determined to be independent, and early found that it took deception and acting to stand up to her mother and the nuns.

When Evelyn was seventeen, she went to England to earn her own living and to be completely independent. She was armed with the military virtues of the Wades. Circumstances existed in order to be overcome. In every situation, the thing to do was to take command of it. Never let the other person stop you with, "Don't bother, it's my problem."

She was an officer's daughter, but the only work she could do was that of a governess, teaching French to British children. A distantly related family, living at Tenby in Wales, employed her. There are two stories about the end of this job: one, that her employer tried to seduce her, and she would not have it; the other, that she wanted to be treated like a relative, not like an upper servant, and he would not do that. It could be that both stories are true. In any case, she went to London, and continued to earn her living by teaching French. She also worked at improving her English; that showed in the clipped tones with which she spoke it all her life. The first evening she and Whitehead spent together, her English was not yet good enough for him to understand just what she was saying; his French was much worse; so they took to looking at pictures.

vi

Alfred Whitehead's feeling for Evelyn Wade was a once-in-a-lifetime-and-for-a-lifetime business, the enduring romantic love that was not quite as exceptional in Victorian times as it is now. She was handsome, gallant, generous. Most important: she was more *alive* than any person he had ever met. Whereas he was quiet and restrained, she was active and outgoing. Married to her, he would have a greater sense that he was fully alive and participating in the real world. In the end, however, romantic love is a mystery. There are no data to explain why he was susceptible to it.

What was her initial feeling for him? She was always a good judge of character, and saw that his was almost saintly; he was the type of man a dominating woman marries. At the same time, he had his own kind of strength; she could respect him, and let him carry on his work on his own terms. Their children would suffer from her possessiveness, but he did not. He needed her support. She once said to their daughter, Jessie, that she accepted his proposal of marriage because "I never saw a man who needed me more." I should not want anyone to suppose that Whitehead had a pathological need for a wife who would dominate him. Not at all. The point is rather that her self-confidence was complete, his was not. Also, he was too wrapped up in abstract thought to tend to his environment and to practical problems; he had little sense about money, but could be terribly anxious about financial security. She would take care of all these things. He satisfied her primary need, which was to be needed, emotionally and practically.

At twenty-four, what could she do that would be better than marrying him? She had no dowry. If she waited for a proposal from some man of the Wade type—she loved military uniforms—it would be unlikely that she could give that man as much as she could give Alfred Whitehead. He would not infringe on her cherished independence. She may have sensed that he had a touch of intellectual genius; I do not know. She saw that he had a good position in the world, even if it was the academic world; she might not like that world, but with her sense of superiority she would not be troubled in the least by being a parvenu in it. Married to him, she would be free of having to earn her keep; she could devote herself to him, to having children, and to trying out her talents at homemaking (these were exceptional, as will be noted in the next chapter).

The marriage was a good bet for both of them.

After it took place a Trinity man who knew them exclaimed,

"Whitehead has made a great mistake. He has married out of his tripos!"[5] Indeed, Whitehead could scarcely have gone farther away from the British intellectual aristocracy, or even from the kinds of women he had been teaching at Girton. Evelyn could never have passed any kind of Cambridge examination. If it were in French irregular verbs, or on some French memoirs or collections of letters, she could do well. But history was boring, mathematics was unintelligible, and no science meant anything to her. She was not interested in knowledge, only in people, and in beautiful things. Truth was not an object of research, but a good story. The primacy given to beauty—over truth and goodness—in her husband's final philosophy is primarily (though not entirely) an effect of his life with Evelyn; so is the special value he gives to "Appearance." She herself did not have learning; she had style.

vii

Whitehead was a thinking romantic: he planned his proposal of marriage. Under the garden of his father's Vicarage there was a large cave, where smuggled brandy and French lace used to be stored. He asked to show it to her, and made his proposal inside it. He was surprised by her acceptance. If he had known her better he would not have been surprised.

This was in the summer of 1890. Whitehead was so pleased by the prospect of life with Evelyn that by the time the Long Vacation ended, he had agreed with Gonville and Caius College to take a 99-year lease on a parcel of land the College owned on the West Road, with a view to building a house there. It was generally felt that this was what a don and his bride should do, if it were financially possible. Whitehead was getting £550 a year from Trinity, as a Fellow and Lecturer; probably he had additional income, stemming from his Buckmaster inheritance.*

Although Whitehead's mother was shocked by the fact of Evelyn's convent schooling, his father heartily approved the marriage, for he was aware of Alfred's insufficiency of self-confidence, thought it would be a pity if he joined a contemplative order, saw that he was madly in love, and supposed that Evelyn might give him what he needed. Professor Sir Thomas Wade was pleased that a Fellow of

*See Chapter II, section vi.

Trinity would be his nephew-in-law, but the other Wades thought that Evelyn was marrying beneath her station.

They were married at the beginning of the first vacation of the next academic year, on December 16—two days before the bride's twenty-fifth birthday. The ceremony was performed in St. Jude's Church, South Kensington, London, by Alfred's brother Henry, assisted by their father. Forsyth was the best man. Evelyn's mother was not present; the two were not on speaking terms then or afterward.

The day of the wedding was an unusually cold one: fog above and snow below. The Whiteheads went to Windsor for their honeymoon. When they arrived they found a red carpet laid down, and a bouquet was handed to Evelyn. She was delighted, but her husband noticed that an older woman was coming up, a woman of some exalted position for whom the red carpet and flowers were intended. To his wife's annoyance, he took the bouquet from her and gave it to the right woman. Evelyn had the better eyes for beauty, Alfred for facts.

viii

Whitehead's marriage had an immediate effect on his mathematical career. It was shortly before he met Evelyn that he had envisaged the large scheme of work described in the last chapter.* It would begin with a big book, using Grassmann's *Ausdehnungslehre* as point of departure. But he had not made the firm commitment that would be marked by a plunge into writing it. Was he up to the job? When he was finally a famous man Evelyn told his assistant at Harvard, "Altie said to me when we were engaged, 'Now, you know, you're marrying a failure.' "[6] But when the Cambridge University Press published Volume I of his *Treatise on Universal Algebra* (586 pages, quarto) and he presented her with a copy—of which she could understand nothing—he wrote on the flyleaf: "Evelyn for whom this book was written, from Alfred. Feb. 18th, 1898. Began Jan. 1891."

I do not mean to suggest that but for Evelyn this book would never have been written, but only that marriage to her spurred him to getting started on it. The inscription on the flyleaf may be an overstatement if taken literally. Whitehead was smart enough to know that she needed to believe in the absoluteness of his need for her.

Two weeks before the marriage, the Council of Trinity agreed to

*See Chapter VIII, section iii.

enter in its minutes "that Mr. Whitehead has been continuously on the educational staff of the College since October 1884." I wonder why, in his old age, he misremembered the year as 1885, and suspect that at some time he fell into the comforting habit of shortening the length of time that he had been a rather lonely Fellow and Lecturer, working away at mathematics without the personal support that Evelyn gave him.

Thomas Whitehead (1790–1877)

Alfred Whitehead (1827–1898)

Sarah Buckmaster Whitehead (1832–1924)

Bird's-Eye View of Ramsgate, 1861

St. Peter's Parish Church in Thanet

Opposite: Detail from a Map of the Isle of Thanet: Churches in the Vicinity of Ramsgate, ca. 1790

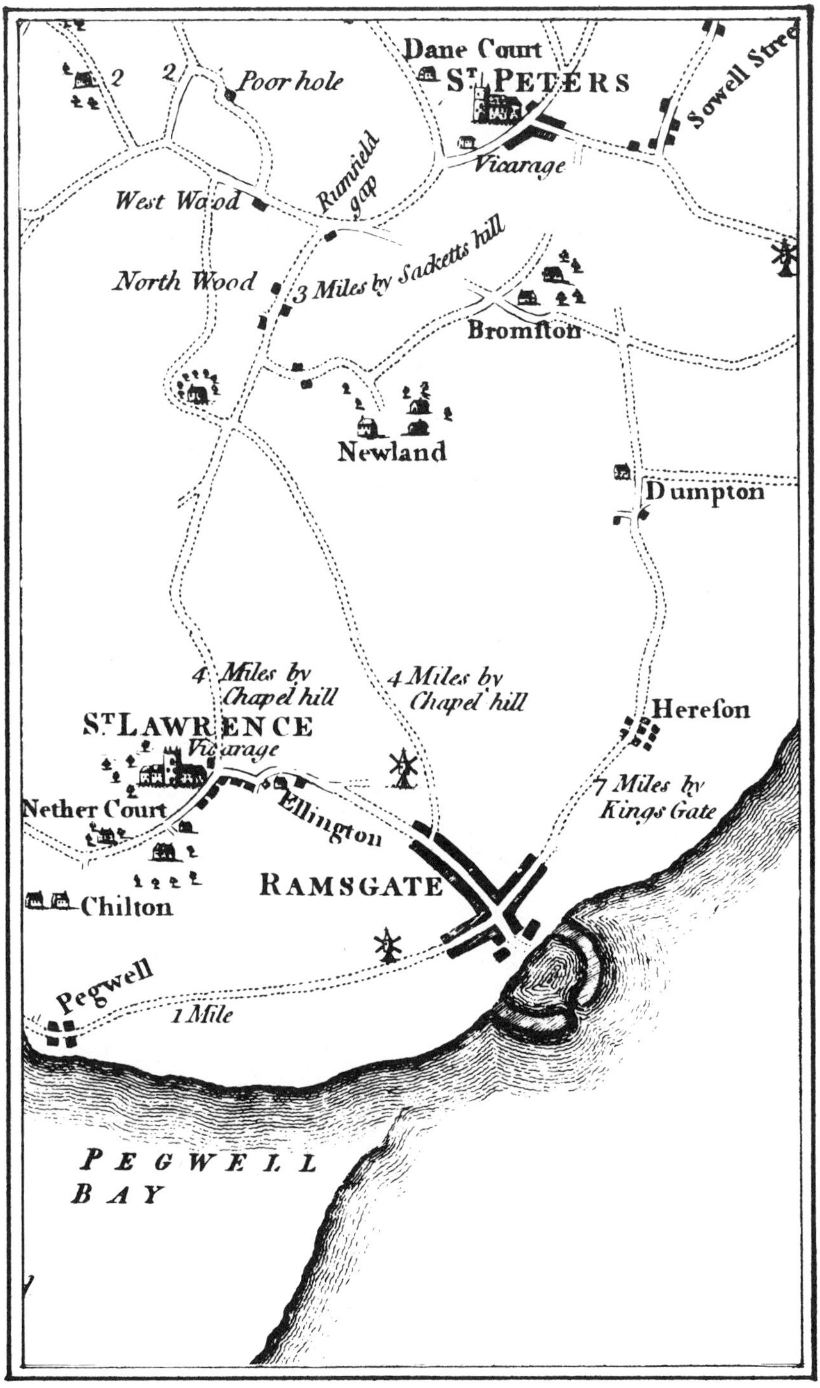
Dane Court
S^T. PETERS
Sowell Stree
Poor hole
Vicarage
Rumfield gap
West Wood
North Wood
3 Miles by Sacketts hill
Bromfton
Newland
Dumpton
4 Miles by Chapel hill
4 Miles by Chapel hill
Herefon
S^T. LAWRENCE
Vicarage
Nether Court
Ellington
7 Miles by Kings Gate
RAMSGATE
Chilton
Pegwell
1 Mile
PEGWELL BAY

Evelyn Whitehead, ca. 1892

Alfred North Whitehead, ca. 1912

Three Generations of Whiteheads, 1898: Alfred, Henry, Alfred North, Evelyn, Jessie, T. North, and Sarah Buckmaster

Evelyn, T. North, Jessie, and Eric, 1899

Bertrand Russell, 1907

Sarah Buckmaster Whitehead and T. North, ca. 1912

Alfred North; his sister, Shirley; his brother Charles; his mother, Sarah; and his brother Henry, ca. 1912

CHAPTER X

The Married Mathematician

i/Birth of the Whiteheads' first child, Thomas North. Pont-Aven. Jessie Whitehead.

ii/Whitehead's private debate with himself: Rome or Canterbury? Growth of historical studies; the Cambridge school. Disintegration of the authority of Newtonian physics. Whitehead's decision. Possible causes, and character, of his agnosticism.

iii/Purpose of Whitehead's *Treatise on Universal Algebra*. Boolean algebra. Grassmann's Calculus of Extension.

iv/The philosophy of mathematics in *Universal Algebra*. Leibniz. Whitehead's attitude toward Platonic metaphysics. His handling of " = ". Reception of *Universal Algebra*. MacColl's review.

v/The Mill House. The Whitehead household. Grantchester village. Houseguests. The children's happy life. Whitehead's Swinburnian poem. Holidays.

vi/Whitehead's encouragement of research by young Cambridge Fellows. Successful opposition to Trinity College scheme to build on "the backs." A reconciler in College Meetings. Senior Lecturer in mathematics.

vii/Whitehead's part in University reforms: the Mathematical Tripos; compulsory Greek. "The Women's War," 1896–97. Whitehead's support of admitting women to Cambridge degrees.

viii/Council member of Newnham. Teachers Training College. Cavendish Laboratory. Sciences at Cambridge.

On arriving in Cambridge after their honeymoon, Alfred and Evelyn took an apartment in Rose Crescent, in the heart of town. Before long they found quieter lodgings at 24 Bateman Street. Their first child, a boy, was born there on the last day of 1891. They named him Thomas North. Both were Christian names in the Whitehead family. Presumably "Thomas" was chosen with Whitehead's paternal grandfather in mind.* "North" came from the Buckmaster side.† The boy was always called North; he signed himself "T. North Whitehead."

The lot on which the Whiteheads were building their house was on the south side of the West Road, next to the one at the corner of Grange Road. On the other side there was a roadway, now Selwyn Gardens. They called the site Pont-Aven, after the village in Brittany where Evelyn had spent the happiest years of her childhood. There was no stream here, only a little ditch; but she was imaginative. She and the architect worked out the plan of the house together, for Evelyn had the taste of a first-rate interior decorator. In a period when many dull little houses were being built, they produced one with individuality and charm. There was not really much space inside its two stories, but the appearance of spaciousness was created on the ground floor. By building close to the road on a lot that lacked width but was quite deep, a long vista of garden was assured. The drawing room, separated from the dining room only by sliding glass doors, had a large recessed window that served as a little conservatory; flowers were essential to Evelyn. The house was of red brick with matching scalloped tiles at the second-story level. The walls, the fireplaces, the hobnailed glass above the front door, were all in the best taste of the time. Evelyn and her architect realized that a butler's pantry was not essential but a ground-floor water closet was worth having. And care went into the arrangement of the kitchen wing and servants' rooms.

---

*See Chapter II, secion i.

†See Chapter II, section vi.

When my late wife (Victoria Lincoln) and I visited this house in 1968,* we saw on the edge of the garden a fine beech tree, older than the house, and three elms that the Whiteheads had planted on the West Road frontage. There was no carriage house; there is no reason to think that the Whiteheads owned a carriage.

Before his marriage, home for Whitehead was his father's vicarage in Broadstairs. It was like a big barn, gloomy and without a touch of graciousness. What a wonderful change of environment Evelyn gave him! The West Road house was the only one they built in England, but Evelyn altered three others that they lived in, always with elegant results. She usually began by knocking out a ground-floor partition; a spacious drawing room was essential, and made it possible for her to have a salon. Designing and decorating their living space, in one house after another, was her way of being creative, and the only way open to her.

For people interested in the Whiteheads' life, the West Road house is still worth a visit. It has been changed but little in the decades since they left it, though the western edge of Cambridge has undergone striking changes.

The house was ready for them early in the spring of 1892.† Whitehead was now working hard on the big treatise he later named *Universal Algebra.* In term, he lectured twice a day—in the morning—at Trinity College. He did not try to take much part in College business. Evelyn's health began to worry him. She was liable to what appeared to be heart attacks; they will be discussed in Chapter XI.

The life a Cambridge don might lead had been greatly changed by the new University Statutes that took effect in 1882. They removed the requirement that a Fellow of a Cambridge college who wished to marry must either give up his fellowship or take holy orders.‡ The academic year 1882–83 was known as "the Year of the Forty Brides," and the resulting baby boom had not yet subsided. Prams were everywhere in the residential area of Cambridge.

The Whitehead's second child, a girl, was born in the West Road house on February 23, 1894. They named her Jessie Marie, after Jessie Wade, an older sister of Evelyn's, the only sister she liked, who had died at eighteen.

---

*The preceding description of the West Road house is based on this visit.

†For the preceding month or two, the Whiteheads lived in the nearby house of a fellow Apostle, James Ward of Trinity College; the Wards were in Europe.

‡See Chapter V, section viii.

Shortly after Jessie Whitehead's birth Sidney and Beatrice Webb were house guests. In later years Jessie liked to tell the story that Beatrice Webb said of the new baby, "I hope she will prove a useful tool." That hope was quite misplaced.

It ecame evident that Jessie was left-handed. This was the first thing about her that, to her mother at least, was not right. A little later she began to stammer. Jessie told me she thought the stammer began when Mummy tried to make her become right-handed. That such an effort may induce stammering is one theory; as I write, the etiology of stammering has not been settled beyond dispute. All her life Jessie was liable to stammer, although there were times—presumably times when she was under no stress—of unafflicted speech. "Now you're trying," her mother would say; whereupon the stammer returned.

ii

In North Whitehead's house at Cambridge, Massachusetts, I have seen among his parents' books a copy of Thomas à Kempis's *Of the Imitation of Christ.* It is not at all remarkable that they should have owned this great classic of Christian piety, popular ever since it was written in the fifteenth century; but this copy was a fine leather pocket edition, inscribed, "Evelyn from Alfred. Dec. 16th 1890 / Feb. 15th 1891." The first date is that of the most important event in their lives; the second, of his thirtieth birthday. I have no idea what her birthday gift to him was. The plausible interpretation of this gift to her is that he was giving her something he prized, with the hope that, as partners in a Christian life, they would often read it together. Perhaps the Christianity she had picked up as a child of Protestant parents in a convent school amounted merely to going through certain motions. He must have seen that she was not as concerned about religion as he was. I dare say he also realized that she was much more worldly than he. The full title of Thomas à Kempis's book is *Of the Imitation of Christ and of Contempt for All Worldly Vanities.*

For about seven years after his marriage Whitehead read much Church history. He was consumed by the question that had led him to visit Newman: whether he should join the Roman Catholic Church. This was not a consuming question for his wife, but she was willing to follow his lead and embrace whatever conclusion he reached; she knew that he alone had the brains to carry out the intellectual and historical analyses which, he thought, a considered evaluation of the Roman and Anglican positions required. Although North and Jessie told me that their father and mother read theology together, I suppose this to mean primarily that he talked over his reading with her. Evelyn was not naturally a student of theology. She would devour novels with a religious message, like Newman's *Loss and Gain* or Mrs. Humphry Ward's *Robert Elsmere,* but I do not think she could have borne much careful reading of the Fathers of the Church and histories of Councils, of Aquinas, of Hooker and other Anglican divines. Whitehead studied such books, and acquired quite a theological library.

It would have been natural for him to discuss this reading and the whole question of conversion with his scholarly brother, Henry. As Henry was now in Calcutta, that was seldom possible; and his heirs have not kept letters from Alfred. Who else did Whitehead talk to on this subject? A little searching by me for evidence of substantial talks or correspondence has yielded nothing. In 1970 the retired Catholic chaplain at Fisher House in Cambridge, Monsignor A. N. Gilbey, possessed no records of Whitehead's keen interest in the Roman faith, and was astonished when I told him of it. This religious debate appears to have been quite private.

The question of religious commitment, when taken with the utmost seriousness, is at once general and intensely personal. As I see Whitehead, it was part of his character to weigh such questions in solitude. I noted earlier, in connection with his membership in the Apostles, that he lacked *close* friends.* And he always enjoyed discussion, but not polemics. The high spirits of the young Apostles made polemics in their meetings fun, but argument for the fun of it was out of place now. In a genuine discussion he would listen, and make remarks that, if recorded, would tell us what considerations he thought important. He was not given to telling a friend that he was consumed by a great personal question, and thrashing it out with him. In this, as in many other respects, he was not like most men. He discussed the religious question with his wife because his decision on it would affect their life together.

A future researcher might find in someone's papers a record of what Whitehead said to him about, say, the doctrine of papal infallibility; I doubt that we can hope for anything more. Whitehead's daughter, reporting what she learned from her mother, told me that this doctrine was the great obstacle in the way of choosing Rome. Newman had accepted it reluctantly. In any case, opportunities for

*See Chapter VII, section iv.

discussions with Roman Catholics were slim. There were fewer than a dozen of them in the entire University. Martriculation required a special episcopal sanction, for in 1883 Newman's adversary, Cardinal Manning, had persuaded the Congregation of Propaganda to admonish the faithful to stay away from Oxford and Cambridge; the ban was not lifted until 1895. The Whiteheads did know (and were fond of) the lonely leader of the few, Baron Anatole von Hügel. He was the Curator of the University Museum of Archaeology and Ethnology, and lived with his wife (daughter of Hurrell Froude) in their neighborhood. But Anatole von Hügel, unlike his famous brother Friedrich, had little interest in theological and philosophical questions; Monsignor Gilbey thought that Whitehead would not have discussed his problems with him.

Lord Acton was a Roman Catholic thinker whom Whitehead admired, as we know from the fact that immediately after his death in 1902 Whitehead proposed a Cambridge University memorial to him.[1] I dare say that Whitehead learned something to his purpose from Acton's troubles with Rome, and that he dropped in on some of his lectures after Acton was appointed the Regius Professor of Modern History in 1895. But I know of no discussions between them.

Whitehead was not going through what Newman and his friends had gone through at Oxford half a century before. Historical studies had intervened. Newman himself made one that Whitehead always admired—the *Essay on the Development of Christian Doctrine,* written in a time of seclusion which ended with his going over to Rome in 1845. It is natural today to call his subject the *evolution* of the dogmas of the Church. Acton thought that Newman's *Essay* did much to make Englishmen think historically, with an eye on the process as well as the result. In the years after 1845, historical studies of various kinds were made in ever-increasing numbers. Leaving German scholarship aside, mention must first be made of the challenge posed to rigid orthodoxy in the Church of England by *Essays and Reviews,* which appeared in 1860. One of its contributors, Mark Pattison (the Oxford tutor, then the lifelong friend of Whitehead's uncle Henry), insisted that "both the [Anglican] Church and the world of today are what they are as the result of the whole of their antecedents." This was Whitehead's position when he finally began, at Harvard, to write about the history of civilization. History began to be one of his major interests in his last year at Sherborne School, and this interest was nourished at Cambridge by discussions with Maitland, Dickinson, and other Apostles. It is therefore reasonable to assume that in his private study of theology he took a position similar to Pattison's. Next we must notice that for two decades preceding 1891, critical-historical studies were especially pursued by Anglican scholars at Cambridge. B. F. Westcott, F. J. Hort (an Apostle in Clerk Maxwell's time), and J. B. Lightfoot comprised "the Cambridge school." Lightfoot and Hort were no longer alive, and Westcott had left Cambridge; but Lightfoot had brought out a great edition of the *Apostolic Fathers;* Westcott and Hort, a new edition of the Greek text of the New Testament. Whitehead's study of theology could not have ignored the publications of the Cambridge school.

In the Anglican communion, the "Christian socialism" of Maurice and of Newman's adversary Kingsley had practically petered out by 1880. A more widespread movement, liberalism, had become very strong. It was mainly a tendency to loosen creedal requirements and to make Christian conduct the prime requisite; the central concern was the alleviation of what was called "the condition of the people," especially the masses of the poor in the great cities. In Chapter VIII we noticed the founding of Toynbee Hall, and Whitehead's support of it. The appeal of the liberal turn was shown by the popularity of *Robert Elsmere,* published early in 1888; the story in effect defended an ethical rather than a mystical interpretation of the Bible. Two years later, the circular announcing the author's own scheme for a settlement house in London (University Hall), appealed to "all those to whom Christianity . . . has become a system of practical conduct, based on faith in God, and on the inspiring memory of a great teacher, rather than a system of dogma based on a unique revelation." To Mrs. Ward (whose father, Thomas Arnold—older brother of Matthew—had been converted to Roman Catholicism under Newman's influence), Newman was "the great Malignant"; and so he was to liberal Anglicans generally. Once more we see how Whitehead was unlike most other people. He had both a concern with the condition of the people and a concern with his own relation to God.

An Anglican who asks himself with the utmost sincerity whether he should go to Rome is assuming that authority is essential for religion, and is seeking that authority. When Whitehead first asked himself this question, the authority of Newtonian concepts in physics was absolute. As the 1890s advanced, it became clouded. Every year brought new discoveries, and the Newtonian framework became more and more strained. New ideas were needed to explain radioactivity. And Fitzgerald suggested that if matter was electrical in nature, a body moving through the electromagnetic ether might contract in a way that would explain the negative result of the Michelson-Morley experiment of 1887. By 1897 it appeared that Newtonian physics was not at all secure. Whitehead later called this crumbling of the scientific rock one of the crucial experiences of his life.[2] It cannot have strengthened his assumption that there was a true traditional authority in religion, however hard to identify. I suppose that this was one cause of the decision he reached in 1897 or 1898.* His decision was not to go to Rome but—as he had put it in that meeting of the Apostles eleven years before—"in the other direction."† He now could find no clear authority in either Rome or Canterbury, no system of dogmas that satisfied him, and no compelling reason for subscribing to any promulgated articles of Christian faith. He had come to ask himself more and more frequently, "How can they be so sure?" The upshot was that Whitehead became an agnostic.

It is a reasonable speculation that almost from the beginning of his private debate he was wondering how he ought to think of God and immortality. If he went to Rome, the answer, in essentials, would be *given* to him. I don't think he liked that, in any area of inquiry. Of course I am reading back from what I know of the mature cast of his mind. The effect on him, at this time, of the disintegration of the Newtonian system that he had *given* to his undergraduate pupils supports the reading back. Also, I suppose that he could not entirely stop his wondering about God and immortality even as he contemplated joining the Roman Catholic Church. Perhaps he saw that an undercurrent of wondering would continue after taking that step.

Whitehead's memory of the clash of religious convictions of intelligent men in meetings of the Apostles had some influence on him. Maybe the freethinkers among the Apostles—Maitland, Sidgwick, McTaggart, all of whom he admired—were right.

Whitehead's agnostic period lasted about a quarter of a century. From 1925 on, he wrote frequently on *his* idea of God. Revulsion against the notion of God as Almighty Power, Maker of heaven and earth, is evident in all these writings. It is a good guess, though only a guess, that the presence of this notion in both the Roman and Anglican creeds was the main reason why he turned his back on both of them. I suspect that his wife was influential in some degree. She took little interest in how the world came into being, but much in eliciting beauty from the world that is given. And I doubt that she thought

---

*I cannot be sure of the year or the month.

†See Chapter VII, section vii.

churches and their creeds as important as he did. She gladly embraced his agnostic conclusion.

In his article on Whitehead in the *Dictionary of National Biography* Sir Edmund T. Whittaker, a former pupil who in 1897–98 was a young Fellow of Trinity, wrote that Whitehead became "an outspoken and even militant" agnostic.* As a good logician, Whitehead realized that it is not possible to prove the nonexistence of God; but the distinction between his agnosticism and atheism was purely theoretical. North and Jessie were told that they need no longer say prayers.

Whitehead's position was by no means unique. The old conflict at Cambridge, between the established Church and Dissent, was no longer of much importance compared with that between religion and disbelief. Whitehead was joining the ever-increasing band of freethinkers, of which Sidgwick and Maitland were charter members.

Nothing could have been farther from the minds of these agnostics than the sensational news that in the absence of God everything is permitted. Was it not in the Fellows Garden of Trinity that George Eliot made her famous pronouncement about God, Immortality, Duty: "How inconceivable the first, how unbelievable the second, and yet how peremptory and absolute the third!"? For Whitehead, the dropping of Christian theism left Christian morality untouched. Its hold on him was so ingrained that he was merely bored by Henry Sidgwick's efforts to secure its basis. One had only to be a gentleman. For North and Jessie, Sunday was to be no different from other days of the week, except that their parents asked them, out of consideration for the neighbors, not to play tennis or be boisterous.

Did not Whitehead and other Cambridge agnostics worship something? Yes; not God, but Truth. James Ward's younger daughter, Olwen, was often in the Whiteheads' house, playing with North and Jessie, when she was a child in the early 1900s. She recalled this episode:[3]

> Once in the course of a childish argument I said that I believed in immortality. The remark caused consternation in the family. . . . The parents of my young friends lamented that I could be so carried away by my imagination, and a few days later they gave me a copy of Wordsworth . . . and inside the book they had written this:—
>
> 'With whom is no variableness, neither shadow of turning.'

*Many years before Whittaker wrote this article, he had himself been converted to Roman Catholicism.

It fortifies my soul to know
That though I perish, Truth is so.
That howso'er I stray and range,
Whate'er I do, thou dost not change.
I steadier step when I recall
That tho' I slip, Thou does not fall.

A. H. Clough

North Whitehead once said to me that there was something a little odd about his father's agnosticism. The tone was a bit like that of a priest celebrating a Black Mass. It was as if he wanted to be religious and was being defiantly atheistic. At least, although the conclusion of the long, private debate was perfectly definite, it was not altogether welcome. The process had been agonizing, nothing one would go through again or like to see one's children go through. Accordingly, most of the Whiteheads' theological library was sold.

iii

The writing of the first and only volume of Whitehead's *Treatise on Universal Algebra,* probably begun in January 1891, went on "amidst other avocations" (as he put it) for five years.* He announced his purpose in the opening sentence of the Preface: "to present a thorough investigation of the various systems of Symbolic Reasoning allied to ordinary Algebra." "The chief examples of such systems," he continued, "are Hamilton's Quaternions, Grassmann's Calculus of Extension, and Boole's Symbolic Logic." Only the last two were investigated in this volume. Quaternions, along with Matrices and the general theory of Linear Algebras (among which Whitehead would notice Gibbs's Vector Analysis) were to be taken up in Volume II. That volume was to include a detailed comparative study of the symbolic structures of all the algebras, which ought to throw light on the nature of symbolic reasoning in mathematics. As it turned out, this

---

*In the Preface (p. x) Whitehead said, "The writing of this volume has been continued, amidst other avocations, since the year 1890." In his "Autobiographical Notes," as well as on the flyleaf of Evelyn's copy of *UA,* he dates the beginning "Jan. 1891"; but see the penultimate paragraph of Chapter IX above.

In the "Note on Grassmann," which brings *UA* to an end, Whitehead wrote, "The present work has been many years in composition and already nearly two years in the press." This may be assumed to have been written at the same time as the Preface, which is dated "December, 1897."

second volume was never finished. *Principia Mathematica,* written with Bertrand Russell, appeared instead.

Whitehead chose "Universal Algebra" as his title only as he was finishing Volume I.* In 1884 Cayley's friend, the eminent mathematician J. J. Sylvester, had published a paper on matrices with the grand title "Lectures on the Principles of Universal Algebra"; Whitehead mentions this paper in his Preface. But whatever Sylvester intended, "Universal Algebra" appeared only in his *title.* The wide scope of the work Whitehead planned made "Universal Algebra" seem an appropriate name for it. In fact, his real subject was the relatively new algebras, which he also (since he took ordinary algebra for granted) called "extraordinary algebras." They commanded his interest by their bold advances beyond the traditional restriction of variables to symbols for numbers.† As to what positive features the algebras shared: all, regardless of subject matter, dealt with syntheses of terms that were not necessarily numerical (or as Whitehead also says, operations on them) and employed two general types of synthesis, called addition and multiplication. Whitehead observed that the commutative and associative laws for addition hold in all algebras, and the distributive law (as symbolized in ordinary algebra, the law that $a\,(b + c) = ab + ac$) also; but the commutative law for multiplication is not universal, and Boole's algebra is characterized by a special law for addition, namely, $a + a = a$.

Whitehead does not present either Boole's algebra or Grassmann's in strict axiomatic form. In the 1890s the axiomatic method had not yet become customary outside geometry. (The first axiomatization of Boolean algebra—by E. V. Huntington—appeared in 1904.)

We should not consider Whitehead's *Universal Algebra* a part of the branch of mathematics that is called Universal Algebra today. That has a better right to the name than Whitehead had, for it requires only the notion of a set (of any kind) and operations on it, and applies identical methods to the study of many algebraic systems such as groups, rings, vector spaces, and lattices. But this science was not founded until 1935, by Garrett Birkhoff. The way to it was not prepared by Whitehead's *Treatise on Universal Algebra* but by later work that was quite different.

To give the full title of Whitehead's treatise we must add the phrase

---

*The minute of the Syndics of the Cambridge University Press, February 14, 1896, recording agreement to publish the work, names it "Generalised Algebra."

†See Chapter VIII, section iii.

with Applications. But we must not for a moment expect to find applications of the new algebras he studied presented separately in an appendix. He carries on his development of an algebra in close connection with important applications. He does this by introducing, early on, a simple but general interpretation of the algebraic symbolism. Thereby the development of the algebra is facilitated, and its utility is suggested. Whitehead was convinced that the new algebras were not only interesting abstract generalizations but "useful engines of investigation."[4] He always held that the validity of an algebra's principles and theorems was independent of their interpretations, but he never showed much interest in uninterpreted calculi. His constant theme was the broad *power* a mathematical system has by virtue of its abstract generality.

Whitehead was more interested in geometry than in arithmetic; that is why the fourth volume of *Principia Mathematica,* which was to deal with geometry and follow the three volumes he wrote with Russell, was to be his responsibility.[5] He had earlier published two tracts on the axioms of geometry. Still earlier (in 1905) he wrote a memoir in which he explored various ways of relating geometrical entities to time and to matter; this will be discussed in Chapter XIV. In the *Universal Algebra,* with which we are concerned now, he expressed his belief that a spatial interpretation would be possible for the symbolism of each of the algebras he proposed to study.[6] Thus he envisaged an investigation and comparative study of the newer algebras, brought together both formally (through the specification of abstract laws) and concretely (through uniformity of interpretation). Finding a spatial interpretation was not a problem in Volume I. Grassmann's *Ausdehnungslehre* was itself a geometrical calculus. Boolean algebra applied to the intersection or nonintersection of regions of space as readily as to logical classes and to the logical relations between propositions; in fact, Whitehead reserves the application of this algebra to logic for the last two of his five chapters on it.

Volume I of the *Universal Algebra* is divided into seven books. The first is introductory; I shall shortly consider some points in it. Book II, "The Algebra of Symbolic Logic," comprises Whitehead's investigation of Boole's algebra. Boole's "logical equations" are intrepreted as concerning regions of space; this enables Whitehead to avoid "the difficulties peculiar to Symbolic Logic" (Sec. 22)—about which logicians were quarreling—and to treat solution of the equations as a mathematical problem. The results, he says in his Preface, are not necessary for any later books of the volume. Those books all use

Grassmann's work as point of departure. Thus there is no organic connection, in this volume, between what Whitehead does with Boole and what he does with Grassmann.

It was natural to take up the Boolean algebra in the first volume of Whitehead's treatise on extraordinary algebras, for it is the one in which *no* symbols representing number or quantity are required. Although no other algebra of this type had been developed, Whitehead saw no reason why others "should not be developed to receive interpretations in fields of science where strict demonstrative reasoning without relation to number and quantity is required" (Sec. 22). (Boole himself retained the use of numerals, but this had been shown to be unnecessary by Ernst Schröder in 1877, and ten years earlier by Charles Peirce.)

I do not know when Whitehead first became interested in Boolean algebra. It is entirely possible that this occurred before he was seized by Grassmann's *Ausdehnungslehre.* John Venn, Fellow at Gonville and Caius College in Cambridge throughout Whitehead's years there, published his famous *Symbolic Logic* in 1881. In it Boole's algebra was made consistent and coherent; also, the logical relations between classes were represented by the intersection or nonintersection of regions of space.

In the second chapter of Book I Whitehead had dealt with the general idea of a *manifold.* Let me say, roughly, that the manifold of a property is an aggregate of elements, each a distinguishable instantiation of the property. The concrete fact from which the manifold of elements is abstracted is a class of things marked out by possession of the property. If the property is spatiality, the points of space, or of any portion of space, comprise manifolds. Whitehead offers, as another example, the manifold of musical notes conceived as representing every note so far as it differs in pitch and quality from every other. But the important kind of manifold for Whitehead is the one he investigates in Book III, "Positional Manifolds." A positional manifold is any one that admits of a process symbolized by addition of the numerical (i.e., non-Boolean) type. To make this idea "more simple and concrete without any loss of generality," he identifies it "with the general idea of space of any arbitrarily assigned number of dimensions, but excluding all metrical spatial ideas" (Sec. 22). It was characteristic of Whitehead, in all his writings on geometry, to go as far as he could before introducing a metric.

Book IV is devoted to the principles of the Calculus of Extension. Book V applies them to a nonmetrical theory of forces (which may be called line geometry or a contribution to pure dynamics, depending on one's interest). Book VI, "Theory of Metrics," applies the *Ausdehnungslehre* to elliptic and hyperbolic geometry, following the post-Grassmannian work of Cayley and Klein. Book VII (the last) applies Grassmann's calculus to ordinary Euclidean space of three dimensions. Much of it deals with vectors; many of the results were new, but anticipated by a member of Peano's school, C. Burali-Forti, while the *Universal Algebra* was in press. The last chapter of Whitehead's volume was "written to shew that formulae and methods which have been developed by Hamilton and Tait for Quaternions are equally applicable to the Calculus of Extension."*

Grassmann carried out one of Leibniz's ideas, that of an algebra in which symbols representing geometrical entities such as points, lines, and planes are manipulated according to a few fixed rules. He worked out his theory for *n*-dimensional space, not the mere three dimensions that we perceive, to which Hamilton and others limited themselves. Using Hamilton's term "vector," we may say that Grassmann created *n*-dimensional vectors. It is commonly noted that tensor analysis is implicit in the *Ausdehnungslehre*. In another aspect, the Calculus of Extension was a generalization of Möbius's barycentric calculus of 1827; Möbius did not dare reject the commutative law for multiplication; Grassmann did. But this topic would take us too far afield.

I should be sorry if what I have thus far written about Volume I of *Universal Algebra* gave the false impression that it consists of *discussions* of Grassmann's and Boole's discoveries. Those discoveries are simply the points of departure for Whitehead's own mathematical constructions, which form continuous arguments. He extends their work. In his development of Grassmann's ideas he seeks unity rather than completeness; he does not hesitate to pass over parts of the two *Ausdehnungslehre* which do not contribute to what *we* might almost call the Grassmann-Whitehead Calculus of Extension.

iv

I come now to the philosophy of mathematics that appears in the *Universal Algebra*. To show the importance of his general subject,

---

**UA*, "Note," p. 573. This Note continues by remarking "some affinity" between some of Whitehead's formulae and those of Gibbs's Vector Analysis, and promising that his "very interesting algebra . . . will be noticed in its place among the Linear Algebras."

Whitehead in his Preface briefly discusses "the fundamental nature of Mathematics" and offers this definition: "Mathematics in its widest signification is the development of all types of formal, necessary, deductive reasoning."[7] This suggests that many mathematical sciences await discovery and exploration.

He continues: "The reasoning is formal in the sense that the meaning of propositions forms no part of the investigation. The sole concern of mathematics is the inference of proposition from proposition." The reasoning is *necessary* in the sense that it has followed the rules of inference; their justification "is not properly part of mathematics; it is the business of experience or of philosophy." This situation became complex when Whitehead, in *Principia Mathematica,* undertook "the mathematical treatment of the principles of mathematics" (as its subject was described in its first sentence) in collaboration with a philosopher. That will concern us in Chapters XII and XIII. But it should be said here that when *Universal Algebra* was written no one except Frege, whose work was unknown, had advanced toward the thesis of *Principia,* namely, that all mathematics is an extension of deductive logic.*

To return to the Preface to *Universal Algebra.* Whitehead explains that mathematical reasoning is *deductive* in the sense that its validity needs only the test of self-consistency. This is what Grassmann claimed for his new "formal" science.† Whitehead uses different words for the same purpose. He distinguishes two types of mathematical definitions. A definition with an existential import abstracts from observation, and forms the starting point of an applied mathematical science. A definition without an existential import "sets before the mind by an act of imagination a set of things with fully defined self-consistent types of relation." Whitehead calls such a definition "conventional," and asserts that "qua pure mathematics, mathematical definitions must be conventional."[8] To this he attaches a footnote, "Cf. Grassmann, *Ausdehnungslehre von* 1844, Einleitung."

Whitebead's broad notion of mathematics is expressed also in terms of a calculus:

> The ideal of mathematics should be to erect a calculus to facilitate reasoning in connection with every province of thought, or of

---

*Bertrand Russell wrote to me (June 18, 1941) that in 1898 (when *UA* was published) his ideas "were still in confusion."

†See Chapter VIII, section iii.

> external experience, in which the succession of thoughts, or of events can be definitely ascertained and precisely stated. So that all serious thought which is not philosophy, or inductive reasoning, or imaginative literature, shall be mathematics developed by means of a calculus.[9]

This passage does not contradict Whitehead's view of mathematical reasoning as deductive, from definitions that may have been created by an act of imagination. The imaginative leap—say, to an *n*-dimensional manifold—by transcending the world sets up no barriers against returning to it, does not push the world permanently out of mind. A fortunate leap creates a more powerful instrument for this return, in Whitehead's view.

The quoted passage calls to mind Leibniz's vision of a universal calculus of reasoning. But Whitehead does not subscribe to Leibniz's doctrine, revived in our own century, that "all our reasoning is nothing but the relating and substituting of characters, whether these characters be words or marks or images."[10] That is done by means of a calculus, and the business of a calculus, according to Whitehead, is to *facilitate* reasoning, not to do it for the mathematician. Whitehead equates the difference between the two which Leibniz (sometimes, at least) wished to identify, explicitly with F. H. Bradley's distinction between "inference" and "external demonstration." "Inference" Whitehead describes as "an ideal combination or construction within the mind of the reasoner which results in the intuitive evidence of a new fact or relation between the data."[11] "External demonstration" is a combination performed with marks on paper in accordance with fixed rules, and this manipulation throws up the result for sense perception and subsequent interpretation. "The signs of a Mathematical Calculus," Whitehead writes, "are substitutive signs"; they are not, like words, instruments for thinking about the meanings expressed, but rather, as G. F. Stout had roughly put it, "means of not thinking" about the meanings symbolized.[12] Whitehead knows that the situation is not so simple: "No sooner has a substitutive scheme of manipulable symbols been devised to assist in the investigation of any originals, than the imagination begins to use the originals to assist in the investigation of the substitutive scheme."[13] A simple truth this, doubtless not original with him. A logician might say that it was of merely psychological interest. What is it doing in a mathematical treatise? It shows that this mathematician notices mental processes as well as schematic forms. A mind that will remark the psychological interplay that occurs in the consideration of signs and things signified is not so likely to take the pains always to distinguish the two in the sharp manner required by modern semantics. Such a mind will instinctively go ahead and develop its subject, allowing the analyses that add precision and fix the theory of the subject to constitute a second stage. By then that type of mind will probably have passed on to other subjects, leaving the original field to specialists.

In many of Whitehead's later books an interest in the workings of the mind shows itself. He never abandons his distinction between thinking and using a calculus. In several books he gives a high place to the invention and organization of mechanical aids to thinking. John Dewey quoted with emphatic approval the declaration in *Science and the Modern World* that "the reason why we are on a higher imaginative level is not because we have finer imagination, but because we have better instruments."[14] The instruments referred to are physical instruments, but their relation to discovery in physical science is analogous to the function of a calculus in pure mathematics. A calculus either makes further thought unnecessary or gives it a new vantage point. Instruments and calculi are like railroads, which make walks in distant mountains possible.

Whitehead's insistence on this distinction which Leibniz wished to rub out may be symptomatic of a difference in the cast of mind of two men who are often compared. If Leibniz had been Whitehead's contemporary, he could have written the *Universal Algebra* and the categoreal framework of *Process and Reality,* and probably he would have made them shorter and more elegant. If we imagine universities at the low level to which they had sunk in the late seventeenth century, we can easily conceive Leibniz's service of a prince replaced by the service of a large corporation, Leibniz in the capacity of chief accountant. We can see him devising some beautiful systems, with much of the work done by machines of his own design. But Whitehead as an accountant is not conceivable.

What evidence is there in this first book of Whitehead's about his attitude toward Platonic metaphysics? There are a fair number of philosophical remarks in the opening chapters; none hint that mathematical objects have a status superior to things. Quite the contrary. Whitehead notes that the importance of a logical scheme derives from its interpretation as "representing" properties of "the world of existing things," and he takes care to state that his logically primitive conception, that of a manifold of elements, is abstracted from a "scheme of things," which is first explained.[15] If Whitehead when younger was ever attracted to an extreme Platonism, the doctrine of evolution and the discovery of alternative geometries—both of which suggest that Nature is patient of many patterns of order—had taken him away from it.

His remarks in the *Universal Algebra* show sobriety, a reflective mind, and freedom from single-minded faith. He has profited from his years in the Apostles. Consider Whitehead's discussion of the fundamental idea expressed by " = " in an algebraic calculus. He calls it *equivalence;* the equality sign denotes that "the signs or groups of signs on either side of it are equivalent, and therefore symbolize things which are so far equivalent." He explains that two things (of any kind) are equivalent when *for some purpose* they can be used indifferently: "Thus the equivalence of distinct things implies a certain defined purpose in view, a certain limitation of thought or of action."[16] He carefully distinguishes equivalence from identity. Although he refers to Bradley's *Principles of Logic* when he writes, "No investigations which proceed by the aid of propositions merely asserting identities such as *A* is *A*, can ever result in anything but barren identities," Whitehead summarizes his position in a mathematician's way: "Equivalence implies non-identity as its general case. Identity may be conceived as a special limiting case of equivalence."[17]

Taking as an example the equation $2 + 3 = 3 + 2$, he points out: "The fact that both sides represent a common five-ness of number is not even mentioned explicitly. The sole direct statement is that the two *different* things $3 + 2$ and $2 + 3$ are in point of number equivalent."[18] Whitehead has here given a sufficient answer to the criticism made by W. V. Quine in 1941.[19] Quine read the equation as asserting that the order of summands is immaterial, and suggested that this could be displayed by "simply superimposing '*x*' and '*y*' in the manner of a monogram." He continued: "Now the one objection to this procedure is the expense of casting monograms; and thus it is that we revert to a linear notation which imposes an arbitrary notational order on summands. The printer foists upon us a redundant notation, issuing in synonyms." But in *Universal Algebra* "3 + 2" is not a synonym for "2 + 3". When Whitehead introduced the concept of addition, he maintained, as Grassmann had, that the symbols symbolize a "process of forming a synthesis between two things [concrete or abstract, material things or merely ideas of relations between other things] . . . and then of considering . . . [them], thus united, as a third resultant thing."[20] Whitehead was concerned with the process as well as the result; Quine was not.

The breadth of Whitehead's definition of mathematics should be thought of as the response of a mathematician who reflected on the nature of his science to the great advances it made in the nineteenth century. Besides the creation of the extraordinary algebras, there was the discovery and investigation of the non-Euclidean geometries. Also, fundamental work had been done in projective geometry, which involves no reference to size, distance, or measurement, and so seems more general than any metric geometry. To a mind that appreciated these advances, there was no reason to think that the growth of mathematics had come to an end. Hence Whitehead's definition: "Mathematics . . . is the development of all types of formal, necessary, deductive reasoning."

In the Preface to *Universal Algebra* he tells us that once the subject of his volume had fully shaped itself in his mind, the various parts were systematically deduced, "with hardly any aid from other works." He continues, "This procedure was necessary, if any unity of idea was to be preserved, owing to the bewildering variety of methods and points of view adopted by writers on the various subjects of this volume." Such a procedure was natural to a thinker who, as I noted earlier, was a loner.* He followed it in all his longer books, mathematical and philosophical.

At the end of his concluding "Note on Grassmann" Whitehead, after paying tribute to the "important papers" of Buchheim and Cox, says that E. Lasker's "An Essay on the Geometrical Calculus," published in *Proceedings of the London Mathematical Society* in 1896 and 1897, "contains results which I should like to have used, if I had seen it in time." Some readers will be interested in knowing that "E. Lasker" is the Emanuel Lasker who had recently become world chess champion.

For help in the final revision of *Universal Algebra,* Whitehead in his Preface thanks three Cambridge friends. Besides Forsyth, they are Arthur Berry (Fellow of King's), who corrected *all* the proofs; W. E. Johnson, who "generously placed at my disposal some work of his own on Symbolic Logic"; and "the Hon. B. Russell, Fellow of Trinity College, who has read many of the proofs, especially in the parts connected with Non-Euclidean Geometry."

I shall come to the history of the relation between Whitehead and Russell in the next chapter. At this point I want to notice that in *My Philosophical Development* Russell recalled being "greatly excited" by the *Universal Algebra,* and lamented that "Cambridge paid it much less

*See Chapter VII, section iv.

attention than it deserved."[21] In this respect Cambridge was like the rest of the mathematical world, apart from Peano's Italian school. Russell wrote in a letter of 1941, "The book was very little noticed, and had not much influence."[22] Among the few reviewers, G. B. Mathews (Senior Wrangler when Whitehead took the Tripos), writing in *Nature,* described it as brilliant: the wealth of applications removed the last excuse for ignoring Grassmann; although it was too early to say whether a great instrument of discovery had been presented, the power of Grassmann's calculus as a means of expression and organization was amply demonstrated; and the author included substantial contributions of his own to mathematical research. The publication of *Universal Algebra* naturally led to Whitehead's election to the Royal Society. After Whitehead's death, Sir Edmund Whittaker wrote that the book "was acclaimed on all sides as a splendid piece of learning and research."[23] Obituaries abound in such statements; if this one is correct, it does not follow that Whitehead's treatise had much influence on mathematicians, even in England. Symbolic logicians read Book II only; algebraists, unless already enamored of queer algebras, neglected this long, queer volume to work at the theory of equations or something equally traditional; geometers gathered, at second hand, that some of the discoveries of Cayley, Klein, and Clifford about the relations between geometries had been restated and discussed from a slightly different point of view. Some mathematicians may have been set against *Universal Algebra* by its Preface and first three chapters; like the *Einleitung* in Grassmann's *Ausdehnungslehre* of 1844 to the mathematicians of that time, they were "too philosophical."

I should like to end my account of Whitehead's first book—an account which, as I am not a mathematician, is subject to correction—by noticing a review of it by a symbolic logician, Hugh MacColl, that was published in *Mind* (*the* English philosophical quarterly) in 1899.[24] For twenty years MacColl had been putting out papers on symbolic logic; independently of Peirce and even of Boole, he had invented a calculus of propositions similar to Peirce's; and he was shortly to produce another algebra of logic distinguished by the general features of C. I. Lewis's system of "strict implication."[25]

MacColl's review opens with a statement that is often a routine one: "In consenting to review this important volume for the readers of *Mind* I fear that I have undertaken a task for which I am but indifferently qualified." After some explanations, he considers the progress of Whitehead's book. "His opening chapter, 'On the Nature of a

Calculus,' is very interesting, and may be understood by anyone of ordinary education and intelligence." To be sure,

> If the reader knows something of common algebra he will grasp the author's meaning more easily; but, for much of this chapter, even this modicum of preliminary knowledge is not absolutely indispensable.
>
> When, however, we enter upon the second chapter, which treats of *Manifolds*, we find ourselves on very different territory. A reader previously unacquainted with the subject cannot read this straight through, as he would a novel or a paragraph in a newspaper; he will have to make frequent halts, and sometimes very long halts, in order to reflect.

The next sentence falls inevitably in place: "This is not altogether the author's fault." He was grappling with a subject at once formidable and elusive:

> The truth is that the subject of manifolds is extremely difficult to understand, and still more difficult to explain. The meaning of the word *manifold*, as defined by its inventor, Riemann, is so very general, not to say vague and attenuated, that it may be called the *ether* of mathematical conceptions.

But could we not have been let off a little easier?

> Mr Whitehead might, I think, with advantage have restricted his discussion to the general characteristics of the manifolds which enter into his compared algebras, and he should have illustrated these more copiously with simple and concrete examples.

Nevertheless, the task which the author set himself, "I consider him to have accomplished with rare ability." To this conclusion a qualification is attached, namely, "judging of the whole from my knowledge of a part." For of course MacColl had been able to read but one-fifth of the *Universal Algebra*. He had not anticipated a large mathematical treatise, but a manageable book on the general principles of symbolic reasoning, "with occasional appeals to mathematics and geometrical diagrams by way of illustrations."

V

During all but the first year of the writing of Volume I of *Universal Algebra*, the Whiteheads lived in the West Road. After half a dozen years there, Evelyn began to look for another house. A third child was expected; it may be that more room was wanted. Why not something quite different, a house in the country? I remember that when the Whiteheads lived in the American Cambridge, after a few years in a fine apartment Evelyn wanted a cottage in the country, and had one built for them in the Blue Hills south of Boston. In Cambridgeshire in 1898, country was close at hand. The Whiteheads' friends the Shuckburghs lived in the rural village of Grantchester, and told them about the Mill House there.* It was centuries old and romantically situated. They were delighted, and took it.

Grantchester was on the river Granta, as the Cam is called upstream from Cambridge. It had been a camp in Roman times. Cambridge was only three miles away, but the inhabitants of the village could not have been more indifferent to the University. As the number of faculty and students was small by modern standards and the automobile was still a toy, Cambridge had not invaded the village. The country around it was flat; sunsets in the big sky were the only exciting things in a peaceful landscape—except in May, when the meadows were ablaze with the gold of marsh marigolds.

There had been a mill at Grantchester since 1280. The one beside this house was not that old, but it was ancient, and still turning grain into flour, to the presumed profit of its owner, Merton College of Oxford University. The Whiteheads' house stood at the edge of the mill pool, immediately below the mill. It was roomier than the one in the West Road, but modest. It is still charming, though the mill burned down in 1928. Whitehead wrote in "Autobiographical Notes":

> Some parts of our house were very old, probably from the sixteenth century. . . . The whole spot was intrinsically beautiful and was filled with reminiscences, from Chaucer to Byron and Wordsworth. . . . The house had a lovely garden, with flowering creepers over the house, and a yew tree which Chaucer might have planted. In the spring nightingales kept us awake, and kingfishers haunted the river.[26]

After Wordsworth, he could have mentioned Tennyson; the old Mill

---

*Evelyn Shuckburgh, Fellow and Librarian of Emmanuel College, was one of the foremost classical scholars at Cambridge. Whitehead, in "Autobiographical Notes," identifies him as the translator of Cicero's letters; this was but one item in a large number of translations, commentaries, and histories (including a biography of Augustus), in which he made Roman civilization alive to Whitehead's generation. He was a good deal older than Whitehead, but a dear friend.

House was said to have inspired him to write "The Miller's Daughter." Whitehead was right about the age of the house; Elizabethan coins were found under floorboards. Inside the walls, rats could sometimes be heard having "terrific fights."*

While Evelyn was seeing to alterations in the Mill House, the Whiteheads lived nearby, in the historic Old Vicarage. Later, Rupert Brooke wrote a famous poem about it; he considered "The Old Vicarage, Grantchester" his masterpiece.†

Whitehead cycled to his daily work in Cambridge. Because of Evelyn's heart trouble, she traveled out of the house in what they called their rickshaw—a basketlike trailer attached to Alfred's bicycle. Someone, seeing this, said Whitehead led "a dog's life," but I am sure he did not mind it; he probably thought it gave him good exercise. His only other exercise was walking. He enjoyed twelve-mile walks with friends, but did not play tennis or cricket or indulge in any hobbies, not even in fishing.

The Whiteheads lived happily in the Mill House for eight years, from about the beginning of 1899. North and Jessie remember it with pleasure as their childhood home: he was seven when they moved in, she five. Eric Alfred Whitehead had been born November 27, 1898. Evelyn had been delivered of a stillborn child, and had had several miscarriages. She was not an easy breeder.

North described his childhood in the first chapter of his unpublished autobiography and devoted the second to Grantchester as it was then. Both are charming. In connection with the following passage, I suggest that the many servants in the house made it rather easier for an academic man to carry on his work than it would be today, with the servants replaced by appliances. North wrote:

> We had living in the house, besides my mother and father and the three children, a cook, a maid, a children's nurse and, for much of the time, a governess. But we also had some help from the village—a daily girl, a "tweeny", who seemed to do most of the work, chased around by the others, who were "learning her". We also had a full-time gardener, who usually turned up drunk. He pushed the lawn mower all over the gravel paths to have something

*The Whiteheads told this "with huge gusto" to Lucien Price in 1938; "one got the idea that life there was exciting" (Price, *Dialogues,* XII [April 28, 1938]).

†Grantchester was Rupert Brooke's retreat from the social life of Cambridge for three years following his graduation in 1909; an earlier name he gave this poem was "The Sentimental Exile."

> to cling to. Then an adolescent boy, "the boots", used to clean our boots every morning, and also clean the table knives by turning a mysterious contraption. . . . I think that is all, except that we had Mother Heffer once or twice a week. She was less than five feet in height, but she made up for her lack of inches in other directions. She was almost exactly spherical. Mother Heffer did the washing in the wash room beyond the kitchen, and a little rough cleaning around the house.
>
> So, in addition to the family, we had four people living in the house, and four more on full or part time who slept in the village. Perhaps I should add the donkey that pulled our governess' cart, or was saddled for the children. Once in an expansive moment we rose to a pony with a mysterious disease known as "the staggers". He sat down suddenly in the middle of the road from time to time, and only got up again when he felt like it.[27]

Most of the villagers were tenant farmers. In addition to them and their hands, there were the miller and his men, thatchers, pubkeepers, a blacksmith, a cobbler, and a carpenter. All lived pretty much as their ancestors had in the eighteenth century. Grantchester was considered one of the more drunken and depressed villages in the Fen country. As the Whiteheads recalled, the villagers lived lives of "Chaucerian looseness . . . ; a bastard child was merely 'a little mistyke.'" And they "relied instinctively on the gentry, as they had done for centuries."[28]

For Whitehead, the major contact with them occurred when campaigns for Parliament were under way. The villagers wanted to hear the local gentlemen debate the issues. He wrote in "Autobiographical Notes":

> During our residence at Grantchester, I did a considerable amount of political speaking in Grantchester and in the country villages of the district. The meetings were in the parish schoolrooms, during the evening. It was exciting work, as the whole village attended and expressed itself vigorously. English villages have no use for regular party agents. . . . Rotten eggs and oranges were effective party weapons, and I have often been covered with them. But they were indications of vigour, rather than of bad feeling.[29]

Whitehead's politics were Liberal, but (unlike those of many Fellows of Trinity) not Unionist; he and Evelyn believed in Home Rule for Ireland. Evelyn was quite pro-Irish. She and the children sang "The Wearing of the Green." Jessie thinks her mother did not really like the English, nor the Scots. Later, the Whiteheads were sympathetic to the Boers.

Grantchester was close enough to Cambridge for Evelyn to hold a little salon. The Mill House was also ideal for houseguests. She started an album in which they signed their names. These included Forsyth, Walter Raleigh and his wife, G. E. Moore, Henry Head, the French mathematician Emile Borel, Elie Halévy, and two former pupils of Whitehead's, Mary and Alan Beeton (children of an interesting, wealthy stockbroker). Whitehead's mother and his brother Henry (still unmarried) were there in April 1899; two months earlier, Henry had been appointed Bishop of Madras. The album entries stop after a year and a half; perhaps Evelyn thought it pretentious to keep a register. I rely on North's memory in reporting that Henri Poincaré was once a guest, and that as he had neglected to pack shirts, after a few days Evelyn offered to lend him one of Alfred's. Bertrand—"Bertie"—and Alys Russell were in residence for long periods, which I shall tell about in my next chapter.

The Mill House became one of the centers of the University's intellectual life. Whitehead's talk was often about general ideas, of a semiphilosophical character. Hours spent with him and Evelyn were never dull. But guests might have to do without simple comforts. E. T. (later Sir Edmund) Whittaker had been a pupil, and was now a colleague on Trinity's mathematical staff; Lady Whittaker remembered: "When my husband and I went to lunch at the Mill House we had to be content with hard chairs outside the kitchen window and to have our plates of suet pudding passed out to us. When the meal was finished the three children were made to rest in hammocks for the sake of their digestion but we were not offered any hammocks."[30] Dinner, of course, was formal; a dressing gong was sounded twenty or thirty minutes before the dinner bell.

The Whiteheads filled the house with books; Jessie and her mother were voracious readers. The center of North's world was a small carpenter's shop, built for him at one end of the stables. "I was given tools and an allowance of wood; also a carpenter came weekly to teach me how to use them, for my father had no manual skills and my mother, who did have strong artistic abilities and interests, was no carpenter."[31] North's permanent interest in designing mechanisms began here. He made toy steamers and a three-wheeled motor car (he had seen only one car, and it had three wheels); the car was driven by springs bought from a clockmaker and ran the length of the stable yard. Out of spectacle lenses he made telescopes with which he could see what Galileo first saw.

The children never seemed to lack anything. When Jessie wanted a puppy dog the family solicitor, Crompton Davies, was told, and brought her one. Every summer morning Whitehead rowed the children to a pool in the Granta near Cambridge for bathing, and sometimes bathed with them. North wrote that his childhood was "conspicuously happy." Jessie's memory was the same. Whitehead was a good, loving father. Jessie could have been Daddy's girl, if Evelyn had been willing to let anyone be Daddy's girl.

Jessie recalled for me a few further facts about her childhood in the Mill House. Letters of the alphabet were posted on the door of her father's study; when he emerged he would tell the children a few things about each letter. He could be shocked into anger; one such occasion was his discovery that their governess had not taught them the names of the days of the week and the months of the year. At times he would be depressed and irritable. Then everybody had to mind his or her p's and q's. (His spells of depression will be dealt with in the next chapter.) For many years he suffered from insomnia, but he never tried to catch up on his sleep. The constant sound of the mill water was not soporific to him; he read, often in Sarpi's *History of the Council of Trent;* when Jessie was wakened by the thump of that big book on the floor, she knew that Daddy had at last fallen asleep.

North told me that his father never seemed hurried. He seldom failed to play with Jessie and him—if he noticed them. The force of this "if" is seen in Russell's recollection:

> His capacity for concentration on work was quite extraordinary. One hot summer's day, when I was staying with him at Grantchester, our friend Crompton Davies arrived and I took him into the garden to say how-do-you-do to his host. Whitehead was sitting writing mathematics. Davies and I stood in front of him at a distance of no more than a yard and watched him covering page after page with symbols. He never saw us, and after a time we went away with a feeling of awe.[32]

Later in the same year that the Whiteheads came to Grantchester, they were joined as residents by a third Cambridge family, that of the distinguished naturalist William Bateson, who was an exact contemporary of Whitehead's. Bateson had enough land around his house for his experiments on plants and fowl; they led him to become the British champion of Mendel's new science when it was rediscovered in 1900. The first six years of his son Gregory's life were spent in Grantchester. In a letter, Gregory wrote, "We had a sort of family myth which lent a crystal halo to Cambridge and especially Grantchester."[33] The Batesons left in 1910. As Gregory recalled: "I grew up feeling in a gentle way that we had been turned out of, or perhaps had deserted, the Garden of Eden. . . . The Whiteheads were the most conspicuous tenants with whom the Garden had been shared."[34] From North's and Jessie's memories, and from the loving way Whitehead in "Autobiographical Notes" wrote about Grantchester, I gather that to them, too, it was a Garden of Eden.

The Mill House was a place of gaiety to almost everyone except Bertie Russell. There was tension in his relations with Alfred and Evelyn. This subject will be taken up in sections v and vi of the next chapter; we may say here that Russell was unhappy because he was coming to hate his wife, and also because he was not getting anywhere in trying to resolve the contradiction he had discovered in the logical foundations of arithmetic. But Bertie was always wonderful with children. He was a perfect intellectual companion to North and Jessie. He would think about what a child said, then put it clearly; they talked philosophy on equal terms. With the gifts of Socrates and Lewis Carroll, he was a fine teacher to children. And Jessie remembers, "Bertie and Daddy used to go out cycling on a bicycle built for two, and put me astride in between them."

It was in one of these years that the Whiteheads had their first ride in an automobile. I mention this because it was the occasion of the only rhymed composition that I am sure Whitehead wrote. The car, chauffeur driven, was that of Mary Beeton's father, an early enthusiast for automobiles. Whitehead wrote to her:

> Dear Mary
>
> Mr. Beeton told me that he admired Swinburne's Songs before Sunrise. I feel sure that he will find the same beauties in the following lyrical effusion—
>
> We moved in the merciless motor,
> We blew on our bugle a blast,
> Silly sheep shorn of life lay behind us,
> And pigs turned to pork as we passed.
> Rustics ran out to arrest us
> And consigned us to suitable ends;
> But withal what most deeply impressed us
> Was the pleasure produced for your friends.

Whitehead's father died in March 1898. Visits to Broadstairs were less frequent thereafter. August and September were times for traveling, even though there was no money to spare. North wrote in his autobiography:

> Our summer holidays were varied and interesting. Sometimes we

> went to the Continent. On two occasions my paternal grandmother took us all to Switzerland. Apart from that we traveled in the simplest possible style, third class in England and fourth class on the French railways. . . . We carried our luggage in straw hand cases with waterproof coverings and stayed at modest inns. All this was presented to us, not as a hardship, but as part of the fun, and so we took it.[35]

I doubt that Germany was visited, for Evelyn had acquired a hatred of that country when she was a child in Brittany at the time of the Franco-Prussian War; during their train's passage through Alsace-Lorraine on the way to Switzerland she kept the window shade down. (North told me that his father liked the Germans of that time better than the French, but I doubt that he let Evelyn know this.)

Occasionally Alfred and Evelyn left the children at home. The most important trip of this sort was the one made with the Russells to a philosophical congress at Paris in 1900; it will be described in Chapter XI.

In August and September 1901 Russell took his wife and the Whiteheads on a five-week holiday to Italy. After taking a train to Venice, the four boarded a tramp steamer that stopped at every port from Venice to Genoa. Whitehead was the one with the Baedeker; he read it before they reached a town, and could tell the others what was to be seen there. Russell says in his *Autobiography*[36] that he once went by tramp steamer from Venice to Genoa; he does not say who went with him. When I asked him about this holiday in 1967, he told me how he had got the Captain to stop using rancid butter and old oil. Of more value was his memory that in the course of a terrific storm Whitehead had quoted from Sydney Smith, "Fate cannot harm me, I have dined today." It was one of Whitehead's delightful habits to relieve tension by coming up with just the right line of poetry.

Ronald W. Clark's detailed *Life of Bertrand Russell* makes no mention of this Italian holiday with the Whiteheads. I wonder what happened that Russell did not want to remember.

vi

As we saw in Chapter VI, when Whitehead was an undergraduate several members of Trinity's mathematical staff were not research men. That was no longer true in the 1890s. And the young mathematicians who were elected to Trinity fellowships toward the end of the decade—Russell, Hardy, Jeans, E. T. Whittaker, and E. W. Barnes—

were keen on research. Their warmest encouragement came from Forsyth—and Whitehead. Whitehead had both an unbounded enthusiasm for new mathematical sciences and an unstinting hospitality to potential genius in the young. These new Fellows had shown their promise in dissertations. One or more of them might create a new science about which a future mathematician could say what I should wager he said to himself when he first saw what Grassmann had created, and what Boole had created: "How wonderful that is! It must be developed."

Whitehead saw himself as a conscientious student and developer of new mathematical sciences rather than a creator. His admiration for the new creations was objective; he was no one's disciple. Hamilton had high hopes for his quaternion theory, which Tait championed and improved. In 1897, when Tait was an old man, the International Society for Promoting the Study of Quarternions was founded. The name was very soon broadened by adding "and Allied Systems of Mathematics" in order to include Grassmann's *Ausdehnungslehre.** Nevertheless, Whitehead did not join this Society. Volume I of his *Universal Algebra* was in press; he would stick to his last, the making of a comparative study of all the new algebras. Either he lacked time to be an active member or he wanted as always to eschew polemics (the Society was hostile to some methods), or both, as I think likely.

His work at the University had been increasing. In 1894 he served for the first time as an Examiner on the Mathematical Tripos. This was an onerous task, but one of the facts of life for a Cambridge don. It is mildly surprising that he was a don for ten years before he consented to serve; Whitehead held a low opinion of the existing Tripos system, and of examinations in general, but I think that insufficiency of self-confidence was also a factor.

Most of Whitehead's nonteaching work in Cambridge was done for Trinity College. A few words about the government of a Cambridge college may be desirable here, especially for American readers, who expect a college to be run by an Administration. A Cambridge college was, and still is, a democracy of all the Fellows of the college, who collectively compose its Governing Body. The Master of Trinity was like a constitutional monarch. At this time there were sixty-three Fellows. Decisions affecting the College were made

*Tait was elected the first President of the Society, but was too ill to serve. The Society was truly international. It published an annual bulletin until 1913, when it became inactive.

by a Council consisting of the Master, four ex-officio members, and eight Fellows elected by the Governing Body. The Council made appointments to the teaching staff and to committees, and made recommendations to the Governing Body. The Council handled matters as they arose by usually meeting weekly during terms, oftener if necessary.

In January 1895 Whitehead was elected to the Council for a four-year term. He was immediately put on a small committee to recommend action on an issue that was coming to a head. As the College buildings provided rooms for only about half of its undergraduates, the Master of Trinity (Montagu Butler) and many Fellows were eager to have a new Court built. Rather more than half of the Fellows approved; they included most of Whitehead's Trinity friends. The site that was now favored was the paddocks, on both sides of the avenue of limes which bisects "the backs"—that is, the land between the river and the Queen's Road. So the beauty of "the backs" was threatened; that distressed Whitehead and a sizable minority of the Fellows. He had been outnumbered on the committee. He now wrote a pamphlet for the Master and Fellows. In it he marshaled practical arguments against building there, and then declared, "The present beauty of the 'Backs' is more than of merely sentimental value: it produces a direct effect in helping the College to achieve the objects for which it exists."[37] The Fellows being almost evenly divided, the opinion of old Trinity men not in residence was sought. When this turned out to be largely hostile to the scheme, the Master, who did not think the Court need spoil the appearance of "the backs," dropped the plan in the interest of College unity. This was in 1897. Building on the paddocks was next proposed after the First World War, but not carried. It was voted down again, overwhelmingly, in 1965.

Among the permanent committees at Trinity the most important was the twelve-man Education Committee, in which questions of curriculum and educational policy were debated. And appointments to the teaching staff were practically in its hands, since it sent its recommendations on appointments to the Council for official action. Whitehead was a member of this committee for several years.

There were other committees for him, but not many. Many Fellows thought the Wine Committee as important as any, but no one in his right mind would have thought of asking Whitehead to serve on it. He was known to be abstemious, which was contrary to Trinity tradition. A good many Feasts dotted the College calendar, and they were, really, feasts. He did not go to them if he could help it. Besides, he could not stand so many people for so long a time.

On the issues that divide academic men in official meetings, which often turn a meeting into a battle, Whitehead's inclination was to study the whole situation and seek such reconciliation of the opposed positions as conditions permitted. When the conditions make reconciliation impossible, restoring calm until they do is something of an achievement. Whitehead became an adept strategist. In meetings of Trinity's Governing Body, called College Meetings, he was generally a calming influence.[38] I have been told of one meeting in which the tempers of opponents were running unusually high. Whitehead took the floor and held it for quite a while, uttering nothing but platitudes. When he finished, everyone had had time to calm down.[39]

The life of a Cambridge don was much simpler then than it is now, when trips to London to be one of the Government's fact-finders or advisers on this or that problem are frequent—not to mention activities in the administration of big science and as a consultant to industry, work for cultural agencies, and appearances on the mass media. The don in Whitehead's days at Cambridge could concentrate on college and university work. (At Trinity, examining for scholarships and examining both poll men and honors men on the frequent College tests of undergraduates was time-consuming.) As for business ventures by dons, they were considered infra dig.

So far as I have been able to find out, Whitehead had no substantial reliable income from outside sources. And his entire academic income came from Trinity College: Lecturer's stipend and fellowship dividend. It was fortunate that he was at Trinity; although the dividends varied from year to year, the variation was not too great; his friends at King's and St. John's received vastly reduced dividends for several years, as a great agricultural depression hit the sources of those colleges' income hard. One Kingsman, in a University play given at Christmas 1895, depicted the Forty Thieves as Fellows of King's who had been driven to this new way of life.

Routh and a few other mathematicians got the greater part of their academic income from coaching, but Whitehead never undertook to be a coach. He was enough of a workaholic to do coaching as well as college lecturing, but he would have had that much less time for his research.

In 1883 the University made a small step toward the present system, in which most lectureships are University appointments. It established University lectureships that gave the holder a small stipend in addition to the stipend from his college. In mathematics five of these were established. Whitehead never received one, although he was at Cambridge until 1910. The only University appointments he ever held were in connection with the Tripos, as Examiner or Moderator.

In June 1903, however, Trinity College gave him a new appointment. This was a result of the growing desire of the College to foster research. There had been talk about instituting a new post, that of Senior Lecturer, and one had already been appointed in Classics. Now, on recommendation of the Education Committee, the Council laid down the responsibilities of a Senior Lecturer and appointed Whitehead for Mathematics, from October 1, 1903, to May 11, 1913. The position, Council said, "should be used for the provision of additional teaching of an advanced type and based on original work."[40] Accordingly, the Senior Lecturer was required to give in each academic year two courses (open to students of all colleges) on the subject of his researches, in addition to the number of courses normally given by Lecturers. He would be paid £100 more than the £400 paid a Lecturer. In 1903–4 Whitehead (who had begun his collaboration with Russell) fulfilled this requirement by giving a course (described as a conversation class) on "The Principles of Mathematics" and one on "Cardinal Numbers" (the subject of a paper he had published a year earlier).

vii

The University had a General Board of Studies, with responsibility for all the triposes and for the coordination of the lecture-offerings in all departments of study recognized by the University. For every department there was a Special Board of Studies which submitted recommendations to the General Board. Whitehead served on the Special Board for Mathematics from 1895 to 1904. There were 24 mathematical lecturers or professors on this board. The topic that agitated it most often was reform of Cambridge's most hallowed institution, the Mathematical Tripos, with its order of rank among the Wranglers.* Whitehead favored abolition of the order of rank. What, no Senior Wrangler? That was unthinkable to the great mathematical coach of the day, Routh, and to many others. Whitehead thought it

*See Chapter V, section iv.

ridiculous that an honors student should have to spend his first three years on what was essentially elementary work in order to write out solutions to tricky problems fast enough to become a top Wrangler when he took Part I of the Tripos, leaving only one year to study the living, growing parts of mathematics, on which he would be examined in Part II. He also thought that the Tripos provided an unsatisfactory mental discipline, and gave too little encouragement to students who wanted to study mathematics for the sake of practical applications. In January 1900 the Special Board for Mathematics recommended abolition of the order of rank and proposed several other reforms. The University was obliged to print twenty-three fly sheets pro and con. The recommendations were approved by the General Board and sent to the Senate of the University for final decision. Every Cambridge graduate who has acquired the M.A. degree (simply by paying a fee two years after receiving the B.A.) is a "member of the University," entitled to a vote in its legislative body, the Senate. This makes it difficult to drop any traditional way of doing things, and so it proved in this case. The Senate approved all the proposed reforms except the one about which feeling ran high: abolition of the order of merit. The Senate did not agree to this until 1907.

Two other University changes that had been brewing for years were held up even longer by the descent of armies of alumni upon Cambridge to cast negative votes. One was the abolition of compulsory Greek; the other was the admission of women to Cambridge degrees.

Examinations in Greek and Latin were parts of the "Previous Examination," or "Little Go," which every undergraduate was required to pass in his first year.* A scheme to end compulsory Greek in this examination was defeated in 1891. In 1904 a proposal was made to require only one ancient language. This appealed to dons in the sciences, but the Senate voted it down by a large majority. I have found no record of Whitehead's position on this issue, and must surmise that he saw some valid reasoning on both sides of it. He had profited from his early training in Latin *and* Greek. Yet this requirement in the "Little Go" in effect made knowledge of these languages a condition for matriculation, and so tended to keep out of Cambridge gifted young men who had not had his training.†

---

*See Chapter V, section iii.

†J. J. Thomson, who came to Cambridge ignorant of Greek, considered the time he

Half a century of agitation preceded the granting of equality with men to women students at Cambridge. The situation throughout Whitehead's years there seems so strange today that it will be wise to give a rather full account and to use many of his own words in telling of the part he played.

Girton and Newnham were not, like women's college in the United States, degree-granting institutions, and they were not colleges *of Cambridge University*. February 24, 1881, was a day annually celebrated at Girton; that was when the University Senate opened the triposes (and the "Little Go") to women. But no woman could be a Wrangler; her sole reward was the announcement that her score equaled that of the nth Wrangler. As for a Cambridge degree, Heaven forbid! In 1880 the first syndicate (as University committees are called at Cambridge) to study the question had had its recommendations defeated in the Senate. The issue arose again in December 1887, after a large number of memorials were submitted to the Council of the Senate. The Council, rightly perceiving that the preponderance of Cambridge opinion was against any change, declined to recommend any action to the Senate.

The women did enjoy the privilege of using some University libraries, and attending the lectures of those dons who permitted their presence. Many dons did so, and many did not. Whitehead always granted permission. There is a story that, having assigned A. E. H. Love's *Elasticity* as textbook in one course of lectures, he said to the men when he saw women students in the back row, "I see you've brought your Loves with you."

"The Women's War" is the name given by the undergraduates to the campaign of 1896–97. It began when over 2,000 members of the Senate asked that the question of degrees for women be reopened. (Whitehead was not among them, but neither was the women's greatest friend, Henry Sidgwick; perhaps the time was not right.) The Council named a syndicate, but the membership of this syndicate was rejected by the Senate in a large vote.

So matters stood when the *Cambridge Review,* in its issue of May 14, 1896, printed an article, "On Ideals: with Reference to the Controversy Concerning the Admission of Women to Degrees in the University." It was signed "W." So Whitehead had signed his two stories in the *Review* ten years before; but they were identified as his by his

---

spent on Greek for the "Little Go" in his first year "utterly wasted" (Thomson, *R&R*, p. 35).

daughter, Jessie, whereas Jessie could only say "probably" about this article. The *Review* has kept no records about it. Its content, I think, makes the probability that some other "W" wrote it almost nil, so I rule that out.

The author leaves no doubt that he favors the admission of women to degrees. He starts out by observing that only its opponents claim to justify their position "on ideal and elevated grounds." He is referring to their favorite counterproposal: the establishment of a separate university for women. This, under the name of The Queen's University of the British Empire, was the butt of a famous witty speech later given by F. W. Maitland. The natural place for such a university, he told the Senate, was the waiting room at the Bletchley railroad station, and its name "the Bletchley Junction Academy. . . . You wait there; but you do not wait there always. You change for Oxford and Cambridge." (In 1896–97 the issue of degrees for women was being debated at Oxford as well as Cambridge. Of the ten universities in Great Britain, these were the only ones in which women could not receive degrees.)

Another name for the proposed women's university was the Imperial University. Whitehead writes:

> To these advocates of an Imperial University there appears nothing ideal in the persistent, irritating, and stupid clinging of women to the University of Cambridge—the desire to have the same degrees as men. . . .
>
> In various tones, sometimes impatiently, but for the most part kindly and politely, those who oppose the admission of women to the University say:—"We speak in your highest, your most sacred interests, when we tell you that it is much better for you not to be admitted to the University. Please start a nice University of your own somewhere else. . . . We will not cast you off entirely; we will help you; you shall have our advice in determining what is most suitable for women's education; and we think it probable you will be a great success.
>
> "Besides, to admit you to the University would be most inconvenient to us in many ways;—most injudicious, and probably very dangerous. Please, please go away."
>
> To this the other side replies:—
>
> "We, too, have our ideal. In clinging to you . . . , in desiring to be admitted as members of your University, we desire to share your great ideal, your high and imperishable standards of knowledge, of scholarship, and of research, without which no higher education, whether of men or of women, is possible."

We today do not care for this lofty rhetoric, but it was customary at the time. The remainder, and longer part, of this article continues the stress on the great ideal of Cambridge University, which the author distinguishes from the pursuit of knowlege for practical use. The women distrust a scheme of education based on considerations of their "highest interests." The only argument they could accept, Whitehead represents them as saying, is this: "During the twenty years we have been within your gates, we have shown, beyond a doubt, that we have no love of knowledge, no desire to pursue truth for its own sake, no capacities and no aspiration, that in admitting us you are traitors to your sacred and inherited trust."

In June 1896 a second syndicate, almost entirely new, undertook to study the whole question. Whitehead was one of its fourteen members; Maitland, reluctantly, was another. A rift in the syndicate's membership soon developed. The following February a report, signed by nine of the fourteen, was issued. It proposed that women who had satisfied the University's requirements be granted the *titles* of degrees. This was not full equality of status with men, for it denied them the opportunity to become members of the University, entitled to a vote in its government.[41] But the proposal did require the University to issue a diploma saying that the woman had earned a Cambridge degree.

In the protracted debate in the Senate, Whitehead defended the syndicate's proposal.[42] It was, he said, conservative. It removed the women's grievance,* and did nothing more. Some opponents of degrees for women raised the specter of a disenfranchised class in the University; Whitehead pointed out that fewer than half of the men who had taken the M.A. in the past year had bothered to keep their names on the University's books, thus showing that what they valued was the title of the degree, not the vote in University government. Whitehead also restated ("in a dull manner," he said) what Maitland had wittily said against the alternative of a separate university for women. This alternative, he added, lacked the support of a single leading woman in the educational world. "The Syndicate," Whitehead noted, "have been criticised for not introducing into their Report any general principles concerning women's education. It would have

*Whitehead said, "If you are going to make changes at all, for heaven's sake satisfy somebody." He was putting to use his reading of Burke's speeches on the American Revolution, as he would again when he came to write his own sociopolitical philosophy, more than thirty years later; compare *AI,* p. 56.

been magnificient if we had done so, but certainly not wise. Our business was with practical proposals." He found the syndicate's proposal "cautious and safe." On committees, Whitehead was a sensible pragmatist.

Maitland, writing to Henry Sidgwick after a nearly five-hour meeting of the syndicate, said, "Whitehead was a real God-send."[43]

All Cambridge was in arms over degrees for women. Many of the most respected dons and professors were with Whitehead, Maitland, and Sidgwick, but others—Forsyth, for one—were not. A poll of the undergraduates, taken by the *Review* at the time it published Whitehead's article, showed them to be almost four to one against degrees for women. When the syndicate's proposal was published, the Union threw open its debating hall to all undergraduates; a motion condemning the proposal was carried by an almost ten-to-one margin. On voting day the undergraduates had themselves a ball, with banners, fireworks, and effigies of women hung over Trinity Street.

The Senate voted against degrees for women, 1,707 to 661. No further effort in this cause was made until after the First World War; it was successful. But competition for University scholarships and prizes was not opened to women until 1928.

viii

As we saw in Chapter VIII, Whitehead began to teach mathematics to Girton students, as well as to men at Trinity College, in the autumn of 1888.* According to Girton's records, he continued this extra teaching, except for the academic year 1890–91, until the end of 1913. Since he moved to London in 1910, the implication is that for three years he took the train to Cambridge at least one day a week to teach women.

In 1901 Whitehead became associated with Newnham College,† and in 1907 was appointed to its Council. He attended its meetings regularly until February 1911. But he never taught at Newnham.

In 1891 a new outcrop of the University, the Teachers Training College, began its work with the ebullient Oscar Browning as Principal. This was the outcome of a movement, initiated in the seventies

---

*See Chapter VIII, section v.

†Newnham came into existence in the 1870s, thanks to many friends of women's education, first of all Henry Sidgwick, and then his wife, Eleanor. It was located in Cambridge, just west of the Queen's Road.

by him and the quiet Henry Sidgwick, to put trained Cambridge men into the nation's elementary schools.* Whitehead was greatly concerned about elementary education, but his efforts to improve it were made after he left Cambridge. There was no occasion for a Cambridge mathematician to lecture in the Teachers Training College. Whitehead may have been given an opportunity to teach mathematics to workmen and other nonacademic persons in the University Extension lectures that James Stuart had led many Apostles to support, but I have no record of Whitehead's giving such lectures. His life was quite full as it was.

When Whitehead came to Trinity College as a freshman in 1880, the Cavendish was the only laboratory building in Cambridge. Ten years later, the University had a chemistry building and a laboratory for anatomy and physiology, and new buildings for medicine, law, and engineering were being planned. New professorships were created at the same time. Another change was the admission of graduates of other universities who wished to do research at Cambridge. In 1895 this got through the Senate with difficulty. J. J. Thomson, who fought for it, was rewarded. That autumn the first "research student" to arrive at the newly expanded Cavendish Laboratory was Ernest Rutherford, from New Zealand. The nineties and the next decade were a great age for the sciences at Cambridge, most of all for physiology. The human environment was good for Whitehead, who would become a philosopher of science.

Ever since he was elected to the Apostles, he had thought about philosophy. In the mid nineties Lowes Dickinson, McTaggart, and others formed what they called an Eranos for the discussion of philosophical problems. Whitehead was a member; so, at a later date, was Bertrand Russell. According to Dickinson,

> This society was somewhat amateurish and perfunctory, and I am not aware that any records of it have been preserved. . . . I do not even know when the society expired. Presumably it died in the end of inanition, but it continued for some years, and was both stimulating and amusing. Jack [McTaggart], of course, was its principal inspiration.[44]

In the Eranos, Whitehead was exposed to, perhaps eventually influenced by, McTaggart's unorthodox version of Hegelian idealism.†

*The Teachers Training College was the predecessor of the University's present Department of Education.

†See Chapter VII, section v.

When Whitehead married, he changed the rooms given him by Trinity College, moving from a large, high-ceilinged room (C.2) in Nevile's Court to a modest one there. From 1902 on, he was content with a very modest, low-ceilinged ground-floor room (G.2) in New Court.

CHAPTER XI

Bertrand Russell

i/Whitehead and Russell's scholarship admission to Trinity College. Russell's undergraduate years.

ii/Russell's mind. His Trinity fellowship dissertation. His need for Whitehead's encouragement.

iii/The 1900 Congress of Philosophy at Paris. Peano's mathematical logic, and Russell's response to it. Russell's discussions with Whitehead on the extension of Peano's logic.

iv/Russell's logicist thesis about the foundations of mathematics. His *Principles of Mathematics*.

v/Russell's emotional conversion on seeing Evelyn Whitehead in pain. The facts about her condition. Russell's surreptitious gifts of money.

vi/Russell's love for Evelyn. Russell and his wife, Alys. "The Free Man's Worship."

The history of the relation between Whitehead and Bertrand Russell begins with, and for a long time is little more than, an account of the help that Whitehead gave to the younger man. He did not get Russell into Trinity College, but in December 1889 he was one of the readers of the examination papers for scholarships written by prospective freshmen. The Honorable (brother of an earl) Bertrand Arthur William Russell, grandson of Lord John Russell, was one of them. He was then seventeen, and in love with mathematics; but his preparation for the examination, at a crammer's school, was not the best. Whitehead thought that his papers showed more ability than was indicated by a summing of the marks he made in them. He persuaded his fellow examiners, who were his old tutor, H. M. Taylor, and Forsyth, that Russell should be awarded a £75 minor scholarship instead of the £50 one which his marks earned, and which Whitehead himself had been happy to accept ten years earlier. What was at stake was not financial aid, of which Russell had no need, but prestige, and the extra impetus that the larger scholarship provided at the beginning of undergraduate studies. If Russell's performance had been outstanding, he could have been given a major scholarship.

Recounting this episode in his *Autobiography,* Russell says he later learned—from whom, he does not say—that Whitehead burned the marks before the examiners met. Perhaps he did, but it is just as likely that this is a dramatic embellishment due to his wife. From Whitehead's later life we know that he believed the end sometimes justified the means, though not as often as Evelyn maintained; but we also know that he was a persuasive man in committee meetings, and usually got his way by patience and shrewdness, without destroying the evidence. We simply do not know what he did in this case.

The more important effect of Whitehead's reading of these examination papers was that when Russell came up to Trinity in October 1890 he found himself called upon within the first week by several of the brighter undergraduates, men who became his close friends. There was no period of loneliness, for Whitehead had spoken to them about him. Otherwise, Russell would have had a hard time, for on his own account, he was extremely shy, and also quite a prig.

According to Russell, his first personal contact with Whitehead occurred that same term, when he attended Whitehead's lectures on statics. "He told the class to study article 35 in the textbook. Then he turned to me and said, 'You needn't study it, because you know it already.' I had quoted it by number in the scholarship examination ten months earlier. He won my heart by remembering this fact."

Talking to me in 1965, Russell could not definitely recall what other course of lectures by Whitehead he attended in his three years as an undergraduate in mathematics. Probably he attended one or two others, but possibly none, as Trinity's staff was large. Whitehead did not teach the infinitesimal calculus. Those who did, offended Russell by their lack of logical rigor concerning its foundations. His Trinity teachers and his coach* so disgusted him with mathematics, presented as a bag of tricks to use in examination problems, that as soon as he passed Part I of the Tripos in June 1893 (with the rank of Seventh Wrangler) he sold all his mathematical books and swore that he would never look at one again. Unlike Whitehead, Russell never wanted to be a professional mathematician.

As a young man he had diverse large interests, primarily philosophical. He wanted to know what ultimate reality is like, and how anything can be known to be certainly true. His interest in mathematics was not in the fruitfulness of mathematical ideas but in the justification for believing in the certain truth claimed for them. When he was eleven he was disturbed by the insistence of his brother, Frank, instructing him in geometry, that he accept Euclid's axioms without question. Boy and man, Russell demanded proof and was obsessed by the search for certainty. (We shall have occasions later to compare his conceptions of mathematics with Whitehead's.) Young Russell was also keenly interested in ethics, social philosophy, and politics (his grandmother Russell, who brought him up, envisaged him as a Prime Minister). Instead of spending his fourth year at Cambridge studying for the advanced Part of the Mathematical Tripos, he officially entered with joy the open world of philosophy, by studying for the Moral Sciences Tripos, chiefly under Sidgwick, Ward, and Stout. In 1894 he won a First Class in that Tripos, and was granted "distinction" in ethics and metaphysics, and in the history of philosophy. The switch from mathematics caused him to miss by a term Whitehead's first

*Russell's coach was R. R. Webb of St. John's College. Routh had retired.

offering of a lecture course on non-Euclidean geometry, then considered an out-of-the-way subject. It would have appealed to him.

It would be a mistake to think that the teacher-pupil relation between Whitehead and Russell rapidly changed into that of friends, colleagues, and then collaborators. Russell's friends were men within three years or so of his own age—Crompton and Theodore Llewelyn Davies, Charles Sanger, the three Trevelyan brothers, and G. E. Moore. "Whitehead," he wrote, "was amazingly kind, but was too much my senior to be a close personal friend until some years later." During Russell's undergraduate years, no evidence suggests advance beyond their relation in 1890. They became colleagues, in the sense of both being Fellows of Trinity, in 1895. The best year to assign to the beginning of their collaboration on *Principia Mathematica,* we shall see, is 1901. It was presaged by their trip together, in the summer of 1900, to a philosophical congress at Paris, which I shall shortly describe.

In dealing with the years before that trip we must make a constant effort to keep chronology in mind, and not slip into the assumptions that Whitehead was already a philosopher, and Russell already a symbolic logician. Whitehead knew Kant's first *Critique,* and the *Universal Algebra* shows that he had studied authorities in logic, but he had no command of any other branch of philosophy; his knowledge was not much beyond what is suggested in Chapter VII, on the Apostles. He was a mathematician who included the algebra of logic in his investigations. Russell, however, was a man who had done much serious philosophical reading—he did not wait until he officially began to study philosophy at Cambridge to join the Moral Sciences Club, but did so in his second term there—and was turning to the *philosophy* of mathematics. What first brought the young philosopher and the mathematician together was Russell's idea, when he was finishing his second Tripos, that the epistemological foundations of geometry would be a promising subject for the dissertation with which he hoped to win a Trinity fellowship. Kant had posed the question, How is geometry possible? That was before non-Euclidean geometries were discovered, so Russell wished to study Kant's problem anew. Early in June 1894 he went to James Ward, who was a Kant scholar. Ward gave him initial reading advice on the philosophical part of his project and sent him to Whitehead for the mathematical part. If he had attended the advanced courses Whitehead offered in 1892 and 1893, which were beyond the range of subjects covered in the Tripos for which he was then studying, instead of spreading out his surplus intellectual energy, Russell would already have been aware of Grassmann and some other

Continental mathematicians he now read at Whitehead's suggestion, and his love of mathematics would not have suffered a year's interruption. Whitehead did not know how to make mathematics, when studied for its own sake, uninteresting; but Russell did not at that time know Whitehead well enough to realize what he was missing. His vow to abjure the subject was a misplaced early instance of his passionate condemnation of whatever he heartily disliked, a characteristic that would always be with him. He was, as he often insisted, a man of passion.

He now made up for lost time in short order. He read the *Ausdehnungslehre,* projective geometry, many writers on non-Euclidean geometry, several *Cours d'Analyse,* and much more—in a year when he was working as attaché at the British Embassy in Paris, marrying Alys Pearsall Smith, studying economics at the University of Berlin, and enjoying spring in Italy with Alys. What a man!

ii

The first thing that should be said about young Bertrand Russell is that he was a superb intellectual athlete. Grasping his reading with lightning speed, seeing logical relations, writing out an analysis of a problem, were as easy for him as breathing. In searching for the solution of a problem he had his unproductive days, but he did not let himself be distracted. He lost little time in waste motions, and was never slowed up by fear of making a mistake; if he made one, he could correct it in a subsequent publication. The second thing that should be said is that his quick mind was highly, perhaps completely, verbalized. When friends gave him an analogy-spotting test in which shapes of increasing complexity were to be matched, he did the simpler problems rapidly, but gave up before the end, saying, "I hadn't got any names for the shapes."[1] (That he was in his seventies when he took this test is irrelevant; when I interviewed him at ninety-three, I noticed no sign of senility.) And Whitehead told me that whereas he himself would be seized by an idea and then grope for the best words to express it, Russell admitted that he thought entirely in words. This was a great advantage so far as productivity was concerned, but it often kept him on the surface of his subject. While Russell wrote much more than Whitehead did, he was not often as penetrating a thinker. Certainly he was by nature a writer, whereas Whitehead was not. Russell's wonderfully clean style has been taken as a model by many who have had to sweat hard and long to produce paragraphs a bit like his; at twenty-eight, he wrote his *Principles of Mathematics* at the rate of ten pages a day.

When Russell's subject was not mathematical, he was inclined to simplify alternatives. He was a definite yes-or-no man. If the alternatives were simple, a little exaggeration enabled his incorrigible wit to score. In talk, everyone except the loser—but including the loser if it was Whitehead—shared his enjoyment of making a clever point. He was great company.

Russell's intellectual ambitions were large. At twenty-one he received a legacy from his father's estate, big enough to permit him to set aside all thought of a profession and devote himself to writing. In the *Autobiography* he says of a day in 1895, when he and his wife of three months were in Berlin:

> I remember a cold, bright day in early spring when I walked by myself in the Tiergarten, and made projects of future work. I thought I would write one series of books on the philosophy of the sciences from pure mathematics to physiology, and another series of books on social questions. I hoped that the two series might ultimately meet in a synthesis at once scientific and practical. My scheme was largely inspired by Hegelian ideas. Nevertheless, I have to some extent followed it in later years, as much at any rate as could have been expected.[2]

The point of the "nevertheless" is that Russell's Hegelian period ended three years later. The first philosophy to which Russell adhered was Mill's. He had imbibed Hegel's, which seemed much more profound, partly from Stout's lectures and partly from reading F. H. Bradley with admiration, but mostly from constant talk with Hegel's apostle at Cambridge, McTaggart. McTaggart was an intimate friend from his freshman year until the War came, and, being six years older and almost as brilliant, was the first living person to influence him strongly in philosophical matters.

Russell admired McTaggart very much, and hoped that some day he would do work as good as his. Unfortunately Jack had no interest in the philosophy of mathematics, and Russell found that what Hegel wrote on it was worthless. Back to Kant! And to the writing of the fellowship dissertation, Russell's first piece of serious original work. When he finished it early in August 1895, he believed that he "had solved all philosophical questions connected with the foundations of geometry."

The manuscript he submitted does not, I believe, exist today. Instead we have the elaboration of it which the Cambridge University

Press published in 1897, as *An Essay on the Foundations of Geometry*.[3] (It is dedicated to McTaggart.) What Russell brought to this subject and no philosophers before him had brought, was knowledge of projective geometry. The relation between it and metrical geometries is the key with which he unlocked the problem of foundations. In finding this key he had learned much from Whitehead; indeed, desire to learn from Whitehead on mathematical matters was characteristic of Russell for decades. It was a firm conviction of Whitehead's, held all his life, that since projective geometry makes no reference to lengths or to sizes of angles—in short, to extensive quantity—it is more general than metrical geometry. Cayley had declared that projective geometry is all of geometry, and Felix Klein had illuminated its mathematical relationships to the non-Euclidean geometries. Then Russell, accepting the Kantian doctrine that the intuition of a form of externality is necessary to sense-experience, argued that projective geometry expresses this form of externality as systematic a priori knowledge, and that when the addition of a distance function launches us into metrical geometry, an empirical element is introduced for actual space, and for the alternatives concerning its structure—Euclidean, Lobachevskian, or Riemannian. For the present purpose, this is all that need be said about Russell's complex construction of the foundations of geometry. Of course he repudiated it in a few years. But it was a remarkable piece of work for so young a man—even allowing that, as he said in the Preface to the book, he received from Whitehead "constant criticism and suggestion throughout the course of construction, especially as regards the philosophical importance of projective geometry."

When Russell turned in his dissertation it was given to Whitehead for evaluation from a mathematical point of view, and to Henry Sidgwick for philosophical evaluation. A month later, however, Sidgwick asked Ward to take over from him; he had too long been out of touch with controversies relating to space. He had read the manuscript, would send Ward a preliminary report to save him a little trouble, and said that both he and Whitehead thought it "decidedly able."[4] So it was to Whitehead and then to Ward that Russell, anxious to get some idea of his chances for a fellowship, went on October 9. Neither one, of course, could tell him whether they would recommend his election on the basis of the dissertation. Ward allowed himself to praise it, but Whitehead did not. That evening Russell wrote to his wife that he had had "a rather depressing talk with Whitehead. . . . He wouldn't say a word directly as to my merits or demerits, but only criticized definite points. He said however that Ward and Sidgwick were more dissatisfied than he was: so I drew the blackest inferences." If he was turned down, he would have two more years in which to try for a fellowship; but Russell was not that long on patience. More important, he had made up his mind that if Ward and Whitehead thought highly of his dissertation, he would devote his life to philosophy, and first to the philosophy of mathematics. If they thought ill of it, he would study economics in order to prepare for the career that would satisfy his practical interests, one in public life; that was also the natural career for the grandson of Lord John Russell, and was strongly urged upon him by his family.[5]

The next day he was elected a Fellow of Trinity, and Mrs. Whitehead sent him this note:

> Dear Mr. Russell
> Please accept our warmest congratulations on your election. Were you doubtful yesterday?
>
> Sincerely yours,
> Evelyn Whitehead

Russell wrote on the note, sometime before it was sent to the Bertrand Russell Archives, "I learned afterward that Mrs. Whitehead had scolded Whitehead for giving me a bad time." In his fullest published account, Russell says that Whitehead

> defended himself vigorously. He said it was the last occasion on which he would be able to speak to me as to a pupil, and that, after the praise that he knew I was to get, I might have paid too little attention to his entirely justifiable criticisms.[6]

That the criticisms were entirely justifiable is Russell's later opinion. After the fellowship election, Whitehead gave him all the praise that he had withheld in the interview.

A basic element in Russell's character was an abiding need for approval of his work and for constant encouragement from people he respected. I do not know the underlying cause of this need, but suppose that his lonely boyhood is part of the cause, perhaps along with his being regularly, until he came to Cambridge, subjected to opinions he could not respect. Whatever the causes, and however strange a need for reassurance may seem in a formidable intellectual athlete, there is no room for doubt that it was there. It did not disappear as he grew older. At seventy-five, when he was writing *Human Knowledge,* he would take a chapter or two to his friend and neighbor Crawshay-Williams to be read aloud, and presumably discussed. But when praise was not forthcoming, "Bertie would begin diffidently to fish. 'I rather like that bit', he would say, Or 'Do you think that point is sufficiently clear?' "[7]

Whitehead could give the most brilliant of his pupils the encouragement he needed, and did so, year after year—without holding back honest criticism of points he considered crucial. He would have done the same for any other pupil; my point is that Russell had a special need for this, and thrived on it. Whitehead's repeated encouragement is what most strikes a reader of the hundred-odd technical letters[8] that he wrote to Russell in the course of their long collaboration on *Principia Mathematica.* In the interview of October 9, 1895, which so depressed Russell (formally their last teacher-pupil interview), Whitehead did relieve Russell's depression by telling him that "certain parts of the dissertation, more or less as they stood, were well worth publishing." Whenever Russell was baffled by the mathematical side of a problem, he looked to Whitehead for help. For example, at a time early in 1897 when he was working on the relation between number and quantity, he wrote to Alys: "utterly baffled in all attempts to get on with quantity and look forward to suggestions from Whitehead." He was doing the same thing twenty years later. Whitehead had then been led by the work of Einstein and Minkowski to turn from the composition of Volume IV of *Principia Mathematica,* on geometry, the notes for which he had been sending to Russell, to that of his first book on the philosophy of physics. He now declined to send Russell his notes, explaining that he worked slowly and did not yet want his ideas set out by anyone—as had been done (with due acknowledgment) in 1914 in some parts of Russell's *Our Knowledge of the External World.* Whitehead wrote to him on January 8, 1917:

> I am sorry that you do not feel able to get to work except by the help of these notes—but I am sure that you must be mistaken in this, and there must be the whole of the remaining field of thought for you to get to work on—though naturally it would be easier for you to get into harness with some formed notes to go on.*

The fact that Russell greatly enjoyed arguing with his friends does not conflict with his having the characteristic just described. His mind

*In telling me of this letter, Russell naturally saw it only as an expression of Whitehead's feeling that he had been plagiarized in 1914. The letter reminds me of Russell's need for help from Whitehead, and shows Whitehead's recognition that Russell would go dancing away with Whitehead's ideas before Whitehead himself had put them in order.

was so much quicker than any opponent's—save Maynard Keynes's—that he usually won the argument, though not always fairly. The personal feeling between him and an adversary who persisted though defeated could be unhappy; his relation to G. E. Moore is an example. On mathematical points he assumed that Whitehead, because of his greater knowledge, was more likely to be right than he. In all his writings, mathematical or philosophical, he was admirably ready to admit that his ideas had been mistaken. But he was also happy to follow Moore in abandoning the idealism they had shared; he allowed Santayana to sweep him out of an early absolutism in moral theory, and Wittgenstein to convince him that mathematics consists of tautologies; when he became acquainted with American behaviorism he was unduly taken with it. He tenaciously retained some philosophical beliefs of his maturity as long as he lived. I shall not discuss them here, as Russell's philosophy is not the subject of my book.

iii

By the spring of 1898 Russell was clear that he wanted to write a big book on the foundations of all of mathematics. He would write it in a small sixteenth-century workman's cottage near the village of Fernhurst, 40 miles west of London on the hilly border between Surrey and Sussex—a region much favored by intellectuals. Millhangar, as he and Alys named the cottage, was ideal for her because of its proximity to the Pearsall Smith estate on Friday's Hill. After they had added two rooms, the Whiteheads would be among those who came for visits; the men were now close friends, Evelyn Whitehead enjoyed witty aristocrats, and Bertie's American wife, who would be called a feminist today, was interesting—for a time.

At the end of July 1900 the four went to Paris together. The First International Congress of Philosophy was being held there; it would be followed by the Second International Congress of Mathematics. At the former, Russell would read a paper, "L'Idée d'ordre et la position absolue dans l'espace et le temps," and Alys would read one entitled "L'Education des femmes." Whitehead read no papers at either Congress, but he was the only Englishman among the five Secretaries for the mathematical one. The great Exposition was a further attraction of Paris that summer.

Russell always said that his attendance at the philosophical Congress of 1900 was a turning point in his intellectual life. He had already made a philosophical turn*about*. In 1898, soon after his first false starts on the big book, he rebelled against the type of philosophy that was then fairly dominant. He threw overboard the idealistic monism to which he had been attached for four years. He stopped believing that everything is mind-dependent, that particular things are but appearances to finite minds and that everything in the universe so involves every other thing that there are no finite self-sufficient truths, nothing that can satisfy our reason short of the system of the whole, which Hegel and Bradley called the Absolute. These doctrines seemed a hindrance in dealing with the principles of mathematics, and he became dissatisfied with them. G. E. Moore spurred him on to reject them entirely, and for good. Russell moved to a position diametrically opposite, and thereafter indulged his natural preference for sharp outlines. Every fact, he declared, is independent of every other fact, and every thing, whether existent or ideal, of every other. The world, as he liked to put it later, is more like a heap of shot than a pot of treacle. Space is an infinite collection of points; it is not relative to the things "in" space. Nor is time, an infinite series of instants, relative to the things "in" time. Russell at this time pictured mathematical entities like numbers and triangles as subsisting eternally in Plato's heaven. He utterly rejected the dialectical logic that Hegel, Bradley, and McTaggart had used to force philosophers to a monistic conclusion. To his delight, he had become acquainted with a powerful instrument of analytic thought, symbolic logic, through the works of Boole and later logicians, and Book II of Whitehead's *Universal Algebra*.

Idealism fared no better than monism in Russell's turnabout. He saw that there were no valid arguments for any dependence of the facts of external nature on the mind. Grass is really green. Space and time are not subjective, as Kant had persuaded almost everyone to believe; geometry and arithmetic were in the clear.

Within a few months of Russell's adoption of his extreme pluralism and realism, circumstances led him to study the philosophy of Leibniz. McTaggart wanted to visit his family in New Zealand early in 1899. He was to lecture on Leibniz in the Lent term, and asked Trinity College if Russell might do so instead; the College approved. Russell at first found Leibniz's monadology as puzzling as most commentators had. Then he saw that he could explain it as the outcome of Leibniz's strictness in adhering to the twin traditional doctrines that every proposition attributes a predicate to a subject and every fact consists in a substance having a property. Leibniz drew the consequence that there can be no direct relation between separate substances; that is why his monads do not interact. Russell's book *A*

Critical Exposition of the Philosophy of Leibniz, based on his lectures, was published by the Cambridge University Press in 1900. The next year, his interpretation was strengthened when Louis Couturat, who had discovered many additional manuscripts at Hanover, published *La Logique de Leibniz.* It and Russell's book, which Whitehead admired,* were the basis of Whitehead's knowledge of Leibniz. The idea that adherence to the subject-predicate logic makes it impossible to explain the togetherness of things in our experience became Whitehead's permanent belief. There is, however, no reason to assume that Whitehead was wedded to that logic until Russell wrote on Leibniz.

When Russell went to Paris in 1900, he was already known through his writings, which included a controversy with the great Poincaré on the interpretation of the axioms of geometry. But he was not happy with what he had written for his book on the principles of mathematics. He was satisfied with his philosophy, but he had not found a satisfactory logical technique for getting at the foundations of arithmetic.

The chief attraction of the Congress for both Russell and Whitehead was the opportunity it afforded them to hear Giuseppe Peano, Professor of Mathematics at Turin, who was then at the height of his fame, though only three years older than Whitehead. Peano was at home in all branches of mathematics and had made valuable contributions to analysis. And everyone knew his set of five axioms for the arithmetic of the positive integers. As that arithmetic, extended to other types of numbers, had been shown to provide a sufficient foundation for analysis, Peano's axioms were the current starting point for discussions of the foundations of mathematics. Something else would turn out to be even more important for Russell and Whitehead: Peano had for some years been showing how the foundations should be presented—not in the terms of ordinary language plus the symbols of ordinary algebra, but solely in the rigorous language he used. This language was composed of symbols he accepted from the current symbolic logic (with some modifications) plus others that he originated, such as "ϵ" for an individual's membership in a class. Precision as well as brevity was achieved by this handling of the ideas vaguely expressed by such words as "the" in the singular and in the plural, "is included in," "and," "either," "there exists," and "if-then." In this way he made it impossible to fall into such prevalent confusions as

*This is shown by the footnote recommendations of it in Whitehead's writings from 1906 to 1925.

that between the properties of a class and the properties of its members. Peano did not mean to be a professional logician; his purpose was to effect a logical reformation of mathematics. He was the leader of a school, and with Italian collaborators had published several volumes of a *Formulaire de mathématique,* in which the principles and main results of the various branches of pure mathematics were being recast in his ideographic symbols so as to free them of covert reliance on what *seems* to be intuitively obvious. He also founded a journal, *Rivista di matematica* [*Revue des mathématique*], which was now in its ninth year.

Both Russell and Whitehead had heard of all this, but Russell tells us that he had not troubled to master Peano's symbolism, and I have no reason to think that Whitehead had done so. Whitehead, however, already had a high opinion of Peano, for he knew some of his work outside mathematical logic. (That name, now more commonly used than "symbolic logic," is due to Peano.) In 1888, when Whitehead was first lecturing at Cambridge on the *Ausdehnungslehre,* Peano had published a brilliant monograph on it.[9] He had studied the *Ausdehnungslehre* of 1844, which was the one Whitehead preferred, and after finding Grassmann's exposition an impediment, had discovered the extraordinary power of this calculus by examining its applications. Working backwards, he reconstructed it, and expressed it much more clearly. His monograph is one of the half-dozen "important and interesting works" on Grassmann's calculus that Whitehead named in his *Universal Algebra.* Another was a book on it by a member of Peano's school, C. Burali-Forti; it appeared the year before *Universal Algebra* and anticipated one of its contributions.

Whitehead's treatise was making only a slight impact on British mathematicians; at Paris he found that the members of the Italian school admired it and were pleased to meet its author. The historically important event at the Congress, however, was the impact Peano made on Russell and Whitehead. They saw that in the discussions he showed more precision and logical rigor than anyone else, including the German authority on symbolic logic, Ernst Schröder. Peano, Russell wrote, "invariably got the better of any argument upon which he embarked. As the days went by, I decided that this must be owing to his mathematical logic." He therefore went to Peano and said, "I wish to read all your works. Have you got copies with you?" Peano had copies, and was glad to oblige. This was the beginning of Russell's great career as a mathematical logician, and of the passing of leadership in that field from Italy to England. Peano soon shifted his interest to linguistics, and became dedicated to advocating the adoption of an international auxiliary language he invented—interlingua, which is Latin without grammar.

iv

On returning to Millhangar in August, Russell mastered Peano's work with his customary rapidity. Peano had applied his methods to the symbolic logic of propositions and of classes, but not to that of relations. The importance of relations was much on Russell's mind. He had once accepted Bradley's famous argument that relations between things are self-contradictory if taken as finally real; he renounced it when he became an ardent pluralist. He also renounced a Bradleian doctrine which had found more favor than the argument against relations, and was indeed generally accepted. This was what Russell called "the doctrine of internal relations." According to it, every relation between two things is grounded in the nature of the related things. This is plausible for relations like "father of," but is not obvious for mathematical relations like "greater than." The doctrine of internal relations was but the current, metaphysically useful, form of the subject-predicate doctrine of propositions. Leibniz had noticed a crucial exception to that doctrine, only to set it aside. In a passage which Russell in his *The Philosophy of Leibniz* deemed of capital importance, Leibniz observed that we may consider the relation between the lengths of two lines, *L* and *M,* abstractly, that is, without pitching upon either *L* or *M* as subject. He said that the relation, so considered, is "*out of* the subjects; but being neither a substance, nor an accident, it must be a mere ideal thing, the consideration of which is nevertheless useful."[10] Is the "ideal thing" a qualification of the mind that contemplates the relation? Instead of taking that Kantian line, thought Russell, stop insisting that relations be reduced to adjectives, and treat them as in their own right part of the ultimate furniture of the world. It was clear to him that propositions which propose that a particular relation holds between two terms, or between three, or between four (in projective geometry, the fundamental notion of cross ratio is a relation between four points on a line), or between more than four, are not, in general, formally reducible to the subject-predicate pattern. Attempts at reduction assume the meaning of the relation, in independence of the particular terms related. At least, this is true of mathematical relations.

The theory of relational propositions, as investigated in the nineteenth century, led to the systematic study of types of relations, and so to a symbolic logic of relations. This logic was initiated by De Morgan and developed by Charles S. Peirce; its current exposition was by Schröder. Whitehead knew what all three had done, but had not tried to include the logic of relations in the compass of his *Universal Algebra*. Peano, as we said, had done nothing with it. Russell got the idea of applying Peano's precise notation to this field, and spent September doing so. This, he has said, was intellectually the highest point of his life. In his *Autobiography* he wrote:

> It seems to me in retrospect that, through that month, every day was warm and sunny. The Whiteheads stayed with us at Fernhurst, and I explained my new ideas to him. Every evening the discussion ended with some difficulty, and every morning I found that the difficulty of the previous evening had solved itself while I slept. The time was one of intellectual intoxication. My sensations resembled those one has after climbing a mountain in a mist, when, on reaching the summit, the mist suddenly clears, and the country becomes visible for forty miles in every direction.[11]

In October, Russell wrote the first draft of a paper on the logic of relations. He did not complete it to his satisfaction and send it off (to Peano, in French, for publication in his journal) until five months later.* In the interval he profited from further discussions with Whitehead. On November 16, 1900, Whitehead wrote:

> Dear Bertie,
>
> I have been studying Peano's arithmetic—*Formulaire,* 1899. He seems to me to have prematurely identified his symbols with those of ordinary mathematics. The result is that he is led into some inconsistencies. . . . My belief is that the development according to his own principles is much more lengthy than the one he gives. I have been working at it in my intervals of leisure.
>
> Yours ever
> A. N. W.

In the portion of this letter not printed above, Whitehead collected four formulae from separate sections of the *Formulaire* and deduced an unacceptable conclusion from their juxtaposition. In the first paragraph of his published paper, Russell, who had corrected Whitehead's careless numbering of the formulae, cited substantially the same de-

*On March 19, 1901, Peano acknowledged its receipt.

duction of the same consequence to show the necessity of modifying Peano's notation.

With "Sur la logique des relations avec des applications à la theorie des series," Russell made a stunning debut in his chosen field.[12] He argued that the necessary changes in Peano's notation required the addition of *relation* as a new primitive—that is, undefined—idea, and that this was in any case required for the completion of Peano's logic. "It is . . . the logic of relations," he said, "which must serve as a foundation for mathematics, since it is always types of relations which are considered in symbolic reasoning; . . . for example, transitive and asymmetrical relations, or one-one relations."[13] Assuming a knowledge of Peano's notation (as was natural in *Rivista di matematica*), but also using a bit of Schröder's, Russell set out the general logic of relations, giving this a simpler and more precise form than that in which he found it. But he did much more than that in this paper. He presented his now familiar definition of "cardinal number": two classes are *similar* if the members of one can be put into one-one correspondence with the members of the other, and the cardinal number of a class *u* is the class of all classes similar to *u*. The second, longer half of the paper was devoted to ordinal numbers and the relations that generate series.

Preparation of the published text of "Sur la logique des relations" was but a small part of the work Russell did after he and Whitehead returned from the Congress at Paris. In the last three months of 1900, while Whitehead was teaching at Cambridge, Russell sat down and wrote the first draft of his big *Principles of Mathematics*. Its thesis was that pure mathematics is *entirely* built on formal logic, and needs no concepts not definable in terms of a small number of logical ones like implication and the inclusion of one class in another, nor any premises not provable from a handful of logical ones. Pure mathematics—he never concerned himself with applied mathematics—could be built up from the formal logic of propositions, of classes, and of relations, without making any appeals to intuition, let alone perception. Peano's axioms for arithmetic could be proved in the terms of symbolic logic alone. So could geometry as a pure a priori science, not as the study of actual space.

Russell did not get his thesis—commonly called the logicist thesis about the foundations of mathematics—from anyone. Gottlob Frege had put it forward—for arithmetic and analysis, though not for geometry—many years before, but his work was almost unknown. Peano knew of Frege's work, but rejected the logicist thesis. Russell learned of Frege from Peano, but did not read his work until after he had independently set out that thesis. There is no reason to think that Whitehead influenced Russell toward it. The young philosopher was tilted in that direction by his renunciation of Kantian philosophy, and then by his study of Leibniz, who conceived the demonstrability of all mathematical propositions by means of a logical calculus. Russell himself was dedicated to arguing the logicist thesis as soon as Peano's logical reforms suggested to him how he could begin to make out the case for it.

Having set his logical compass in the long first Part of *The Principles of Mathematics,* he carried out his great adventure in the six successive Parts with analyses of the mathematical concepts of number, series, infinity and continuity, space, and motion. The analyses, accompanied by sharp criticisms of earlier, vague views, were fairly exhaustive and fairly rigorous; nothing that could compare with this had been done before. Whitehead's help was invaluable. In his Preface Russell wrote, "At every stage of my work, I have been assisted more than I can express by the suggestions, the criticisms, and the generous encouragement of Mr. A. N. Whitehead." As we saw in Chapter X, the author of *Universal Algebra* had a keen interest in the fundamental nature of mathematics. He could not but encourage Russell, for he felt that his former pupil was on the right track. By January 1901 he was convinced of this.

In *The Principles* Russell made a strong case for his claim that pure mathematics is entirely a prolongation of formal logic, but he did not demonstrate its truth. Demonstration would require the precise statement of the axioms of logic in Peanesque symbols, and chains of deductions from them to the principles of arithmetic and geometry. Russell proposed to do this in a second volume, which would be addressed exclusively to mathematicians.

On its philosophical side, *The Principles of Mathematics* embodied the extreme realism and atomistic pluralism that I described a few pages back. In his Preface Russell, with excessive modesty, said that he derived his position from Moore. What did Whitehead think of it? On this I must say what I have said of so many factual questions that the reader must by now find the refrain tedious: there is, to my knowledge, no record written at the time. That should not, in this case, be surprising. As a working mathematician, Whitehead was not, and did not need to be, concerned with the issues that agitated Moore and Russell. His philosophical activity began some fifteen years later, when he joined the Aristotelian Society; his views then will concern me in Volume II of this biography. So will his doctrine (which he announced a few years later) that nature, as studied by science, is independent of the knowing mind—an epistemological realism which was not drawn from Russell or Moore. In any case, the philosphical doctrine that was most important in *The Principles of Mathematics* was that of the independence of every term from every other, with emphasis on the externality of relations to their relata.

V

The first few months of 1901 were of prime importance for the developing role of Bertrand Russell in Whitehead's life, both intellectually and emotionally; intellectually, for the beginning of the collaboration that issued in *Principia Mathematica;* emotionally, for Russell's feeling about humanity in general and Evelyn Whitehead in particular. The intellectual aspect is reserved for my next chapter.

Russell at this time was anxious for Whitehead's company, for he was revising some chapters and rewriting others of the manuscript of his *Principles of Mathematics.* Whitehead, admiring Russell's originality, especially admired this work. His own continuing occupation was the composition of Volume II of the *Universal Algebra,* but he was also making further investigations of the algebra of symbolic logic. As the Russells since 1898 had been spending one term of each year in Cambridge, they and the Whiteheads agreed to share their friend Maitland's house in Downing College for the Lent term. (Maitland's health had been failing for several years, and he had begun to spend the winters in Madeira.) The arrangement was welcome to Evelyn, perhaps was her idea. Her husband was very happy in the Mill House at Granchester, but Evelyn was always restless and loved nothing more than a change of house, unless it was the company of a brilliant aristocrat.*

On February 10 Bertrand Russell had a signal emotional experience. He and Alys had gone to Newnham College in the evening to hear Gilbert Murray read part of his translation, not yet published, of Euripides' *Hippolytus,* and Russell was deeply moved by it. On their return to Downing College, they found Evelyn suffering intense pain, apparently from angina pectoris. Russell knew, or thought he knew, that Evelyn had a weak heart. He had seen her having chest pains

*Possibly the Mill House was undergoing some of the alterations Evelyn made in it.

before, and with increasing frequency this winter. As he recalled the event in his *Autobiography,* this was

> an unusually severe bout of pain. She seemed cut off from everyone and everything by walls of agony, and the sense of the solitude of each human soul suddenly overwhelmed me. Ever since my marriage, my emotional life had been calm and superficial. I had forgotten all the deeper issues, and had been content with flippant cleverness. Suddenly the ground seemed to give way beneath me, and I found myself in quite another region.[14]

Within five minutes, Russell tells us, he became a completely different person. He saw that nothing can penetrate the loneliness of the human soul "except the highest intensity of the sort of love that religious teachers have preached." Actions not inspired by love are not good. The root of evil is force, and war is wrong. "Having been an Imperialist, I became during those five minutes a pro-Boer and a Pacifist." After this apparently mystic illumination, Russell became for a time a very gentle man. He felt that he was in close touch with all of his friends and many of his acquaintances. He began to love children, and found that they loved him.

> The Whiteheads' youngest boy, aged three, was in the room. I had previously taken no notice of him, nor he of me. He had to be prevented from troubling his mother in the middle of her paroxyms of pain. I took his hand and led him away. He came willingly, and felt at home with me.[15]

There is an obvious orthodox Freudian explanation of this experience of Russell's, and at least one such explanation has been published.[16] According to its authors, Dr. and Mrs. Bennett Simon, the sight of Mrs. Whitehead in paroxysms of pain led Russell to identify with Eric Whitehead and revived in him memories of and feelings about the death of his own mother when he was roughly the same age. (The correspondence in age was closer than the phrase, "aged three," in Russell's account suggests. Eric Whitehead was in fact two years, two and one-half months old; Russell, when his mother died, was two years and one month old.) Bertrand, like Eric, was the younger son. His father was already depressed, soon became ill, and died before Bertrand was four. Russell in 1901 had seen signs of depression in Whitehead, and had clearly perceived how much his mental health depended on Evelyn. He now had an absolute need for Whitehead's help in his work. The child Bertrand had experienced fear and loneliness, and fears and fantasies that angry feelings for parents could lead to their disappearance. To a Freudian, the mystic feeling of union with others, experienced by Russell now, was compensatory. Elsewhere in the *Autobiography* he tells of almost strangling a friend when he was an adolescent, and of almost strangling Alys in a nightmare; he had his share of murderously aggressive feeling. His "conversion" (Russell's word) to pacifism is explained as a defense against that feeling. The Simons, who note that his pacifism was never absolute and show considerable sympathy with it, are not reductionists, and insist on the importance of conscious motives and of social and historical forces. Their construction of the psychodynamics of Russell's sudden change of attitude seems to me quite credible. Whether it is so or not, we may hardly doubt that when Russell saw Evelyn Whitehead in danger of death he was overcome by a traumatic residue of his own childhood loss and loneliness.

To be sure, he had misperceived Evelyn's physical state. She was nowhere near dying. She continued to have "heart trouble," with palpitations and rapid breathing, until she died—sixty years later. The severe pain that Russell witnessed was caused not by angina pectoris but by the simulation of it which is termed false angina, or pseudoangina. In saying this, I am relying on the opinion of Norah Schuster, M.B. (later Mrs. Norah Nicholls), who lived with the Whiteheads in London for fourteen months in 1920–22, when she was Assistant Pathologist at St. George's Hospital. Norah was one of the daughters of Sir Arthur Schuster; another daughter, Margaret, married North Whitehead about this time. Norah was present during one heart attack, called in a consultant, and herself gave Evelyn a general physical examination. The result was a good bill of health. Evelyn supposed that she had had rheumatic fever at some time when she was working as a governess before her marriage, but no effects of it were noticeable. When I asked Norah Nicholls (in 1967) about Evelyn's heart attacks, she called my attention to the description of pseudoangina in a British *Synopsis of Medicine:* strictly functional, with no organic lesions but with a neurotic and hysterical factor present; also, it is more common in females and is never fatal. She was devotedly fond of Evelyn, but "hysterical" was, inevitably, the word she applied to the heart attacks. In short, Evelyn Whitehead was a noncardiac cardiac.

In 1921, when Norah Nicholls lived with her, Evelyn was considered an invalid (or semi-invalid)—a "sofa lady," though not of the Victorian type that went in for fainting spells and smelling salts; her false angina was much more persuasive. I am not saying that it was an act; she was conscious only of the real pain. False angina was the way in which her passionate, hysterical nature found a vent. She was a sofa lady who always had just enough strength to be wonderful. Unconsciously she used her illness to get what she wanted, but never as an excuse for inactivity.

After the episode that so moved Russell, Evelyn's apparent heart disease continued, and everything was done to make her life easy. There was a nanny for the children—Jessie was seven years old, North nine—but she spent most of her time, Jessie told me, "taking care of Mummy." By March, Alys Russell, who spent much of her time doing the nanny's proper work, was telling her mother that Mrs. Whitehead was a regular invalid, with a trained nurse. But physical care was not the remedy Evelyn sought. Russell, on that February evening, thought he saw Evelyn suffering not only a heart attack but also the pain of loneliness and neglect. In a letter written ten years later to Lady Ottoline Morrell, with whom he was then much in love, he told of this episode without mentioning the occasion or naming any names.

> I came to know suddenly (what it was not intended I should know) that a woman whom I liked greatly had a life of utter loneliness, filled with intense tragedy & pain of which she could never speak. I was not free to tell my sympathy, which was so intense as to change my life. I turned to all the ways there might be of alleviating her trouble without seeming to know it & so I went on in thought to loneliness in general, and how only love bridges the chasm—how force is the evil thing, and strife is the root of all evil and gentleness the only balm. I became infinitely gentle for a time. . . . I resolved to bring some good and some hope into her life. All this happened in about five minutes. In spite of many faults and many backslidings I succeeded on the whole in what I undertook then. When I told her of you, she bade me remember that she is permanently better and happier owing to me. But it took me rather more than a year to acquiesce in her pain and to learn to love the cause of it, tho' he deserves much love.[17]

In short, Alfred was to blame for Evelyn's spiritual anguish.

Russell was generally naïve about people; this was Bertie at his most naïve. How he helped Evelyn will be taken up shortly. At this point, observe that for a woman who did not want him to think she was lonely, she succeeded very well in revealing that she was. Alfred adored her, and was always tender, but he was not a demonstrative person, which was what she wanted. She was also very possessive.

Although North and Jessie resisted this, her husband did not. The only freedom he desired was freedom to do his work. She gave him that, and in every house they had in England arranged a place where he could work uninterruptedly. Unfortunately for her, it was hard for him to stop working. If they spent an academic vacation in the country or traveling on the Continent, his work was always on his mind. Mathematics was his mistress, and I cannot believe that Evelyn entirely overcame jealousy of her.

In this marriage between two strong characters, both obstinate, there were conflicts that could not always be kept from the children. Jessie has a memory of her mother screaming in rage at the top of the stairs. Her father seldom expressed anger; he retired into silence. Jessie told me that her mother had an iron will; no one, to my knowledge, dissents. She compared her father's will to steel: he would bend but not break. When the two collided, he would appear on the verge of a nervous breakdown, which in fact never occurred; and she would have one of her heart attacks, which in fact never hospitalized her. I think she got her way more often than he did. When they disagreed about the children, he would allow that this was her province more than his. (That was unfortunate for North and Jessie.) In 1901, none of her intimates could have known that her heart attacks were strictly noncardiac. But Alfred, after ten years of marriage to her, had become used to them, and probably knew that when she had one she was not about to die. Bertie and Alys did not know that. They felt that he was not taking her condition seriously enough. What a strange man he was!

A possessive woman with an iron will, whose heart attacks no longer had as strong an effect as she unconsciously desired, Evelyn naturally often felt that she was not being fully appreciated, and that her husband was neglecting her. There was also a pervasive cause of her feeling unfulfilled. Her place as the wife of a middle-class don, which she outwardly accepted, was not the right place for her. She belonged in the theater, and in the company of aristocrats rather than dons' families.

Was Evelyn Whitehead's life at this time "filled with intense tragedy and pain of which she could never speak"? Certainly not. That was how she meant Bertie to see her. There had been tragic events, but not of an unusual kind: her favorite sister died young, and after Jessie's birth Evelyn's next child was stillborn. She would suffer the greatest blow of her life when Eric, the child she loved most and who never thwarted her, was killed in 1918. But in February 1901, the only pain of which she could never speak to her husband was the intense feeling that his extreme self-control could break at any moment, bringing him to the edge of insanity. In his *Autobiography* Russell wrote of Whitehead:

> He used to frighten Mrs. Whitehead and her servants by mutterings in which he addressed injurious objurgations to himself. At times he would be completely silent for some days, saying nothing whatever to anybody in the house. Mrs. Whitehead was in perpetual fear that he would go mad. . . . I think, in retrospect, that she exaggerated the danger, for she tended to be melodramatic in her outlook. But the danger was certainly real, even if not as great as she believed. She spoke of him to me with the utmost frankness, and I found myself in an alliance with her to keep him sane.[18]

Since Whitehead never reached the point of being unable to lecture or to carry on his mathematical research, one wonders whether Russell called the danger "certainly real" in order to shut out the possibility that he could have been misled. Certainly Evelyn Whitehead "exaggerated the danger." But Russell did her less than justice in explaining that she "tended to be melodramatic in her outlook." A more accurate description would be that she *was* dramatic, even theatrical. Jessie Whitehead agrees with this description. The reason "melodramatic" does not fit Evelyn is that she had *taste*. She habitually talked of situations in a way that turned them into good theater.*

When their time in Maitland's house was up and the Whiteheads went back to Grantchester, the Russells went with them. Alys was very helpful to the still-ailing Evelyn and with the children, and Bertie was working with Alfred. The Russells did not return to Fernhurst until May. In July, Bertie was back at the Mill House. On the thirteenth he wrote to Alys:

> Evelyn and I had an exceedingly frank talk last night on the subject of Alfred. She saw all that we felt about him when she was ill, in a way that was really uncanny. But at the time she was too much inclined to think herself neglected by him to argue with us. However, I was able to say last night whatever was necessary to her happiness. She was much relieved to have talked it out.[19]

In a postscript added the next day he said, "She has cured herself of the feeling of being neglected." He did not inform Alys that money was

---

*This description of Evelyn Whitehead was given me by my late wife, Victoria Lincoln, who knew her.

talked about, and that he had undertaken financial help. In his *Autobiography* he set down as a fact that

> Whitehead was obsessed by fear of lack of money, and he did not meet this fear in a reasonable way, but by spending recklessly in the hope of persuading himself that he could afford to do so. . . . Mrs. Whitehead was always discovering that he had run up large bills with Cambridge tradesmen, and she did not dare to tell him that there was no money to pay them for fear of driving him over the edge.[20]

Shortly after this was published, Russell, whom I had visited, wrote me about his information as follows:

> I had much talk with Evelyn Whitehead about the Whiteheads' affairs. . . . The precise "muttering" of which I have written, I heard. It was clear that he was very worried, and I attributed his worry to his financial difficulties largely, I think, because of what she said. I think that she was basically right about it. I did what I could to alleviate the situation. It was very distasteful to me as I by no means liked to provide money of which Whitehead knew nothing. His attitude towards his financial difficulties I learned of from her, but I believed her account because it tallied with my own general observations. I remember no facts or definite impressions that support this general impression.[21]

I suppose that Russell must have observed that Whitehead was sometimes reckless in spending money; he must have seen that when Whitehead came across a book that interested him—which was often—he bought it forthwith. And so, because of what Evelyn told him, he *found himself* "in an alliance with her to keep him sane." His part was to give her money secretly. In the *Autobiography* he wrote that Whitehead

> would have found the humiliation unbearable if he had known of it. But there was his family to be supported and *Principia Mathematica* to be written, and there seemed no other way by which these objects could be achieved. I contributed all that I could realize in the way of capital, and even that partly by borrowing. I hope the end justified the means. Until 1952* I have never mentioned this to anyone.[22]

Russell's generosity continued for several years. In a 1952 draft of the *Autobiography* he gave a total figure, subsequently deleted, of £8,000.

---

*Whitehead died in 1947.

This sum is so huge that I believe Russell, reading the typescript and mistrusting his memory, chose to name no figure rather than try to make a correction.

There is one fact that neither Russell nor his biographer took into account. The family expenditures were in Evelyn's hands. She ran the Mill House, greatly altered and beautified it, hired the servants, and dressed the family. With what Cambridge tradesmen, other than booksellers, did Alfred run up large bills? It was Evelyn who went to the draper's. From the description of the Whiteheads' life at Grantchester given in the last chapter, it appears evident that merely carrying on that life must have pretty well used up Alfred's lecturer's stipend and fellowship dividend from Trinity College, which were their only income.

Not much money—in some periods none—was spent on alcoholic beverages. And Evelyn economized on food, so much so that some of those who came to the Mill House for dinner thought she economized too much. After some years she learned she could not always assume that what you give guests to eat doesn't matter so long as the silver and cutlery are good. She had style, and an eye for beautiful things for their house. And she was shrewd in using her money to the best advantage in buying them; also in dressing the family and doing things for the children. Still, these things took money. There was not nearly as much as she felt she ought to have, and still leave Alfred free to buy books as he pleased.

The story of Whitehead's financial incontinence is repeated as fact by Russell's biographer and by others. So far as I know, the only source of the story is Russell himself. And what he says is completely general, mentioning no examples of reckless spending by Whitehead. When Evelyn gave Bertie an account, not implausible to him, of the cause of the financial difficulties, he would not have pressed for details. He was not hard-headed enough to do so, least of all in this situation. Instead of accepting Russell's account of the circumstances, we must conclude that Evelyn, in order to afford to maintain her beautiful way of living, gave him a picture that contained little more than a grain of truth. The facts, so far as they can be ascertained at this late date,* are that Whitehead spent freely on books, that Evelyn either spent or wanted to spend more than they could afford on other things, and that her husband could not bear to hear that there was not enough money for them to live in the style Evelyn wanted to live in, enough

*I made an unsuccessful effort to find bank records in Cambridge.

for her to have the beautiful things that she wanted and he wanted her to have, while not skimping on anything for the children. She was not above sponging on friends who could afford to help her out; other instances of this will appear in Volume II of this biography.

vi

During the second half of 1901 and into 1902, the Russells lived with the Whiteheads in the Mill House at Grantchester. (By 1902 there was a new bedroom with a view of meadow and woodland—paid for with money Russell had supplied.) The result was that Bertie discovered—early in 1902, according to his account—that he no longer loved Alys. He fell in love with Evelyn Whitehead. The indirect evidence that he did so is quite strong. Of course he could not mention it to anyone, and in his *Autobiography,* where the women he loved at various times are identified, he says nothing about his feeling toward her. When he told Lady Ottoline Morrell about his past life in the letters that began in 1911, he said that between Alys and her, he had loved a woman he did not name. The first stanza of a poem he wrote for Ottoline reads: "Thrice have I loved. Once in the morning dew, / Singing with spring-time birds a careless strain, / Forgetting earth, I soared to heaven, and knew / Joys that forget the burden of man's pain." The second stanza was: "Again I loved. Piercing the prison of flame, / Where one stern soul in lonely anguish burned, / Forgetting earth once more with love I came / Into that hell whence no light hopes returned." How Yeats, had he seen this, would have been enraged by that "Thrice have I loved"! Setting aside the poor Edwardian character of Russell's verse, the point to notice about the second stanza is that Evelyn Whitehead, as he perceived her, fits. Ottoline did not doubt that she had been his second love. Nor did Jane Harrison, who took Alys's side against Bertie; nor did Beatrice Webb, who remained a close friend of both. Russell's biographer does not really doubt it either, though he notes that proof is lacking.

There is no documentary proof because, although Russell wrote letters at the drop of a hat, Evelyn destroyed her letters from him—and, I think, practically all correspondence from her years in England. She was not much more of a keeper than her husband. On his instructions she destroyed his correspondence along with his papers after he died in 1947. And Whitehead in his will directed that his letters to his wife be destroyed; always obsessed with privacy, he would probably have done so even if there had been nothing to conceal. Finally, the small batch of letters from Evelyn to Russell that are in the Russell Archives tell us nothing significant about their relation except that she remained a sympathetic friend even in the tense wartime years.

Despite the absence of documentary evidence, conclusions that are highly probable emerge when we look at the quartet who lived together in the Mill House for many months.

What was Bertie's wife like? Alys was the youngest daughter in the remarkable Philadelphia family of Robert Pearsall Smith and Hannah Whitall Smith, who were evangelists and rather heterodox Quakers; their story has been expertly told by Alys's great-niece, Barbara Strachey.[23] The mother scored a great success with her religious classic, *The Christian's Secret of a Happy Life*. She was the dominant figure. The father's revivalist preaching had found a ready audience in England in the 1870s, and in 1888, with their son, Logan, and daughter Mary (who was committed to living there), they crossed the Atlantic for good. Bertie met Alys when he was taken to visit the family in the summer of 1889. He was seventeen, and she twenty-two. For him, it was love at first sight. Alys was beautiful, and of high moral character—something very important to him; in fact, both were earnest, puritanical prigs. But apart from following her mother in fanatical devotion to the cause of Temperance and that of equality for women, Alys, like Logan and Mary, engaged in forthright discussion of all subjects. This Apostolic intellectual freedom was unaccustomed fresh air to Bertie, who had not yet entered Cambridge. (Alys had recently graduated from Bryn Mawr College.) Her love for Bertie developed during his undergraduate years and never wavered. When he had finished his triposes, they became engaged. Defying the objections of his grandmother, Russell married her in December 1894. It was a very happy marriage for more than six years.

Although Russell's outlook broadened as he approached his thirtieth year, Alys's did not. She remained, as his biographer nicely says, "a good woman in the worst sense of the word." She quite lacked subtlety, so her good intentions appeared to be general and superficial. She tried, hopelessly, to be the sort of wife she thought Bertie wanted; but she was out of her element in Cambridge. Three years after he stopped loving her, he wrote in his Journal, "Stupidity is really the main thing in her character."

When the Russells were living with the Whiteheads, the great conversational trouble was Alys's literal-mindedness. Alfred and Evelyn came to dislike her, and Bertie found Evelyn an infinitely more interesting person. It did not matter that she was older than Alys; the difference was only one and one-half years. It did not matter that she was the uneducated one of the four; she was highly cultivated, witty, and gay, despite her invalidism and loneliness. Her suffering seemed more profoundly significant than Alys's good works. (Alys was always traipsing off to Manchester or some other city to lecture on Temperance, or giving charity to poor girls.) Instead of having to face this contrast between a dull good woman and an elegant one who was fully alive, Bertie ought to have been married to some member of England's intellectual aristocracy.

I do not think that Evelyn tried to seduce him, or even flirted with him deliberately. She did not need to. Her pitying glance, when Alys's remarks were unusually stupid, was quite enough to show her sympathy with him. Did he make love to her? When he knew that his love for Alys was dead, he would probably have tried, had Evelyn not been the wife of his teacher, long-time friend, benefactor, and collaborator. Did she encourage him to make love to her? Again, I think not. He wrote of his feeling for the woman he did not name as the love that "died a gradual death for want of nourishment." Adultery could have been arranged with little trouble, say on her shopping trips to London. Most writers nowadays assume that when there is opportunity for adultery, and the wife is sympathetic to the man in love with her, it takes place.* But the assumption fails when Evelyn Whitehead's character is taken into account. In the code by which she lived, loyalty was one of the highest virtues. She did not need a consummation of Bertie's love for her so much as the money he was supplying. She was quite happy in the knowledge that a brilliant, witty member of the nobility was in love with her. What a lift that must have given her! Why mess up her life with adultery? And so she used her children to hinder Bertie from declaring his passion in romantic circumstances. Yes, she would enjoy doing what the play called for: she would go with him in a boat on the Granta, to read "Sohrab and Rustum" by moonlight. But she took Jessie along, to Bertie's evident distaste. Evelyn's relation to him remained one of affectionate friendship all her life.

After Russell found that he could not abide Alys, they lived together for nine painful years of nominal marriage before he left her to

*When I was in London in 1968, I telephoned the editor of Lady Ottoline Morrell's memoirs, Robert Gathorne-Hardy, to say that I had undertaken a life of Whitehead and to ask if he had any material for it; he said to me, a perfect stranger, "Mrs. Whitehead had an affair with Russell, you know."

be Ottoline's lover. The years were marked by Alys's nervous breakdowns and Bertie's frequent absences. As the strain became perceivable to their friends, rumors spread that Evelyn Whitehead was the cause of it. When Bertie finally walked out, Alys wrote to her sister, Mary: "The Whiteheads have always been the best and truest of friends to us (quite contrary to that silly rumour of her flirting with Bertie 9 years ago which had absolutely no foundation). . . . There was no other woman in the case, least of all Mrs. Whitehead." Barbara Strachey, who devotes three chapters to Alys and Bertie, believes Alys must have sensed that he had fallen in love with Evelyn, but could not admit it to anyone.

There is no evidence concerning Whitehead's awareness of the situation. He could not have been as ignorant of Bertie's feeling toward Evelyn as of his financial support. I suppose that he sometimes worried about the former, and then fell into silence "for some days," as Russell later said. He hoped no climax would occur. So far as we know, none did. The situation was easier after 1902, for Russell refused to live in Cambridge. He told Alys that she got on his nerves when they were with Cambridge people. That was true, but it was also true that Bertie realized the woman he loved was not for him. The Russells tried living elsewhere. (The Millhangar was impossible, as he had come to think Alys's mother, whom she adored, a very wicked woman.) They tried London and a couple of places in the country before they built a house—they had become resigned to living together—in Bagley Wood, outside Oxford. Thus the greater part of the collaboration on *Principia Mathematica,* by two men who were pure Cantabrigians, was carried on by mail. They met in Cambridge when they had to talk over their work.

Russell's exaggerated perception of Evelyn Whitehead's suffering, on that February evening in 1901, stayed with him. It, and his own loneliness, led him to write "The Free Man's Worship." As he recalled in one of his late autobiographical essays, "Early in 1901 I became suddenly and vividly aware of the loneliness in which most people live. . . . I expressed the outcome of this experience in 'The Free Man's Worship.'" He began it at Christmas 1902 in I Tatti, the Bernard Berensons' beautiful villa in the hills above Florence. Bertie and Alys, miserable with each other, were visiting them. Ten years before, Alys's sister, Mary, had left her husband, Frank Costelloe, and two daughters to run away with Berenson. Costelloe having died, they were now a happily married couple. How poignant, then, was the necessity of enduring life with Alys! To the few who knew Bertie's situation, it was significant that the essay opened with Mephistopheles' history of the Creation, as told to the Dr. Faustus who was willing to sell his soul for a second youth.

In the Prologue to "The Free Man's Worship," God, grown weary of the endless praises of the choirs of angels, amuses himself by arranging the formation of our solar system and the emergence of Man,

> with the power of thought, the knowledge of good and evil, and the cruel thirst for worship. . . . God smiled; and when he saw that Man had become perfect in renunciation and worship, he sent another sun through the sky, which crashed into Man's sun; and all returned again to nebula.

Russell had long since shed all theistic belief; his point is that the universe revealed by modern science is even more void of meaning. He assumes that this universe is entirely mechanistic, and all of man's fears, desires, and acts are "but the outcome of accidental collocations of atoms." Only in thought and aspiration are we free, free in imagination to create, free to reject the worship of Power and to worship Goodness. "Shall our God exist and be evil, or shall he be recognized as the creation of our own conscience?" Man, in short, is to worship his own ideals. Russell, in his naïveté does not realize that bad ideals are as numerous as good ones. He would build a temple for the ideals of gentleness, of truth as it is pursued in "the untroubled kingdom of reason," of the beauty of music and architecture, or the beauty that "shines and glows" in "the golden sunset magic of lyrics" and triumphs in Tragedy, the proudest of the arts. Resignation to one's personal fate is essential. "To abandon the struggle for private happiness, to expel all eagerness of temporary desire, to burn with passion for eternal things—this is emancipation, and this is the free man's worship."[24]

Back in England, Russell completed the essay. (It is only ten pages long.) The penultimate paragraph, on conscientious sympathy for our fellows, was written as exhortation to himself to treat Alys decently. She did not like it; what she wanted was the love he had once passionately given her.

"The Free Man's Worship" is the most famous of the many popular essays Russell wrote in his long life, and justly so. Its lyrical prose, briefly shown in the bits I have quoted, was perfect for his purpose; Joseph Conrad greatly admired it. Whitehead, who at this time was as atheistic as Russell, must have been in delighted agreement with the rejection of God conceived as Almighty Power. But when his Harvard students in the late 1920s tried to get Whitehead to say what he thought of "The Free Man's Worship," all they could get out of him was, "That's how Bertie felt at the time." (By then Whitehead had his own, nonmechanistic conception of the world, and his own new conception of God as supremely good.) Indeed, Bertie felt differently ten years later, when he had found Ottoline. To her he wrote that this essay was "only for people in great unhappiness." He was pleased to republish it time and again for their benefit.*

On the intellectual side, "The Free Man's Worship" was much too simple to be true. It set out an almost Manichean opposition between matter and mind, with no hint of how "the lofty thoughts that ennoble" a human life can arise in a human brain. The resolution of this dualism was to be one object of Russell's powers of philosophical analysis in the two decades following completion of his part of *Principia Mathematica*.

*A count made by the Russell bibliographer Harry Ruja showed that when Russell died in 1972 "The Free Man's Worship" had been republished in five of his own books and in thirty-six others. Between the 1910 and 1918 reprintings in his books, Russell changed the title of the essay to the more modest "A Free Man's Worship."

CHAPTER XII

Principia Mathematica

i/When and how the Whitehead-Russell collaboration began.

ii/Whitehead's papers for the *American Journal of Mathematics*.

iii/Whitehead's criticisms of Russell's first efforts. The collaboration procedure. Whitehead's attitude toward Russell's philosophy; philosophical assumptions in *Principia*.

iv/Axiomatization of the premises of symbolic logic. Russell's "material implication." Russell's "propositional function." The six Parts of *Principia Mathematica*.

In 1910, 1912, and 1913 the Cambridge University Press published the three large volumes of Whitehead and Russell's *Principia Mathematica.* Since that time so much research has been done on the foundations of mathematics that the assessment of their work must be left to the mathematicians and mathematical logicians who are familiar with these developments. I can, however, give some account of what the authors tried to do, of the problems they wrestled with (so far as this can be done without becoming unduly technical), and of the personal aspects of their work together.

Russell's biographer, Ronald W. Clark, thought it impossible to date the beginning of the collaboration with Whitehead; he wrote, "after the autumn of 1900 and before the summer of 1902."[1] Let me survey the evidence.

In the preceding chapter, I pointed out that Russell proposed to carry out a rigorous demonstration of the thesis of his *Principles of Mathematics* in a second volume of that work. In his Preface (dated December 1902) he wrote,

> The second volume, in which I have had the great good fortune to secure the collaboration of Mr A. N. Whitehead, will be addressed exclusively to mathematicians; it will contain chains of deductions, from the premisses of symbolic logic through Arithmetic, finite and infinite, to Geometry.

After Whitehead's death in 1947, Russell wrote a short article, "Whitehead and *Principia Mathematica,*" in which he said: "The project of deducing mathematics from logic appealed to him, and to my great joy he agreed to collaborate. I knew that my mathematical capacity was not equal to accomplishing this task unaided."[2]

From Russell's language in these two statements I conclude that he asked Whitehead to collaborate with him; but I should not absolutely rule out the possibility that the idea was first mentioned by Whitehead, nor the possibility that at some point they simply found themselves working on what could only become a joint volume. In an interview with Russell on June 4, 1965, I asked if he could recall any details about how the collaboration began; he could not, and told me, "It grew up." Upon being pressed to give in rough terms a date of beginning, he said, "late in 1900 or early in 1901." He allowed that I was right in believing that he was the one who proposed the collaboration. (In this interview, Russell's memory seemed good for a man just turned ninety-three, being questioned about matters that no longer interested him.)

Among earlier sources of information from Russell, there is his correspondence with Philip Jourdain.[3] On April 15, 1910, Russell, responding to Jourdain's request for information about Peano's influence on him, told of his conversion to Peano's symbolic method at the Paris Congress in the summer of 1900 and his application of it to the logic of relations in September.* He added that he tested his logic of relations "by its applicability to Cantor." To this he attached a footnote about Whitehead: "I was with Whitehead when he was writing an article for AJM [*American Journal of Mathematics*] on applications of Cantor to Logical Equations, and I persuaded him that my logic of relations would greatly facilitate his work. This produced his conversion (Jan. 1901)."[4]

Another source is a 1911 letter to Lady Ottoline Morrell, in which Russell said that he persuaded Whitehead to think as well of Peano as he did, and that this was the beginning of their "formal cooperation."[5] A third source is a letter of 1932 to G. Revesz, in which Russell said that when he interested Whitehead in Peano's logic as extended by himself (to relations), Whitehead "perceived that it gave the correct symbolism for the foundations of arithmetic, finite and infinite, on which subject he published a paper." This paper is either the one Russell referred to in his letter to Jourdain, or one on cardinal numbers that Whitehead wrote and published (acknowledging that Russell had written one of its five Sections) a little earlier. I shall shortly discuss both papers.

In these letters Russell seems, characteristically, to be thinking only of what he did, and forgetting that, as I noted in Chapter XI, he had profited from discussing with Whitehead his extension of Peano's logic to relations in September 1900 and from Whitehead's critical study of Peano in November.†

---

*There is a fuller account in Chapter XI, section iv.

†See Chapter XI, section iii.

The autumn of 1900 is mentioned in the application to the Royal Society for a grant in aid of publication of *Principia;*[6] this was written by Whitehead and checked by Russell in December 1909; it says, "Apart from their preparatory studies both authors have worked at it since the autumn of the year 1900." In the Preface to *Principia* we find: "The present work was originally intended by us to be comprised in a second volume of *The Principles of Mathematics.* With that object in view, the writing of it was begun in 1900." They seem to have been thinking of the paper "Sur la logique des relations," of which Russell wrote a draft for Peano's journal "early in October (1900)"[7] after discussing his ideas with Whitehead. That seems to have been the first piece of work that should be counted as work for what became *Principia Mathematica.*

In July 1943 Russell wrote a short autobiographical essay, "My Mental Development," for the *Library of Living Philosophers.* In it, he told of his enthusiasm for Peano after the Congress at Paris in the summer of 1900, and then said:

> Basing myself on him, I invented a notation for relations. Whitehead, fortunately, agreed as to the importance of the method, and in a very short time we worked out together such matters as the definitions of series, cardinals, and ordinals, and the reduction of arithmetic to logic. For nearly a year, we had a rapid series of quick successes. Much of the work had already been done by Frege, but at first we did not know this.[8]

In June 1901 this period of honeymoon delight came to an end. Russell had discovered a contradiction in the theory of classes, on which the definitions of cardinal and ordinal numbers were based.*

According to the account just quoted, Russell and Whitehead worked together from the autumn of 1900. Of what they produced by June 1901, only Russell's paper for Peano and Whitehead's "On Cardinal Numbers" were printed.

All that Whitehead wrote for publication about his collaboration with Russell is a short passage in "Autobiographical Notes." After discussing his *Universal Algebra* and noting that its Volume II was "in preparation" until 1903 but never published, he wrote:

> In 1903 Bertrand Russell published *The Principles of Mathematics.*

---

*In "How Bertrand Russell Discovered His Paradox" (*Historia Mathematica* 5 [1978]: 135), I. Grattan-Guinness examined the diverse dates Russell gave for this discovery in his books and letters, and concluded that June 1901 was "the most likely date."

> This was also a "first volume". We then discovered that our projected second volumes were practically on identical topics, so we coalesced to produce a joint work. We hoped that a short period of one year or so would complete the job. Then our horizon extended and, in the course of eight or nine years, *Principia Mathematica* was produced.[9]

When Whitehead wrote this he was eighty, older than Russell was when writing any of his statements about the beginning of the collaboration; hence I give more weight to Russell's written statements.

Attention to chronology will suggest how Whitehead's statements should be corrected and interpreted. Recall that Russell finished the first draft of his *Principles of Mathematics* on the last day of 1900. He began an extensive, troublesome revision of it in January, when the Russells and Whiteheads were sharing Maitland's house for the Lent term.* In June he discovered the contradiction. After vainly seeking a solution for the rest of 1901, he decided to finish getting his *Principles of Mathematics* ready for publication without including more than an appendix on a possible solution. He finished the job (except for changes made in the proof!) at the Mill House on May 23, 1902. In the summer he drafted the first Section of the second volume and sent it to Whitehead. In 1903 Whitehead decided to give up work on the second volume of his *Universal Algebra* as a project separate from the joint work with Russell.

Whitehead hinged his account of the beginning of the collaboration to the date of publication of Russell's *Principles*. That date is irrelevant, because he was familiar with the book while Russell was rewriting it, often in the same house with him.

My conclusions are as follows. Mutual discussions that contributed to *Principia Mathematica* via Russell's "Sur la logique des relations" began in September 1900. In January 1901 Whitehead came into agreement with Russell on the indispensability of Peano's symbolism for work on the logical foundations of mathematics. As the sharing of this conviction made it possible for them to write a book together, it appears probable that within a month Russell, who did not like to postpone anything, asked for and got Whitehead's agreement to be the mathematical collaborator he badly wanted for the second volume of his *Principles of Mathematics*. In the spring Whitehead finished his article "On Cardinal Numbers," which incorporated some results of their discussions; and Russell wrote one of the five Sections of this

---

*See Chapter XI, section v.

article. The year after Russell's June discovery of the contradiction went first into unsuccessful attempts to solve it, then into rewriting *The Principles of Mathematics,* while Whitehead continued work on the second volume of his *Universal Algebra.* In the summer of 1902 Russell drafted a beginning for their joint volume. Sometime in 1903, Whitehead gave up work on his second volume.

The above chronology seems to me the likeliest. But I would not rule out the possibility that Russell did not get Whitehead's agreement to collaborate so quickly. It may not have come until later in 1901, or even sometime in the first five months of 1902. Clark's opinion was the safe one to hold.

ii

Whitehead's second volume would presumably have been about as long as the first, which ran to 586 pages, quarto. No manuscript pages are extant.* The one paper that he published in the years spent on the volume was not meant to be part of it. "On the Geodesic Geometry of Surfaces in non-Euclidean Space" was read to the London Mathematical Society on March 10, 1898, and published in Volume 29 of their *Proceedings.* It investigated the properties of those surfaces in any non-Euclidean space, elliptic or hyperbolic, which are such that on them the geometry of geodesics (i.e., lines of equal distance) is that of straight lines in a Euclidean plane. Whitehead used Grassmann's Calculus of Extension to prove his theorems most easily, and shortened his paper by referring the reader to the first volume of *Universal Algebra.* An older friend, William Burnside (a former Fellow of Pembroke College, now Professor of Mathematics at the Royal Naval College, Greenwich), read this paper in manuscript; at Whitehead's request a three-page Appendix by him, written from a synthetical point of view, was attached to the published paper.

Early in 1899 Forsyth, who was a Fellow of the Royal Society, communicated to it a paper by Whitehead, "Sets of Operations in Relation to Groups of Finite Order." An abstract was published,[10] but not the paper. Burnside is relevant to this work of Whitehead's also; his *Theory of Groups* (1897) was the first treatise on the subject in English. Groups of finite order are simply those with a finite number

*And Whitehead was not in the habit of writing letters about his work while it was in progress. He just might have written to Forsyth about it; but most of Forsyth's correspondence was destroyed by bombs in the Second World War.

of elements. Whitehead's purpose was to place the theory of such groups in relation to a special algebra, which he calls the Algebra of Groups of Finite Order. He compares it with the Boolean Algebra of Symbolic Logic that he had investigated in his *Treatise*. They have "many affinities," he says. This is not surprising, since both algebras are nonnumerical; in each, the law of addition is $a + a = a$, and numbers as factors do not appear. Recall that in *Universal Algebra* Whitehead remarked that although the Algebra of Symbolic Logic was the only algebra so far developed in which no symbols representing number or quantity are required, he saw no reason why others should not be developed.[11] We may say that in this paper's purely algebraic treatment of the theory of groups of finite order he himself exhibited an algebra that was nonnumerical in the same way as the Algebra of Symbolic Logic.* He did not thereby create a new branch of mathematics, but took one already in existence and fitted an unusual type of algebra to it.

According to the abstract of this paper, Whitehead in it took the distinct elements of a group as so many operations; and since a group is a special type of set, he made some study of the general properties of sets. If the paper had been published, Whitehead just might have acquired a somewhat higher reputation among British mathematicians than the middling one he actually gained in his years at Cambridge. But perhaps not. To judge from the abstract, in this paper he was once again the solitary thinker, not adding to a line of thought others were pursuing, but placing what they had done in a new perspective of his own.

Whitehead was aware that a mathematician could do much more with Boolean algebra than the logicians had done with it, and more than he himself had accomplished in the *Universal Algebra*. When he and Russell went to the Congress at Paris in the summer of 1900, Whitehead had completed a two-part paper, "Memoir on the Algebra of Symbolic Logic," and sent it to the *American Journal of Mathematics*, then edited by his friend Frank Morley.[12] A logician, even a symbolic logician, would have little interest in this memoir; I suppose he might say that it dealt with the higher mathematics of Boolean algebra. Whitehead wrote it to show that this algebra had many interesting mathematical properties, unnoticed in the literature because its purely mathematical study had languished while attention was concentrated

---

*"Boolean algebra" did not become the customary name until H. M. Sheffer used it in 1913.

on applying it to logical operations. To Whitehead, taking Boolean algebra as pure mathematics meant investigating the consequences of its laws in abstraction from all subject matter—in other words, as a department of what came to be called abstract algebra. The memoir deals with such things as the invariants of functions of independent variables in Boolean algebra, the factoring of functions, their transformations, and the conditions under which the transformations constitute a group.[13]

In Whitehead's published works, the effect of the Paris Congress on him first appears in the paper "On Cardinal Numbers," which he also sent to Morley for publication.[14] This may be the paper he was writing when the Whiteheads and Russells were sharing Maitland's house in the early months of 1901. When Whitehead wrote *Universal Algebra* he was aware of Cantor's general theory of sets; he now takes it for granted. Of the five Sections of the paper, the first explains Peano's logical symbolism; the second, Russell's symbolism for the logic of relations. These Sections are preceded by a Preface, in which Whitehead expresses unrestrained enthusiasm for them:

> I believe that these two methods are almost indispensable for the development of the theory of Cardinal numbers. . . . I believe that the invention of the Peano and Russell symbolism, used here, forms an epoch in mathematical reasoning.

Whitehead was convinced that sufficient clarity can never be reached by ordinary language, supplemented by algebraic devices; Peano's ideographic symbols are needed.* Whitehead's position provoked Charles Peirce to write an inaccurate letter to Morley in which he called Whitehead's claim "downright silly." The subsequent history of mathematical writing suggests that Whitehead went too far. However, logical deduction by a strict symbolical method is especially called for when the reasoning concerns infinite numbers; there, imagination cannot guide the choice of words.

Whitehead tried to promote Peano's method in Britain by reading a paper, "Peano's Symbolic Method," to Section A, Mathematical and Physical Science, of the British Association for the Advancement of Science at its 1904 meeting in Cambridge. The paper was not printed in the *Proceedings* of this meeting.† Whitehead's effort accomplished little; Peano's symbolism never took hold in British mathematics.

---

*See Chapter XI, section iii.

†Whitehead was not on the BAAS membership list.

With the third section of "On Cardinal Numbers" a chain of definitions and deductions in the Peano-Russell symbolism begins. Whitehead writes, "Section III is entirely due to Russell, and is written by him throughout." Russell here defines "cardinal number" as in his paper for Peano: the cardinal number of a class is the class of all the classes that can be put into one-one correspondence with that class.* Now, Cantor's work antedated Peano's, and although Peano had put the elements of Cantor's theory of cardinal numbers into his own notation, he had done little else with the theory. Whitehead's task, undertaken in the fourth and fifth Sections of the paper, was to develop it in Peano's precise notation. He defines the sum of any class of cardinals, finite or infinite, and goes on to such familiar operations as the raising of a number to a power. Cantor's definition of multiplication required the number of factors in a product to be finite. Whitehead overcomes this restriction with his definition of a multiplicative class. If *k* is any finite or infinite class of classes, none null and no two having any term in common, we reach the multiplicative class of *k* by choosing *one* term, in all possible ways, from each of the classes composing *k*. The product of the numbers of terms in the classes belonging to *k* is then defined as the number of terms in the multiplicative class. This conception played quite a role in future work on the foundations of mathematics.

Whitehead also proves many familiar theorems, notably the binomial theorem, for infinite as well as finite cardinal numbers.

He says of these concluding Sections that he is "indebted to Russell for great improvements in the symbolism of the proofs and for the correction of many oversights." It seems likely that Whitehead, for his part, helped Russell in the third Section, but not as much as a rigorous text would have made desirable. When W. V. Quine wrote a brief summary of the paper, he commented, "In general the proofs of theorems are sketchy; and in particular I have been unable to decipher the alleged proof that $n \neq n + 1$ for finite n."[15] In fact, the alleged proof occurs in the Section written by Russell. Quine also noted three other specific omissions or defects; these too are all in that Section. The sketchiness of the proofs may have been deliberate; Whitehead wrote in Section V that "the proofs of some propositions practically proved by Russell in Sect. III will be omitted." "Practically proved" is an odd phrase to use in a mathematical paper.

Whitehead followed "On Cardinal Numbers" with a two-page

*See Chapter XI, section iii.

sequel, "Theorems on Cardinal Numbers."[16] Quine's judgment was that despite the weaknesses he had noted, "the two papers were highly significant," for they afforded

> the first considerable development of infinite arithmetic within mathematical logic. Cantor and others had used no logical formalism; Peano had deduced none but a few trivial theorems in infinite arithmetic; and Frege, though deducing the laws of finite arithmetic in rigorous detail, did little with infinite numbers.[17]

Whitehead's "Memoir on the Algebra of Symbolic Logic" was at press when Russell discovered Peano's symbolic method. Whitehead, following his conversion to that method in January 1901, made another investigation of Boolean algebra in the spring. He called his paper "The Logic of Relations, Logical Substitution Groups, and Cardinal Numbers," dated it July 4, 1901, and sent it to the *American Journal of Mathematics* for publication.[18] An adequate explanation of the title would take us into lengthy technicalities. It should be said, however, that Whitehead here returned to the subject of logical equations, which had been central in Book II of the *Universal Algebra* but not in the memoir. He now generalized his theory of Boolean equations in terms of Peano's precise notation and Russell's extension of it to relations. He also, using Cantor's theory of infinite numbers,* extended his generalization to the utmost, so that equations in infinitely many variables became meaningful.[19]

The memoir and this paper show Whitehead making important contributions to Boolean algebra. They are not much noticed in histories of the subject. They came at a time when, because of the apparently final form that Schröder had given it, a decline might have set in. That was the opinion of the Cambridge logician W. E. Johnson, given in 1905 to the Degree Committee of Cambridge's Special Board for Mathematics. Whitehead had applied for a Doctor of Science degree, which required the candidate to submit the research he had done; he submitted the *Treatise on Universal Algebra* and the papers published in the *American Journal of Mathematics*. The committee asked for evaluations from Johnson and from H. F. Baker, then Cayley Lecturer in Mathematics and subsequently Lowndean Professor. Baker confined his report to the *Treatise,* he said, because the papers were written in Peano symbolism. Johnson, reporting on these papers, found that

*To be more precise, Whitehead applied to logical equations the Cantorian theory as he had developed it in "On Cardinal Numbers."

Whitehead had developed a new line of research "with remarkable results and in a manner exhibiting extraordinary power," thereby giving "new life to the study of symbolic logic."[20]

Whitehead received the doctorate. Later, the Harvard mathematician who gave Boolean algebra its classic axiomatic treatment, E. V. Huntington, could speak of that algebra as "originated by Boole, extended by Schröder, and perfected by Whitehead."[21] Whitehead's contributions to the Algebra of Symbolic Logic were such as only a mathematician could have made, not a logician—not even one of Russell's stature. (Whitehead liked to call Bertie the greatest logician since Aristotle.)

iii

The Bertrand Russell Archives at McMaster University contain 102 letters from Whitehead to Russell. Most of them were written in the course of their joint work on *Principia Mathematica*. These are my concern now. Much can be learned from them, despite the absence of Russell's letters to Whitehead.

Bertie and Alys, utterly miserable with each other, spent the summer of 1902 near Broadway, in Worcestershire. Here Russell made his first draft of the beginning of *Principia*, ∗1–∗5, which deals with the logic of propositions. Whitehead sent back a sharp criticism: "Everything, even the object of the book, has been sacrificed to making the proofs look short and neat. It is *essential*, especially in the early parts, that the proofs be written out fully." Russell had not kept this letter, but simply torn out and kept these two sentences from it. I saw them in June 1965, on a visit to him before he sold his papers to McMaster. He had given me a folder containing twenty-seven of his Whitehead letters, told me I might copy what I pleased, and set me up in his spare study with an ancient typewriter. On emerging, I asked him why he had preserved this scrap of a letter. He did so, he said, because "it shows that the fulness of the *Principia* was due to Whitehead." (His retrospective comment, "Whitehead was entirely right," was written on the sheet.)

In his comment on work that Russell sent him later that summer (apparently his draft of ∗9), Whitehead again criticized the handling of the proofs:

> The impression I have from a careful study is of very elaborate definitions which are not used, some proofs very careful; others

> equally important carried out by common sense in the style of Euclid.
>
> The *proofs* are the *only* interest of this subject—no one will read this unless the *proofs* are careful and minute.

Russell did not go into this in Chapter VI, "Principia Mathematica," of the first volume of his highly personal *Autobiography*. There he quoted Whitehead's "Everything, even the object of the book, has been sacrificed to making the proofs look short and neat," only to show how his work suffered from the hopelessness of his life with Alys.

> Pain made me sentimental, and I used to construct phrases such as 'Our hearts build precious shrines for the ashes of dead hopes.' I even descended to reading Maeterlinck. . . . The unreal, insincere, and sentimental frame of mind into which I had allowed myself to fall affected even my mathematical work. . . . This defect in my work was due to a moral defect in my state of mind.[22]

As time went on, Bertie came to accept his situation. In Whitehead's subsequent comments on Russell's drafts, sharp rebukes do not appear. Instead, there are the frequent praise and encouragement that Bertie needed.

In the article "Whitehead and *Principia Mathematica,*" Russell described their collaborative procedure as follows:

> When our work was sufficiently advanced, we parcelled out the topics, each then produced a first draft of whatever was in his assignment, the other then went over it and probably made considerable changes, and then his revised draft was finally revised by the first author. In most parts of the book, there was, in the end, very little for which either had *sole* responsibility.[23]

In a later book, *My Philosophical Development* (1959), Russell enlarged this account somewhat, saying:

> The problems with which we had to contend were of two sorts: philosophical and mathematical. Broadly speaking, Whitehead left the philosophical problems to me. As for the mathematical problems, Whitehead invented most of the notation, except in so far as it was taken over from Peano; I did most of the work concerned with series and Whitehead did most of the rest. But this only applies to first drafts. Every part was done three times over. . . . There is hardly a line in all three volumes which is not a joint product.[24]

In "Whitehead and *Principia Mathematica*," Russell wrote, "Whitehead was more patient and accurate and careful than I was, and saved me often from a hasty and superficial treatment of difficulties that I found uninteresting." I myself have often found Whitehead's mathematics in other books so complicated that I felt he never feared a complication and would rather introduce one than risk an oversight; hence I can readily believe Russell's next sentence: "I, on the other hand, sometimes thought his treatment needlessly complicated, and found ways of simplifying his drafts."

Russell in this paper calls the notation for relations, used in the paper he wrote in 1900 for Peano, "clumsy," and credits Whitehead with starting the notation used for them in *Principia*.

That work may be said to have resulted from the confluence of Russell's determination to reach firm conclusions about the nature of mathematical truth, Whitehead's explorer's interest in widening and unifying mathematics, and Peano's successes with a new symbolism. (To these we need add only Cantor's creation of transfinite arithmetic.)

The fact that Russell was the philosopher, Whitehead the mathematician, is reflected in the collaboration correspondence. Russell brought to the work his philosophical preconceptions and critical weapons. One weapon was Occam's Razor, which shaves off ideas that are not strictly necessary. A Russell proposal to eliminate "class" and "relation" from the list of primitive ideas by defining them in terms of others drew a warning from Whitehead[25] to temper rigor by "practical considerations": mathematicians habitually reason about classes, and we must try to systematize their reasoning even if they could have avoided the notion.

When writing in Chapter XI about Russell's *Principles of Mathematics,** I noted that the general philosophy underlying this book was an atomistic pluralism which, the author said, he had derived from G. E. Moore. Most important was the doctrine that relations and their terms are independent entities. As Whitehead in his mature philosophy (written in America) rejected this doctrine, I asked him in 1941 about his past opinions; he answered that the doctrine was never the foundation of his own thought, and that during the collaboration he neither agreed nor disagreed with Russell about it.

In *My Philosophical Development* Russell mentioned the philosophical assumptions that were implicit in the symbolism of *Principia Math-*

---

*See Chapter XI, section iv.

ematica. "This symbolism assumed that there are 'things' which have properties and have, also, relations to other 'things.' "[26] Whitehead accepted this commonsensical assumption without committing himself to Russell's atomistic pluralism. The discussions in *Principia* also assume that truth is a matter of relation to fact;[27] Whitehead always agreed with Russell on this, although in his late philosophy he interpreted the truth-relation metaphysically. He and Russell said in their Preface that they had "avoided both controversy and general philosophy." These were avoidable in the mathematical treatment of the principles of mathematics; the authors would let their massive logico-mathematical system provide its own justification. There was, then, no need for Whitehead to agree with Russell's attacks on the various views concerning logic and mathematics that Kant, Bradley, Lotze, and others had advocated, attacks that were prominent in *The Principles of Mathematics;* there was no need even to discuss these matters with him. There is no serious discussion of them in the collaboration correspondence.

iv

The first task that Russell and Whitehead faced was to formulate the premises of symbolic logic, for no formulation was at hand. Frege had presented a precise axiomatic treatment of the logic of propositions in his *Begriffsschrift;* it was published in 1879 but not read. The propositional interpretation of Boolean algebra was well known, but not yet axiomatized. They turned to it. (Thus Whitehead's mathematical contributions to Boolean algebra, as described in section ii, were not irrelevant to his collaboration with Russell.) They had to frame a basic apparatus of axioms and undefined ideas; following Peano, they called them primitive propositions and primitive ideas, respectively. The deduction from them of logical propositions and then of mathematical concepts and principles would be carried out in Peanesque notation, suitably modified and enormously expanded. This "official" text of *Principia Mathematica* is accompanied by explanatory English, which Whitehead in his letters to Russell calls "the talk."

The authors symbolize any proposition by p, and introduce its negation, for which they use a tilde ($\sim p$), as a primitive idea. As another primitive idea they choose the disjunction of p and any other proposition q, writing it "$p \vee q$"; this is used when they wish to assert "either p or q or both." It is also called the Logical Sum of p and q.

"Assertion of a proposition" is another primitive idea; the idea and its symbol, $\vdash p$, were taken from Frege.

The connection between two propositions that we express by the words "*p* implies *q*" is the connection that, by enabling theorems to be proved, is the workhorse throughout *Principia*. But the authors do not adopt it as a primitive idea. They write it "$p \supset q$," and define it as an abbreviation for "$\sim p \vee q$." This was Russell's idea, which he called "material implication" and is often called "Russellian implication." Whitehead did not dissent from it, but Russellian implication provoked severe criticism from readers, notably from C. I. Lewis.[28] The criticism was inevitable. As *p* and *q* were any propositions, there was no requirement that they be relevant to each other; the definition required only that either *p* be false or *q* be true (or both). It followed, paradoxically, that a false proposition "implies" any proposition, and a true proposition is "implied by" any proposition. Obviously, material implication is not what we normally mean by logical deduction of one proposition from another. It is in fact the weakest of the various relations that logicians have called implication. Whitehead and Russell were abstracting from the specific meanings of *p* and *q*, and attending only to their being *either* true *or* false. What the proving of theorems demanded of the implicative connection was only that if *p* is true, *q* also is true. As they explain in ∗1 of *Principia*, "If *p* implies *q*, then it cannot be the case that *p* is true and *q* is false, i.e. it must be the case that either *p* is false or *q* is true." They set up what "must be the case" as their general definition of implication.

The proofs in *Principia* are not affected by the strangeness of material implication. A proof begins with a true proposition. If *p* were known to be false, "not-*p* or *q*" would be true, but not used to infer *q*. If *q* were already known to be true, "not-*p* or *q*" would again be true, but not used to infer *q* because no inference is needed. All that needs to be asserted is the disjunction "either *p* is false or *q* is true." It was along these lines that Russell later wrote an answer to Lewis's criticism.[29] Plainly, the trouble resulted from setting aside everything about propositions except their truth-values (T *or* F); this was not peculiar to Whitehead and Russell.

The primitive propositions of *Principia* are small in number, and none are questionable. Who would doubt "If *q* is true, then '*p* or *q*' is true"? I shall not list the primitive propositions. Both their number, and the number of primitive ideas, were much reduced by other writers in the years following the publication of *Principia*. It is worth noting, however, that Whitehead and Russell constantly used, tacitly, three assumptions which H. M. Sheffer called prescripts; they are *rules* for passing from one assertion to another. One was the rule known as "modus ponens": when "if p, then q" is given and p is asserted, then q may be asserted. A second rule is based on the fact that p stands for *any* proposition; when p occurs in a formula, any symbol or string of symbols that represents a proposition may be substituted for it throughout the formula. The third rule authorizes substituting the defined meaning of an expression for that expression, for example, "$p \supset q$" for "$\sim p \vee q$." If it were not for the roughly four hundred definitions in *Principia,* the work would be almost infinitely long.

The great majority of the assertions in the work are not assertions of propositions as I have thus far considered them, but of what Russell called "propositional functions." These are assertions that contain one or more variables, and turn into propositions when values are assigned for all the variables. An example from ordinary speech would be "x is older than y," which yields a proposition when "Harvard" and "Yale" are assigned as values of x and y. It would have been better to call "x is older than y" a propositional form or matrix, but Russell's name got accepted, although it departs from the customary use of "function" in mathematics. An example of the customary use is the statement that the volume of a sphere is a function of its radius; another example is provided in *Principia* itself when the Logical Sum of two propositions is introduced as "a propositional function with two arguments p and q." But Russell also wanted a convenient name for expressions like "ϕx," where ϕ symbolizes a predicate. He and Whitehead wrote:

> Let ϕx be a statement containing a variable x and such that it becomes a proposition when x is given any fixed determined meaning. Then ϕx is called a "propositional function"; it is not a proposition, since owing to the ambiguity of x it really makes no assertion at all. Thus "x is hurt" really makes no assertion at all, till we have settled who x is.[30]

The basic mathematical uses of ϕx are in the assertions that ϕx is true for *all, some* (i.e., at least one), and *no* meaningful values of x. The first and second are primitive ideas, symbolized respectively by "$(x) \cdot \phi x$" and "$(\exists x) \cdot \phi x$." The third is easily expressed by "$(x) \cdot \sim \phi x$." Thus *Principia* makes its start on quantification theory. So far in this section, I have chiefly followed Chapter I of its Introduction. According to Russell, Whitehead wrote most of it, and was responsible for the notation "$(x) \cdot \phi x$."[31]

Part I of *Principia,* "Mathematical Logic," goes on to the logic of classes and concludes with the logic of relations. Part II, "Prolegomena to Cardinal Arithmetic," takes up the second half of Volume I, which ends 634 quarto pages after its beginning. Volume II is still longer. It begins with Part III, "Cardinal Arithmetic"; here the constructions are more careful than in Whitehead's paper "On Cardinal Numbers." Hundreds of theorems are proved, and infinite numbers are not excluded. They were rejected by a vocal group of top mathematicians, but Whitehead and Russell were deaf to them, and continued to admire Cantor's creation.

Whitehead had the primary responsibility for cardinal arithmetic, Russell for the subject of Part IV, "Relation-Arithmetic."[32] Russell later wrote, "From the mathematical point of view this was my most important contribution to the work."[33] Indeed, he created relation-arithmetic. He defined "relation-number" in a way analogous to his definition of "cardinal number": it is the class of all the relations that can be put into one-one correspondence with a given relation. This kind of correspondence is readily illustrated by maps: Edinburgh stands in the relation "north of" to London, and the place marked "Edinburgh" is *above* the place marked "London" on every map, of whatever size, that is conventionally drawn. In *Principia* relations are taken in extension, as ordered couples, triplets, quartets, etc. The sums and products are like those of classes. The arithmetic of relation-numbers is more complicated, especially for infinite relation-numbers. The utility of relation-arithmetic lies in its precise mathematical elaboration of the formal properties that must obtain when two things are said to have the same "structure."

Part V of *Principia* has Series for its subject. In *The Principles of Mathematics* Russell had given the first precise general definition of series. The topic is now treated as an application of relation-arithmetic. Whitehead alone, according to Russell's memory, was responsible for the last Section in Volume II, "On Convergence, and the Limits of Functions."[34] Volume III opens with a continuation of Part V, and concludes with Part VI, "Quantity," which also was Whitehead's responsibility.[35] This Part begins with the arithmetic of signed integers, ratios, and real numbers. It then goes on to vectors, which are treated as repeatable one-one relations. The final topic of *Principia* is measurement. Here no primitive ideas concerning physical objects are introduced. Measurement is taken to consist, mathematically, in determining ratios between vectors in some family of vectors. (Physical measurement requires that the variables in the mathematics be given physical values.) The theory of measurement developed here was meant to be used to introduce coordinates in geometry.[36]

In the Preface to Volume III Whitehead and Russell say: "Geometry we have found it necessary to reserve for a separate final volume." Sometime around 1905 they agreed that Whitehead would take sole responsibility for it, as he was especially interested in geometry, Russell in other parts of their subject. The story of the fourth volume will be told in section vi of the next chapter.

CHAPTER XIII

Principia Mathematica (Continued)

i/Russell's discovery of a contradiction in the logic of classes. The effect of this discovery on Frege. Russell's theory of types.

ii/Russell's theory of descriptions.

iii/Character of the collaboration, and Whitehead's way of thinking, as seen in Whitehead's letters to Russell. Two general differences in their philosophies of mathematics.

iv/Decision to make the joint work independent of Russell's *Principles of Mathematics*. Choice of its title. Whitehead's arrangements with the Cambridge University Press, and with the Royal Society, for publication of *Principia Mathematica*.

v/Short expositions of *Principia Mathematica*. The reception of *Principia*. Reasons for its slight permanent influence on mathematicians. Treatment of *Principia* as Russell's book.

vi/How the collaboration ended. Contrasting philosophical views of Whitehead and Russell. Final comment on the collaboration.

While the first volume of *Principia* was being drafted, Russell was wrestling again with the special problem that he had left unresolved when he sent his *Principles of Mathematics* to the Cambridge University Press. He had to find a satisfactory way of evading the contradiction he had discovered in the logic of sets. (He and Whitehead always called sets "classes"; hereafter I shall use their word when I am referring to them.) In the 1890s Cantor's paradox of the greatest cardinal number and Burali-Forti's paradox of the greatest ordinal number had been discovered. Russell's paradox was more elementary, as it did not involve number. Some classes, he noticed, are not members of themselves, but others are; for example, the class of men is not a man, but the class of nonmen is one of the things that are not men. Consider these two broad classes of classes and form the class *S* of all the classes that are not members of themselves. Is *S* a member of itself or not? Suppose it is; then it must have the defining property of this class, which is to be not a member of itself. Start instead with the hypothesis that *S* is not a member of itself; then it is not not-a-member-of-itself, and so is a member of itself. Each hypothesis entails its own denial.

Russell noted the similarity between his paradox and the ancient one of Epimenides the Cretan.[1] When Epimenides said that all Cretans are (always) liars, was he speaking truly or falsely? Russell was plagued by the feeling that the effort to solve these paradoxes was a trivial occupation. But the contradictions about the greatest cardinal and greatest ordinal numbers showed that the construction of the principles of arithmetic could not be accomplished without solving them.

Here Gottlob Frege should be brought into the story of *Principia Mathematica*. (In recent decades his work has received as much attention as Russell's and much more than Whitehead's.) Recall that Frege axiomatized the deductive logic of propositions and anticipated by many years the logicist thesis of Russell's *Principles of Mathematics*. Russell did not know this until early in 1901, when he read Frege for the first time. He wrote to him on June 16, 1902, disclosing the contradiction he had discovered a year earlier.[2] When the letter came, Frege was getting the second volume of his *Grundgesetze der Arithmetik* ready for the printer. The work would carry out in detail that deduction of arithmetic from logic which he had sketched in an earlier book, and do it in a formal system that used set theory. He replied to Russell, "Die Arithmetik ist ins Schwanken geraten [Arithmetic is tottering]." Frege took the blow nobly. He tried (unsuccessfully, it turned out) to avoid contradiction by weakening one of his axioms in an appendix to his book. But a third volume of the *Grundgesetze,* needed to complete his program, was not written. His creative life was over. Before he died in 1925, he concluded that the attempt to deduce arithmetic from logic was all a mistake, and set theory an aberration in the history of mathematics. He had always exempted geometry from his logicist program, for he held Kantian views about space, whereas Russell, anxious to refute Kant's entire philosophy of mathematics, included all known branches of pure mathematics in his program. All of them were meant to be included in *Principia.* Poor Frege in his last years held that geometry was the basic mathematical science; arithmetic would somehow have to be derived from it.

In June 1901, when Russell informed Whitehead of his paradoxical discovery, the response was a telegram that quoted Browning: "Never glad confident morning again."* He could not have spoken more truly; much of the time Russell spent on the *Principia* for the next half-dozen years was devoted to seeking a complete solution to his contradiction. In the spring of 1903 he sent Whitehead what he thought was a solution; the response was again a telegram: "Heartiest congratulations. Aristoteles Secundus. I am delighted." Russell later wrote this notation on the telegram: "But the solution was wrong."

Russell called the possible solution that he had put forward tentatively in an appendix to his *Principles of Mathematics* "the doctrine of types." He held that every propositional function ϕx has a range of significance, beyond which the assignment of values to x does not yield a meaningful proposition. The values of x fall into types: an individual (i.e., something devoid of complexity) is of the lowest type, a class of individuals is one type higher, a class of classes of individuals is of the next higher type, and so on. Russell believed that this doctrine was along the right lines, and that he had only to resolve

*The line is from "The Lost Leader." On Russell's glad, confident mornings, see the quotation from him in Chapter XI, section iv, above.

subtle technical difficulties. They resisted him, and he often despaired. In December 1905 he abandoned the theory of types and proposed three alternative theories, but each presented difficulties of its own. He then returned to the theory of types. The elaboration of it that he settled on, and that was used in *Principia,* was not expressed until 1907, when he wrote the paper "Mathematical Logic as Based on the Theory of Types."[3]

The purpose of the hierarchy of types was to prevent whatever involves *all* of a collection from being taken as one of the collection. Poincaré and others had suggested that the legitimacy of statements about all of a collection requires, and requires only, that the collection be finite. The requirement that Russell proposed instead was "what may be called *logical homogeneity*": all the terms of a collection must fall within the range of significance of one propositional function.[4] Since the question of whether a particular class is a member of itself presents different types, it is ruled out as meaningless. The Epimenides paradox and the other logical paradoxes are evaded in the same way: the self-reference present in each paradox introduces a second type.

Logical heterogeneity must also be guarded against in the theory of relations. Russell's theory of types in its final form is highly complicated. It is known as the ramified theory of types, for it distinguishes "orders" within types. In order to keep all proofs clear of typical ambiguity, the theory of types was introduced early in the *Principia* (*12), with a new primitive proposition called the axiom of reducibility. This axiom was by no means obvious; it was needed to legitimate many mathematical theorems and to make the hierarchy of types manageable. I am not going to explain it; nor shall I take up any other facets of Russell's theory of types. I shall add only that the theory is heavily used in the Parts of *Principia* that deal with the logical foundations of cardinal arithmetic, and that, on Quine's estimate, the theory is responsible for roughly one-fifth of the page count of the three volumes.[5]

Throughout Russell's work on the theory of types, Whitehead gave encouragement and offered suggestions. For example, on April 23, 1905, he began a ten-page letter with "Dear Bertie, Your work is admirable. I think you are in sight of land." He then made seven comments. "*Firstly* as to a small point . . ." Whitehead often began his criticism in this way. The "small point," and one other, concerned the notation Russell was using. Whitehead ended the letter by saying that he was very busy, but "thought that a few notes might suggest trains of thought to you." Then came the inevitable postscript: "You will want a notation for . . . I suggest . . ."

Whitehead's mature judgment on the theory of types may be found in the beautiful brief account of *Principia* which took up half of "The Organisation of Thought," his presidential address in 1916 to Section A (Mathematics and Physics) of the British Association for the Advancement of Science. He said of the theory of types:

> It cannot be neglected without the introduction of error. Its theory has to be settled at least by some safe hypothesis, even if it does not go to the philosophic basis of the question. This part of the subject is obscure and difficult, and has not been finally elucidated, though Russell's brilliant work has opened out the subject.[6]

Russell himself expressed a similar opinion in 1918, when he wrote his useful *Introduction to Mathematical Philosophy*. The theory of types "emphatically does not belong to the finished and certain part of our subject; much of this theory is still inchoate, confused, and obscure. But the need of *some* doctrine of types is less doubtful than the precise form the doctrine should take."[7]

The exposition of the theory of types in *Principia* is not as guarded as Russell and Whitehead were in the retrospective judgments just quoted. As the function of the theory in their system was entirely negative, nothing in the system would have been invalidated if several more years of work had resulted in a more elegant and philosophically more basic theory of types. Russell had produced a theory which set out with perhaps unnecessary complications a rule of safety that solved the mathematical contradictions, saved the definitions and deductions in *Principia* from typical ambiguity, and enabled him and Whitehead to finish their book.

ii

After the theory of types, the contribution to *Principia* for which Russell is most remembered is his theory of descriptions, particularly his analysis of definite descriptive phrases of the form "the term which etc."[8] The apparatus for dealing with ordinary mathematical functions such as the logarithm of a number or the sine of an angle begins with $(\imath x)$ (ϕx), meaning, "the term x that satisfies the propositional function ϕx." But Russell was equally interested in the descriptive phrases of ordinary discourse; his favorite examples were "the author of *Waverley*" and "the present king of France." He and Whitehead devoted hours on end to discussing these cases.

When a descriptive phrase is used in designating something that exists, it does the same grammatical work as the name of that thing. Logically, however, the two are quite different. A name denotes in its own right. Russell's general idea about a descriptive phrase was that it is like the symbol for the derivative "d/dx" in differential calculus; by itself it has no denotative meaning, and is to be called an "incomplete symbol." Something else must be supplied, so as to define the *use* of the symbol. "Scott is the author of *Waverley*" is defined to mean, "For all values of x, 'x wrote *Waverley*' is equivalent to 'x is Scott.'"

It seemed to Russell that his discovery of incomplete symbols effected a break in his struggle with the horrid contradiction about classes. Did not symbols for classes form another species of incomplete symbols? They are so treated in *Principia Mathematica*.[9] This enabled the authors to avoid assuming that classes have some sort of being in their own right, while providing a notation for classes and defining the meanings of the propositional functions that determine classes. There was thus no quarrel with the habitual use of classes by mathematicians. However, Russell's hope that this procedure would set up a solution of his contradiction proved mistaken: the hierarchy of types was needed for functions as well as for classes. And it appears that in dispensing with classes Russell accepted properties as having some sort of being in their own right, since the natural interpretation of the propositional function ϕx is "x has the property ϕ."

Whitehead was pleased with his collaborator's analysis of descriptive phrases, and thought it fundamental to logic.[10] It soon gained wide acceptance, but did not escape basic criticism.

iii

Instead of attempting further expositions of topics in *Principia Mathematica* that may be too technical for the reader's interest or my competence, I want now to show the *character* of Whitehead's collaboration with his brilliant former pupil as it appears in various letters, including four from Russell to his wife. Another purpose is to show how the future philosopher, Whitehead, was thinking in this period of his life. But I shall not here make comparisons with the views to be found in what he published after he came to Harvard in 1924 as Professor of Philosophy, and must warn philosophers whose primary knowledge is of the later work that this is a very risky business.

I have already drawn on a few of Whitehead's early letters to Russell. The Russell Archives contain more than sixty other collaboration letters.[11] I shall quote from about one-third of these in chronological order. In a few cases the passage will itself show why I selected it for quotation; in most, explanations will be provided.

Whitehead wrote to Russell from "The 'Stores,'"* in London, on September 28, 1900—just after their discussions at Millhangar of Russell's extension of Peano's logic to relations. Whitehead began this letter, "I am studying the symbolism at 'odd' moments, as you will observe. Here are my reflections in the Stores." He then said that two of Russell's notations "offend against the rules of symbolism," but two others "are exactly what are wanted. . . . Nothing else at present. But the more I think of it, the more I am delighted with your work, and am longing for a sketch of it to gloat over."

On May 26, 1903, Bertie wrote to Alys from Trinity College:

> Alfred has worked like a horse since he got my letter, and has done a lot of things that have to be considered: we have to adopt a joint policy before we can go on with the writing out of our book, and that demands discussion. We began at once yesterday, the instant he had had his tea.

From Grantchester on June 19, 1903, Whitehead wrote to Russell about his first draft of *Principia,* Part I, Section C, "Classes and Relations":

> Dear Bertie
>
> I have looked through your ms hastily and approve most highly. It would want indeed a dull phlegmatic soul not to take the keenest pleasure in reading it.
>
> Of course it is only a first draft and thus *minor* explanations do not count at present. All your *major* explanations are admirable. My only remarks on the notation are as follows
>
> I I do not see the advantage of the change from . . . to . . . .
>
> II The notation is still not complete and distinctive, and since everything is defined for all values of variables this is the sin against the Holy Ghost. . . .
>
> By the bye you are sometimes too sparing in the use of brackets. . . . We must pay the penalty for our generality, which is extreme explicitness.

*The Army and Navy Cooperative Society, Ltd.

These criticisms were accompanied by Whitehead's positive suggestions about the notation.

On April 11, 1904, Bertie wrote to Alys, "I read Alfred's recent work, which is very good. He has been most active for some time past; I only wish my part of the book got on equally fast."

On April 25 Whitehead wrote to Russell from Trinity College just before going off to lecture, posing two courses between which, he thought, Russell would have to choose in his initial treatment of relations. Russell chose neither one, and neither appears in *Principia* as published. In a postscript Whitehead added: "I hope that you have found no flaw in your work. I like it immensely." He believed in Russell's genius, and always bore in mind his need for encouragement and praise.

The next day Whitehead wrote from Grantchester, proposing an idea about how to handle "$x \epsilon x$" in working on the theory of types: "Introduce a fundamental type in which x occurs once only." He continued,

> This is very vague, I can only make it precise by introducing the idea of a proposition as a *series* of *ideas* in which x is not repeated. This seems very vague and complicated, . . . Also does God intuit propositions as a *series* of ideas? Perhaps he thinks of $x \epsilon x$ all at once, as a proposition about x.

Douglas P. Lackey, in his article "The Whitehead Correspondence,"[12] quoted Whitehead's last sentence and the preceding question, and commented, "This shows that even at this early date Whitehead was considering the possibility that God might have two aspects, temporal and a-temporal." I do not believe it, and cannot believe that anyone who knew that Whitehead was a confirmed agnostic during his collaboration with Russell would seriously make such a comment. Everybody uses the word "God" all the time without implying anything about God's nature that is not in an English dictionary.

Later the same day, Whitehead wrote, "My ms of this morning indicates more divergence from your point of view than on reflection I feel."

Whitehead's belief in their need for Frege's assertion-sign was wittily expressed on April 30, 1904. Russell had illustrated a point by the proposition that the number of persons at a particular meeting, three thousand, surprised people, and had symbolized this in a way which, Whitehead objected, made the proposition surprise people, "whereas it is the *asserted* proposition which surprised people. 'Who would be afeard at a proposition' as Billy Bottom might say."

On May 25, 1904, Whitehead wrote that Russell's last letter to him "registers a big advance," as it makes possible the statement of a primitive proposition that will exclude the contradiction. "I shall begin again on the Cardinals at the point which I left off at the end of the Easter vacation, but shall not get much continuous work on it till July." This was usually Whitehead's situation. His work as Senior Lecturer in Mathematics, and other obligations at Trinity, prevented him from doing much continuous work on *Principia* except in the academic vacations.

On June 1, 1904, Russell wrote to Alys from Trinity College: "Alf. was much pleased with the MS I brought. He carries it about, and reads it at odd moments. He must have read it ever so many times by now."

Russell to Whitehead, October 27, 1904:

> My dear Alfred, I send you herewith a somewhat rambling MS., containing a mixture of rhetoric and aspiration. In my own feeling, it embodies a distinct advance: I have begun to feel the contradiction to be obvious and just what one might have expected. . . .
>
> I wonder if you could find a proof of any of the propositions in [a section] of my MS? I do not despair of their being demonstrable, but I have failed to prove any of them in spite of many efforts.[13]

Whitehead's letter of April 23, 1905, from which I quoted in section i, contains an important criticism that I did not mention there. Russell, who had the time and energy to write articles, had had an exchange with the Columbia mathematician C. J. Keyser. Keyser had argued that the current theory of transfinite numbers presupposed the existence of at least a denumerably infinite number of individuals, and suggested that this be made explicit in a separate axiom which he called "the axiom of infinity."[14] Russell's response was that the existence of an infinite number could be proved a priori.[15] First prove that there is such a number as 0 (not hard to do), then the existence of the number 1 (as the number of members in the null class), then the existence of the number 2 (as the number of members in the class consisting of the numbers 0 and 1), and so on. The number of terms in the complete set of classes, classes of classes, etc., will be $\aleph_0$. Whitehead was skeptical, and asked Russell, "Does not the theory require the explicit recognition of (at least) an indefinite plurality, if not of the definite cardinal numbers?" Russell himself later compared his demonstration to pulling rabbits out of a hat.[16] The theory of types ruled out this sort of proof, and an explicit "axiom of infinity," equivalent to Keyser's, was adopted in *Principia* (as *120.04 in Volume II).

On September 28, 1905, Whitehead wrote from Sherborne, where he was visiting his sister:

> Dear Bertie
>
> My letters to you have been long overdue. . . . The difficulties of your last letter are very curious. I disbelieve in your lack of technical skill, and rather think that you are on the track of something important. When I get home I will, as you suggest, work at the subject. We cannot safely get on with the other parts of series till this is finished.

In Whitehead's letter of December 4, 1905, he said something that surprises me: "The aleph-series of Cardinals has always bored me extremely and I shall be glad to assist at its funeral." I do not know how to explain this jesting remark.

On the same day he wrote to Russell, "There has been very little spare time during this term. Thus I have not done much more than keep abreast of your work."

On Christmas Day 1905 Whitehead wrote: "I am delighted with your ms. I am afraid that I cannot send it on just yet. I will lend it to Hardy at the beginning of term." Transfinite arithmetic was at this time one of G. H. Hardy's interests; he had published a paper on it. The letter to Russell continued: "There is one slight slip which may possibly affect some of your reasoning, so I correct it at once. . . . All well here, and the children very happy over their Christmas presents."

On April 3, 1906, Russell wrote to Alys from a hotel in Bedfordshire where he was on holiday: "Alfred and I got through an astounding quantity of work considering the shortness of the time, and fortunately we reached an agreement on all the points we had differed about."

Russell was worried about the "existence" of propositions. Whitehead wrote to him May 5, 1906:

> I have read over your ms on propositions three or four times with the greatest care. Of course it is only a sketch, but I hope that you will hang on to the problem, just as you stuck to the Contradiction.
>
> False propositions are a great difficulty to me. You say—and this seems sense—there is only the fact that Caesar is dead, and there is not in addition . . . the true proposition, "Caesar is dead". But then what the devil is there in respect to "Caesar is not

dead"? . . . It seems to me that for the present "Caesar is not dead" must take the place in your mind which was previously occupied by "The king of France is bald". I sincerely hope that my difficulty is trivial, and rests on my ignorance of philosophical logic.

Mrs. Whitehead's health, as her husband perceived it, sometimes affected the collaboration.* For example, on October 1, 1906, Whitehead wrote from the Mill House:

> Evelyn has been very seedy. Don't mention it. A bad heart attack, no pain, but some rather ugly symptoms, such as feelings of faintness etc. She is to rest very carefully for a fortnight. She can see people quietly. When you come on the 17th bring some books, or work, for she ought not to talk too much. Also Crompton† comes on the 20th, so would you mind leaving on the 19th, so that there can be an evening's complete rest in between. Cook‡ is looking after her, and is very emphatic on the necessity of the strictest precautions.

It is most unlikely that Russell objected to these requests.

Whitehead and Russell thought of quantification theory as the heart of mathematical logic. They read "All men are mortal" as "For all values of *x*, if *x* is a man, then *x* is mortal." In an article, "The Philosophy of Mathematics," which Whitehead published in 1910, he wrote of the crucial discovery "of the generalised conception of the variable and of its essential presence in all mathematical reasoning," and attributed this discovery to "the joint and partially independent work" of Frege, Peano, and Russell.[17] Now, Whitehead in all his mathematical work had a sharp eye for tacit assumptions; in the collaboration correspondence he frequently asked Russell to consider whether he was making a tacit assumption. Thus on May 5, 1906, when Russell was trying to handle ambiguity as to type, Whitehead wrote:

> Is there not an omitted, essentially non-symbolical, Pp,§ viz that of

---

*See Chapter XI, section v.

†Crompton Llewelyn Davies was now the youngest son in the Llewelyn Davies family described in Chapter V, section vii. His brother Theodore, a wonderful person to whom he was devoted, had died the year before. Thereafter Crompton was frequently at the Whiteheads' house. Evelyn became very fond of him, and may for a time have been chastely in love with him. Russell was a close friend of Crompton's (Russell, *Abiog.*, Vol. I, pp. 57–59).

‡Evelyn's physician.

§Primitive proposition.

> the *identification of ambiguities,* viz that "the *x*" of one proposition is the same as "the *x*" of another proposition, or of another occurrence. This principle seems to be the ghost of Aristotle's dictum De omni et nullo. It is certainly with us the principle by which Barbara is put on its legs. Our notation obscures it by presupposing it. But your new views by emphasizing ambiguity throw it into relief. The principle is that "the any entity" under consideration can be identified with "that any entity" of a previous consideration.

On June 16, 1907, Whitehead sent Russell an eleven-page letter that will repay our attention and commentary. It began, "After sitting for $6\frac{1}{2}$ hours yesterday on Committees drawing up class lists and bestowing emoluments, I signed my last document of College business and am free today, except for a few trifles of belated unofficial business." There followed a ten-point outline of Russell's current position "as I understand it" on independent and dependent entities, individuals and classes, intension and extension, and the hierarchy of types. Whitehead then put down "some unsystematic reflections"; on the last page of the letter he wrote, "Don't bother to confute me in detail, the above is only written to stir you up, not to waste your time." He believed in Russell's creative power as a philosophical logician, and said now that he could easily acquiesce in his position "in the same way in which liberal Catholics acquiesce in Papal decisions." "But," he continued, "my power of intuition as to ultimate philosophical ideas is so feeble that I attach no importance whatsoever to my acquiescence." Although Whitehead left philosophical analyses and final decisions about them to Russell, he did not hesitate to express doubts, especially about Bertie's destructive rationalism. In this letter he wrote of Russell's current theory:

> My doubts arise as follows:
>
> The theory reposes upon a few ruthless narrow presuppositions, (a) which may or may not be true, (b) which rule out as due to confusion of thought many ideas which naturally arise.
>
> Taking (b) first, Two men are in a room, they form a class of *two* things, they and that class form a class of three things. The latter class, it is true, is a class of three things *in another sense* to that in which the former class is a class of two things. But your theory of types refuses *all sense* to the latter class (Does it not?). This seems to me to be very paradoxical. But this brings us back to (a). . . .
>
> But the keystone of the theory is that there is a definite determinate class of individuals and that all other entities are emanants

> from them. This may be true, but in view of the fate of all metaphysical doctrines in the past, I also observe that it may not be true.

On the basic notion of an entity, Whitehead wrote, "For our purposes we may define an entity as that which in any sense can be amenable to arithmetic ideas." It is this amenability that must be of varying types.

In general, Whitehead wanted to be sure that Russell considered alternatives to a position he was taking. In this letter he asked, "Is it not possible that there is no such thing as a determinate class of individuals at all?" Even a Kantian doctrine that a form of thought is responsible for the determination of a class of individuals is possible. After saying that a technique Russell had suggested was "nearly perfect," Whitehead wrote:

> But I want the explanation to be as non-committal as possible and to point out its adaptability to widely divergent lines of thought. For after all philosophy finds formal logic as a going concern, and different philosophies only differ in giving different explanations of one and the same formal procedure.

On June 27, 1907, Whitehead wrote, "I have got to work again now, and am working at the 'Finite powers of Relations', expanding it into a chapter on [mathematical] Induction." Whitehead was writing on a subject for which Russell had primary responsibility.

The last Section of Volume I in *Principia,* "Inductive Relations," is *90–*97. In it the concept of "ancestral relation," which Frege first defined, is the starting point. Whitehead and Russell use this concept as the basis for their subsequent treatment of finite and infinite and their theory of series. On January 5, 1908, Whitehead wrote to Russell, "I like your list of **90–97 immensely. I cannot imagine that we can improve on them in any substantial way. Your articles 96 and 97 are especially good. The propositions surprised and delighted me."

The next day, Whitehead wrote: "I have been studying you on Types. As far as I understand, I approve highly. But the points are so subtle that I have grave doubts as to whether all the difficulties that you are dodging are really present to my mind." He then made many remarks in the hope that they would "help to stimulate thought," and concluded his letter by saying, "I am in a fog as to where you are."

A week later, Whitehead was able to write:

> Dear Bertie
>
> I have been meditating on the final MSS which you sent me. I am not sending it back just yet, because I want to continue the process of meditation. But I will just jot down my impressions up to now.
>
> All the technical part . . . strikes me as perfectly beautiful. The theory of relative types and of ambiguous symbols has surely reached its last perfection. I would no more think of laying sacrilegious hands on it than of burning down Canterbury Cathedral.
>
> Jan 14th [At this point I was unexpectedly interrupted and twenty four hours have elapsed.]
>
> The hierarchy of types does not—at least as far as its explanation is concerned—appear to me to have reached its final shape.

This letter illustrates a general feature of Whitehead's discourse, both oral and written. Although it was peppered with careful qualifications, hyperbole was likely to appear when he felt strongly, and when he especially wanted to encourage Bertie.

It was a long time before Whitehead was satisfied by his collaborator's written explanation of the theory of types. On May 26, 1909, he wrote:

> Dear Bertie
>
> I am working hard at types and in some respects nearly see my way i.e. as to methods of exposition. But there is one fundamental obscurity in the theory as "exposed" in the *American Journal*[18] article which I do not see through. . . .
>
> Could you pretty soon send me your thoughts on the point? I have a few days now but *not* later. Then I will send you all I have done and you can hack it about.

Some points of difference between Whitehead and Russell have appeared in the foregoing quotations from their correspondence. Two general differences in their philosophies of mathematics, differences not yet noticed, should now be remarked.

1. Although Whitehead accepted Russell's thesis that the concepts and principles of pure mathematics are constructible from those of logic, there was a difference in their interests. To Russell mathematics was only a prolongation of logic; to Whitehead logic was the essential first chapter in the book of mathematics. The first sentence of Russell's *Principles* was "Pure mathematics is the class of all propositions of the form 'p implies q,' where p and q are propositions containing one or more variables, the same in the two propositions, and neither p nor q contains any constants except logical constants." Russell the

logician was preoccupied with the truth of "*p* implies *q*," but Whitehead the mathematician retained an interest in the truth of "*q*" which went beyond the use of *q* as a premise in the chain of deductions constituting pure mathematics. If *q* is true, it may also be a premise in a branch of applied mathematics. The possibility of applications was essential to Whitehead's actual interest in mathematics, and only theoretically to Russell's. When Whitehead wrote out a description of the *Principia* in their application to the Royal Society for a grant in aid of publication, Russell inserted the adjective "pure" wherever Whitehead wrote "mathematics."

2. Although Whitehead was the more expert in devising symbols, did more of it, and frequently criticized Russell for insufficient attention to what strings of symbols a given symbol authorized, there is some evidence that he was less prone than Russell to forget the distinction between a symbol and its object. This was brought out by I. Grattan-Guinness in the course of a detailed criticism of Russell's symbolic device, the "matrix," which Russell tried out in 1906 to do the work of a class.[19] (Whitehead severely criticized Russell's idea; it was not used in *Principia*.) Let me add that Russell with his highly verbalized mind seems to have thought of all mathematical terms except those which denoted individuals as linguistic, although the language was not English but the ideal language invented by Peano, himself, and Whitehead. Russell in his old age wrote:

> I set out with a more or less religious belief in a Platonic eternal world, in which mathematics shone with a beauty like that of the last Cantos of the Paradiso. I came to the conclusion that the eternal world is trivial, and that mathematics is only the art of saying the same thing in different words.[20]

One step in this so-called progress was Russell's acceptance of the view, urged by his pupil Wittgenstein after *Principia* was published, that mathematics consists of tautologies. Whitehead never accepted that view, and attacked it when he came to write philosophy. There is no evidence that he started with Russell's extreme belief, so I think that he was not tempted toward the opposite extreme.

iv

The Preface to *Principia* tells us that as the work on it advanced, the authors found their subject much larger than they had supposed, and they were able to handle satisfactorily many questions that Russell's

Principles of Mathematics left "obscure and doubtful";* hence it became necessary to make their book independent of that volume. Nothing in the collaboration correspondence enables us to date this decision. In letters to his friends Russell for a long time referred to the collaborative work as Volume II of his *Principles;* the last such letter known to me is dated November 10, 1905.† When Whitehead in the autumn of 1906 proposed to the Royal Society that Russell be elected a Fellow, he listed among his books "vol i" of the *Principles,* from which we may infer that Whitehead was still thinking of the joint work as its second volume—unless Whitehead made a slip, which is quite possible. My conjecture is that the authors made their decision in 1907. On March 30, 1908, Russell in a letter to Philip Jourdain wrote, "Whitehead and I are making good progress with *Principia Mathematica.*"[21]

There seems to be no record, in the collaboration correspondence or elsewhere, of the choosing of this new title. Did they have Newton's great work in mind?‡ Not if there is anything in Jessie Whitehead's report to me that her father and Bertie were shocked when they realized that they had appropriated Newton's title. G. E. Moore's *Principia Ethica,* published at Cambridge in 1903, could have given them a precedent. But they need not have had it in mind either. A sufficient reason for choosing "Principia Mathematica" was that it fitted their subject (the encyclopedic character of the work made "Foundations" unsuitable), and the use of Latin distinguished it from *The Principles of Mathematics.*

After Russell in 1907 sent off "Mathematical Logic as Based on the Theory of Types," he began the long, arduous task of writing out the final draft of *Principia.* Whitehead's academic duties left him no time for this. It was dog-work that took about ten hours of Russell's days for eight months of two consecutive years. On March 18, 1908, he wrote to an American friend, Lucy Donnelly, "Since September I have written about 2400 pages of the MS of our book, and I am still only in the 3rd of 8 Parts." He was thinking of geometry, which Whitehead was to work out but never finished, as comprising Parts

---

*Russell became quite dissatisfied with this book as soon as it was published in 1903, and remained dissatisfied.

†Bertrand Russell to his American friend Lucy Donnelly, November 10, 1905. Russell published this letter in the first volume of his *Autobiography.* He wrote, "But the end of Volume II is as far off as ever—the task grows and grows." In an interview on May 7, 1967, Russell assured me that the reference was to Volume II of his *Principles.*

‡*Philosophiae Naturalis Principia Mathematica.*

VII and VIII. Two months later, in a letter to an English friend, Ivy Pretious, he said that since September he had written "about 2,000 pages." Russell was often careless in his use of round numbers, but it may be that 400 or more pages, written in the autumn and winter, were discarded after discussions with Whitehead.

In the Easter vacation of 1909 Russell spent a short holiday in Cornwall with the Whiteheads, then went to his house in Bagley Wood, outside Oxford, to finish the job. In his *Autobiography* he recalled, "Every time that I went out for a walk I used to be afraid that the house would catch fire and the manuscript get burnt up."[22] There was no fire, Russell covered page after page with symbols, and on October 12 Whitehead wrote that he would make a preliminary visit to the Cambridge University Press, and said, "Like the Greek ten thousand we ejaculate 'Printing! Printing!' "

Finally, on October 18, Russell could write Lucy Donnelly that he had

> arrived at a great moment: tomorrow I go to Cambridge, taking with me the MS of this book for the printer. There is a certain amount at the end that is not yet finished, but over 4000 pages are ready, and the rest can be finished easily. I have been working like a black to get the last bits of revision done in time for my visit to Cambridge tomorrow, and now the MS is packed in two large crates.[23]

In the *Autobiography* Russell remembered the manuscript as "so large that we had to hire an old four-wheeler" (instead of a hansom) to transport it to the Cambridge Press.[24]

Whitehead took charge of arrangements with the Press; it was to him that Richard T. Wright, secretary to the Syndics, wrote on October 29. Wright had had an estimate prepared on the cost of printing 750 copies of the manuscript received; it was "alarming." He proposed what had been suggested by "one or two of the Syndics": an application to the Royal Society for a grant to help defray the expense. Whitehead at once passed the suggestion on to Russell, saying that he "will not commit us in any way till we have talked the matter over," and adding, "Do you think that there are parts of the book which can be shortened or omitted without much loss. I doubt it." The next day he wrote again, to say that he would talk to Wright, and to give Russell his own view on shortening the book: "it seems to me improbable that we could save more than 100 printed pages without very seriously diminishing the value of the work, and even that in the end would be bad commercial policy," since the completeness of *Principia* was its "enormous advantage" over tracts or short books that might be published for students of the foundations of particular mathematical sciences. Russell agreed; the manuscript was not cut down.

This *Principia* was about as long as Newton's, and although it contained almost no diagrams (Newton's was full of them), special type would have to be cut for its special symbols. It is not surprising that, according to Whitehead's letter of November 1 to Russell, Wright told him that "a loss of £500 on the book must be allowed for." Whitehead continued, "The Syndicate are anxious to publish, and he [Wright] is under the impression that they would be willing to undertake a *loss* of £250. Leaving £250 to be got from the Royal. I will consult Forsyth as to the best method of application." The making of the application was postponed, as the Syndics now wanted to see a referee's report on the manuscript. W. E. Johnson supplied one that satisfied them, if not the authors. On December 10 the Syndics, after increasing the estimated loss on the book to £600, decided to contribute £300 if another £300 was obtained from the Royal Society or some other source.

Whitehead's next step was to talk with the senior Secretary of the Royal; he was the mathematical physicist Sir Joseph Larmor, who on Stokes's death had succeeded to the Lucasian Professorship at Cambridge, once held by Newton. Whitehead reported to Russell on December 15 that he had had "a satisfactory interview with Larmor—short, and to the point. . . . Larmor was just like Wright, personally most friendly—but non-committal." Larmor would lay the matter before the Council of the Society, and ask the President to refer it to a small committee in advance, so as to secure a decision in January. "Our written application does not much matter but we had better send one in as a matter of form." Whitehead concluded this letter by enclosing his draft of the application. Russell made a few changes in it, added a little, and appended the Table of Contents of the whole work; Whitehead added his signature and took the application to Wright. The archives of the Royal Society do not contain it, so probably it was not formally presented to the Council. From its text, now in the Bertrand Russell Archives, we can see that it describes *Principia Mathematica* as a work of mathematics and pays little attention to its philosophical side; the application concludes with a defense of the symbolism used.

On January 20, 1910, the Council of the Royal Society resolved to pay a subsidy of £200 in two annual instalments to the authors, "the payment of the second instalment being contingent on substantial progress being made with printing of the work."[25] Whitehead and Russell supplied out of their pockets the £100 still required. Russell later wrote, "We thus earned minus £50 each by ten years' work. This beats the record of *Paradise Lost*."[26]

The printing was slow work; Cambridge University Press had only one compositor who could set up the queer symbols. The authors did not expect Volume I to be ready before 1911. But it was published in December 1910. Volume II came out in 1912, and Volume III in 1913.

V

The most well known of the writings to which students turn to learn about *Principia Mathematica* without having to read it is Russell's *Introduction to Mathematical Philosophy,* written in 1918 while its author was in prison. That book, however, is not an account of *Principia;* it is an introduction to mathematical logic and its application to topics in mathematics. It begins with the subject that is most familiar to the reader, the series of natural numbers, and reaches the theory of deduction, with which *Principia* began, only in the last quarter of the book.

Whitehead wrote two short accounts of *Principia* that are not as well known as they should be. The first occurs in his article "Mathematics," written for the eleventh edition of the *Encyclopaedia Britannica.*[27] More than half of the article is devoted to replacing the traditional definition of mathematics as "the science of discrete and continuous magnitude." Whitehead proceeds by raising critical questions, presenting Peano's axioms for arithmetic, and showing how arithmetic is but a subdivision of the general theory of classes and relations. In summarizing the deduction of that theory from the premises of logic, Whitehead explains the contradiction Russell discovered and his theory of types. Whitehead concludes that mathematics in general is the "science concerned with the logical deduction of consequences from the general premises of all reasoning." Although he attached to this definition a footnote that gives Russell the credit for expressing and developing the idea, the words "the general premises of all reasoning" have rather, to my mind, a Whiteheadian ring. At any rate, this is a way of speaking which he continued to use, but which does not jibe with Russell's later belief that mathematics consists of tautologies.

Whitehead's second short account of *Principia* is the latter half of his 1916 essay "The Organisation of Thought." It is a masterpiece of exposition. He briefly summarizes in simple language the successive topics of *Principia*. Attention should be called to his statement of the significance of the book's achievement: "The whole apparatus of special indefinable mathematical concepts, and special *a priori* mathematical premises, respecting number, quantity, and space, has vanished."

We come now to the reception of *Principia Mathematica*. Russell, long afterward, remembered:

> Both Whitehead and I were disappointed that *Principia Mathematica* was only viewed from a philosophical standpoint. People were interested in what was said about the contradictions and in the question whether ordinary mathematics had been validly deduced from purely logical premisses, but they were not interested in the mathematical techniques developed in the course of the work.[28]

Russell gave two dramatic illustrations of ignorance of what had been done in *Principia*. The first sentence quoted from him undoubtedly exaggerates, but it is certain that the most interested readers were logicians, or philosophers curious about the nature of mathematical knowledge, not mathematicians. Warsaw was almost the only place where *Principia* was enthusiastically studied. Most mathematicians elsewhere were bowled over and paid only lip service to it. They could not dip into it and work with it easily. Indeed, it contained few extractable memorable results of fairly immediate use to them. It was interesting to know that separate sets of axioms for separate branches of mathematics were no longer needed, but their research seldom required them to make use of this unification. A typical reaction from a mathematician is the one expressed by Sydney Chapman at the end of his long, distinguished career: "These volumes are full of strange symbols, . . . I bought the first volume, but I found it rather arid reading. I got no stimulus from it, and I didn't continue to buy the later volumes."[29] (Cambridge printed only two-thirds as many copies of each of them as it did of Volume I.)

"One of the great intellectual monuments of all time" is the stock assessment of *Principia Mathematica*. Monuments aren't alive, and the book had little influence on living mathematics. Its presence did have the beneficial effect of putting an end to mysticism about mathematics.

New developments, which began shortly after its publication, had much to do with its lack of influence. The distinction between mathematics and metamathematics became important; but *Principia* contained no metamathematical theory of its system. Constant attention to the distinction between the mention and the use of symbols became *de rigueur;* Russell and Whitehead seldom attended to this. Some logicians and mathematicians who studied *Principia* doubted that it completely proved the derivation of arithmetic and analysis from the premises of logic; so the work was called a great and splendid failure. Russell's theory of types was much discussed; the result was the emergence of "modern" type theories in the work of Tarski, Carnap, Church, and Gödel. And in 1931 Gödel in his famous theorem proved the existence of an undecidable formula in any formal system adequate to number theory, and the impossibility of proving the consistency of such a system within the system. Much earlier, Hilbert's formalist conception of mathematics and Brouwer's intuitionist conception (championed also by Weyl) had become formidable rivals to the logicism of *Principia Mathematica*. In 1937 Russell wrote short retorts in an Introduction to the second edition of his *Principles of Mathematics*. Whitehead published no answer to these competing schools of thought.

Russell began his 1948 article "Whitehead and 'Principia Mathematica'" with "There is in some quarters a tendency to suppose that Whitehead's part in our joint work was less than in fact it was." The tendency has continued; historians and writers of encyclopedia articles now treat *Principia* as Russell's book. There are natural reasons for doing so. The book's thesis, and the controversial topics in it that these writers discuss (such as the theory of types and the theory of descriptions) were Russell's. Also, he defended the thesis and those theories whenever an opportunity arose. At Cambridge, he was the one who lectured on the *Principia* from 1910 to 1916. On April 29, 1910, Whitehead resigned his Senior Lectureship at Trinity preparatory to moving to London. He was a member of the College Council, and on May 27 he was able to tell Russell that the Council had just appointed him Lecturer on Logic and the Principles of Mathematics; this position, he said, "will give you a splendid opportunity to 'expose' the subject—just what is wanted."

In Cambridge it became and remained customary to say of the collaboration on *Principia* that Whitehead was the mathematician who did the dog-work for Bertie. This judgment reflects the almost exclusive interest there in the philosophical aspect of the work. It would have been more accurate if Cambridge had said that Whitehead was the critical mathematician who developed and consolidated the ideas involved in Russell's logicist thesis and (except for the theory of descriptions and the theory of types) took the lead in giving them accurate symbolic expression.

Principia Mathematica increased Russell's fame as a philosopher. It did nothing for Whitehead's standing among mathematicians.

vi

Reading the proofs of Volumes II and III was a long business. Most of it was done by Russell. In January 1911 Whitehead discovered that throughout Part III, "Cardinal Arithmetic," they had not guarded against a particular ambiguity as to type. He stopped the printing of Volume II for four months while he wrote a new Introduction to it and rewrote some of its articles.

After Whitehead moved to London in the summer of 1910 he worked on *Principia* Part VII, "Geometry." Although he showed his drafts to Russell for comment, Whitehead had sole responsibility for this Part. As nearly as Russell could remember in 1965,[30] this decision was made about 1905, and was based on the fact that Whitehead was especially interested in geometry, he in other parts of their subject. Instead of finishing the fourth volume of *Principia*—already delayed by the effect of the outbreak of World War I on his family—Whitehead, never an advocate of the absolute theory of space, decided to take into account the revolutionary ideas of Einstein and Minkowski, and for this purpose wrote *An Enquiry Concerning the Principles of Natural Knowledge*.[31] On January 8, 1917, Whitehead, fearing that Russell would dance away with his ideas before he himself had put them in order, refused to send Russell his notes for this book. Russell in his *Autobiography* says that this put an end to their collaboration.

It had ended almost three years before. Their whole-hearted agreement about the principles of mathematics did not carry over to the philosophy of space. Russell in his "Portrait" of Whitehead (1953) said, "He and I disagreed in philosophy, so that collaboration was no longer possible."[32] Whitehead said it better in his "Autobiographical Notes": "Our fundamental points of view—philosophical and sociological—diverged, and so with different interests our collaboration came to a natural end."[33] I do not believe they could have collaborated on any subject other than the one on which they did.

As I noted in Chapter XI, while they were working on *Principia* Whitehead neither agreed nor disagreed with Russell's doctrine that relations and their terms are independent entities.* Whitehead did not

*See Chapter XI, section iv.

agree with the way Russell saw the world. One of the Autobiographical Talks in Russell's *Portraits from Memory* contains a passage in which he says that after throwing Hegel overboard (that occurred in 1898), he saw the world as consisting of separate things with hard, precise boundaries. Then he writes:

> It was Whitehead who was the serpent in this paradise of Mediterranean clarity. He said to me once: "You think the world is what it looks like in fine weather at noon day; I think it is what it seems like in the early morning when one first wakes from deep sleep." I thought his remark horrid, but could not see how to prove that my bias was any better than his. At last he showed me how to apply the technique of mathematical logic to his vague and higgledy-piggledy world, and dress it up in Sunday clothes that the mathematician could view without being shocked. This technique which I learnt from him delighted me, and I no longer demanded that the naked truth should be as good as the truth in its mathematical Sunday best.[34]

Possibly this is exaggerated. Still, it speaks volumes; especially the phrase, "At last he showed me . . ." *At last.* There were opposed biases, but Whitehead was not in a hurry to demonstrate the superiority of his. What he finally showed Russell was merely a technique that delighted him. I imagine Whitehead knew it would delight him.

In his *Enquiry*, published in 1919, Whitehead gave the name "extensive abstraction" to this technique, and relied on it to define points and instants.

He and Russell held contrasting views of symbolization. On May 31, 1910, Whitehead began a letter to his collaborator, "I think you missed the point of my previous communication which was hurried and slipshod owing to press of engagements." He then emphatically expressed his view of the relation between symbols and their meanings. What underlies a symbol is an idea; otherwise the symbol would be merely a shape on paper. The "logical system" of *Principia* is a network of ideas; it is

> independent of our symbolism, and uses the symbolism to convey ideas. The question as to when the symbolism conveys determinate ideas is outside the system, and is only important because owing to the informity of our minds we have to use symbolism (words or otherwise) formally, without perception of the exact ideas which ought to be conveyed. . . . We *ought* not to want any *theory* of types or any of the philosophical part of logic, we ought to *see* meaning, and "no meaning" is *nothing*.

But we don't perceive meaning with any exactness. The theory of types, as a rule of safety, is needed as "the substitute for the exact perception of meaning."

In section ii of Chapter XI, I remarked the difference between the way Whitehead thought and Russell's verbal way of thinking. Because of this difference, I doubt that Russell could have got inside the view Whitehead maintained in this letter, and would have dismissed it as intuitionistic. The importance of this divergence can hardly be exaggerated.

In the history of men's intellectual products the collaboration that created *Principia Mathematica* stands out. I am not aware of any in the history of mathematics that lasted as long. Of course their project was a big one; after Whitehead committed himself to it, neither he nor Russell wanted to give up. But what were their attitudes toward each other? To Whitehead, Russell was a brilliant young logician toward whom he felt a continuing duty to give help; I think that the teacher-pupil relationship did not wholly disappear so long as the collaboration lasted. To Russell, Whitehead was a solid mathematician who had contributed to symbolic logic and whom he had convinced of the truth of his logicist thesis. There was mutual respect.

The fact that early on, one partner fell in love with the other's wife,* makes their long collaboration yet more unusual. Whitehead could not have been blind to the emotional fact. His self-discipline made it possible for him to act as if it were not so, and to put up with vague feelings of unhappiness about it. I do not think he probed the situation at all. What a wise person he was!

Russell's work on *Principia* was a refuge from his unhappiness with Alys. The decade of collaboration was not a happy period in either his or Whitehead's life.

Joint work, long continued, requires compromises on specific points of disagreement. There is no evidence that this necessity spawned any rancor, though both men were strong individuals. Of course, rancor was foreign to Whitehead's nature. And Russell recognized Whitehead's authority as a mathematician.

A wonderful thing about their collaboration is the perfect preservation of the individuality of each partner, made possible by their mutual respect and affection. In consequence, each was free thereafter to develop his own philosophic views in his own way, not predetermined by their long work together.

---

*See Chapter XI, section vi.

CHAPTER XIV

"On Mathematical Concepts of the Material World"

i/Purpose of the memoir: to define and compare, in terms of mathematical logic, the classical concept of the material world and four alternatives to it. Ultimate entities that are not punctual but linear.

ii/Whitehead's treatment of time. His sophisticated treatment of space; the "Theory of Dimensions" and "Theory of Interpoints."

iii/Further discussion of geometry and mathematical physics.

iv/Lack of influence of this memoir on physics.

v/Its relation to Whitehead's other work. Physics and logic. Geometry and the changing world.

Although geometry was not taken up in the three published volumes of *Principia Mathematica,* Whitehead, in the middle of the decade in which they were written, published a long paper on geometry and the motion of matter. Its title, like those of several of his scientific papers, begins with "On." This sounds old-fashioned, and surely the word is unnecessary. But in Whitehead's case its use is appropriate. He was not in the habit of writing to finish a line of investigation; rather, he wrote to start one on a familiar subject. This is true of all his major works, including the memoir (as he usually referred to it) now to be considered. It was his bad luck that his new methods, however meritorious, were seldom taken up by his peers.

Looking backward thirty years later, when he was a famous philosopher at Harvard, he told me that insofar as he could lay claim to any originality, he thought "On Mathematical Concepts of the Material World" was the most original thing he had done.[1] This memoir was received by the Royal Society of London on September 22, 1905, and published in their *Philosophical Transactions,* Series A, in 1906.[2] It begins:

> The object of this memoir is to initiate the mathematical investigation of various possible ways of conceiving the nature of the material world. In so far as its results are worked out in precise mathematical detail, the memoir is concerned with the possible relations to space of the ultimate entities which (in ordinary language) constitute the "stuff" in space.

Whitehead frames and evaluates the traditional way of conceiving the material world and four alternatives to it. He calls each a "Concept of the Material World." "Precise mathematical detail" is achieved by using the symbolism in which *Principia* was being written. Whitehead says, "Peano's chief symbols are used. The changes and developments from Peano . . . are due to Russell and myself working in collaboration for another purpose." He adds, "It would be impossible to disentangle our various contributions."[3] But in a footnote he calls attention to Russell's paper "Sur la logique des relations." In it Relation was taken as a foundational idea, as it is in this memoir. Indeed, Whitehead views the material world as a set of relations and of the entities forming the "fields" of those relations.[4] Each of the five Concepts he investigates is a complete set of proposed axioms and definitions about the relation that is taken as fundamental and the entities comprising its field.

Concept I is "the classical concept of the material world," embodying the habits of thought that had prevailed since Newton. This concept employs three mutually exclusive classes of entities: points of space, instants of time, and particles of matter. The theory of the motion of matter is superposed on a presupposed independent theory of space and a presupposed independent theory of time. (The classical concept arose in an age when geometry was the only developed science.) The superposition (according to an analysis first suggested by Russell in *The Principles of Mathematics*[5]) requires a class of relations of "occupation of a point at an instant," a new relation being required for each permanent particle. The general laws of dynamics can then be added to the deductive scheme as axioms about the properties of this class of relations.

Whitehead writes:

> Opposed to the classical concept stands Leibniz's theory of the Relativity of Space. . . . This theory, though at present it is nominally the prevailing one, has never been worked out in the form of a precise mathematical concept. It is on this account criticized severely by Bertrand Russell . . . , who, however, has gone further than any of its upholders to give it mathematical precision.[6]

Whitehead now criticizes the classical concept. This is important for students of his philosophical development: he is giving his first criticism of the view that he dubbed "scientific materialism" in *Science and the Modern World*. The criticism is logical, not physical or philosophical. It is that Occam's Razor provides a sufficient reason for trying to reduce the number of relations involved, and to make constructions which do not require three independent classes of entities if fewer will suffice. Also, Whitehead looks upon the dualism of points and particles as a challenge to theoretical thought. He was never happy with final, unrelieved dualisms.

In setting out the classical concept, Whitehead assumes Euclidean geometry, and makes use of Oswald Veblen's recent construction of its nonmetrical part from axioms referring to but one class of entities, "points," and one undefined triadic relation among points, "between."* After formulating Concept I in terms of the precise logic of relations being developed in *Principia*—something in itself well worth doing—he presents two others in which geometry is brought into closer contact with physics. Concept II, based on a suggestion of Russell's,[7] replaces "material particles" by dyadic relations between points and instants. Concept III, which may in part be traced back to Leibniz, replaces the two classes, points and instants, by what may be called either moving points or particles of ether. Whitehead presents two variants of Concept III.

In Concepts IV and V ultimate entities that are *linear* rather than punctual in nature are introduced, resulting in a tremendous difference from the classical concept. These linear or directional ultimates are analogous to lines of force, though endless. The points of space are defined in terms of their properties, and the lines and figures of ordinary geometry are defined as complexes of these *defined* points. Concept IV takes two forms; one is dualistic, that is, requires a class of particles of matter as well as the linear ultimates. The other is monistic, that is, dispenses with particles of matter *as ultimates*. Concept V, the development of which climaxes the memoir, depends upon a new treatment, elaborated in the memoir, of the theory of the dimensions of a space. Concept V is a monistic, Leibnizian concept of the material world, and the most interesting of the five. Whitehead writes: "Concept V, in particular, appears to have great physical possibilities. Indeed, its chief difficulty is the bewildering variety of material which it yields for use in shaping explanations of physical laws."[8]

The memoir concludes with a sketch of a way in which Concept V might be used to make possible a simple formulation of electromagnetic physics. Classical dynamics and, Whitehead hopes, eventually the entire structure of physics, is in this Concept expressible in terms of a single pentadic relation between the members of a single class of ultimate physical entities and the instants of time, and one auxiliary relation (required to determine kinetic axes of reference). The single undefined relation, *R,* is the intersection in order of three linear ultimates by a fourth at an instant of time.

ii

Certain specific features of this work now claim our attention. First, Whitehead's treatment of time. There is here no trace of the

*The properties of points and of betweenness are said to be "defined" solely by the axioms.

novel idea that forms the basis of his later physical constructions and is carried over into his metaphysics: the idea that the instant of time, or, in physics, the "configuration at an instant," should not be assumed as primitive and undefined, but, like the point of space, be derived from physical elements that are epistemologically more primitive, though their logical properties are more complex. Whitehead here asserts, on the contrary, that "time must be composed of *Instants*. . . . *Instants of time* will be found to be included among the ultimate existents of every concept." Again, "In every concept a dyadic serial relation, having for its field the instants of time and these only, is necessary."[9]

Having been educated in the post-Relativity era, we may suppose it to be obvious that instants of time should have been treated as points of space were treated in the latter part of the memoir. Besides committing an anachronism, we should be forgetting that the purposes of the memoir were mathematical, not at all epistemological. And we ought not to expect a man who is busily working a new field with a new instrument (symbolic logic) to think up a new theory of time as well as new theories of matter's relation to space. It is not humanly possible to make progress in any field if one tries to revise all the prevailing concepts at one stroke. Whitehead had for years been a geometer, familiar with the idea of the point as complex.

When and how Whitehead changed his mind about the instant of time is seen in his letter of September 3, 1911, to Russell. Whitehead was then going over proofs of the second volume of *Principia Mathematica,* and had been working on geometry, subject of Volume IV, for a year.

> I hope to post the proofs at the same time as this letter, viz before 5 or 6 o'clock when the mail departs. But last night when I should have finished them, the idea suddenly flashed on me that time could be treated in exactly the same way as I have now got space [which is a picture of beauty, by the bye]. So till the small hours of the morning I was employed in making notes of the various ramifications. The result is a relational theory of time, exactly on four legs with that of space. As far as I can see, it gets over all the old difficulties, and above all abolishes the instant in time. . . . This has always bothered me as much as the "point", but I have had to conceal my dislike from lack of hope. But I have got my knife into it at last.[10]

Interviewing Russell in 1965, I asked whether Whitehead, like himself, favored the absolute theories of space and time early in their collaboration. "No," he answered, and added, "I think he was *born* a relativist."

In the memoir of 1905 Whitehead does not go into the conflict between absolute and relative theories, for he is not there concerned with philosophical questions. But he was keenly interested in the controversy. The memoir makes the absolute theory of space part of the classical concept of the material world, and a source of weakness in it. The concept that he recommends, Concept V, is relativistic.

Whitehead's treatment of points in the memoir next requires comment. He writes that although geometers are used to the idea of the point as complex, they nevertheless assume points as ultimate entities, since the definition of the projective point is introduced subsequent to a geometry of points; in fact, the straight lines referred to in the definition are those concurrent at a point. Whitehead says that his memoir grew out of an attempt to remove this circularity, by finding out how to define a point in terms of lines. He worked out two independent theories for this purpose, the "Theory of Dimensions" and the "Theory of Interpoints" (i.e., intersection points).[11] With their aid he was able to elaborate his linear concepts of the material world. The two theories are unintelligible if one is not acquainted both with advanced geometry and with the symbolism of *Principia*. Instead of making a show of explaining them, I shall set down a few facts about them which are of general interest.

1. We find in them (along with a paper written in 1914, to be noticed in my second volume) the closest existing indication of the manner in which the polyadic relations, essential to geometry, would have been worked out in the unwritten fourth volume of the *Principia*.

2. The Theory of Dimensions is initially presented in its most general form, in which it is a contribution to the theory of sets (Whitehead writes "classes"); then the application to geometry is made.

3. In the Theory of Interpoints Whitehead for the first time uses the procedure that he would push to the limit in his books on the philosophy of physics but abandon in *Process and Reality*, namely, the inclusion of temporal entities (the instant in this case, the "duration" later) in his statement of the point-defining relation.

4. One notices in these theories Whitehead's first employment of certain terms, notably "primeness" in the Theory of Dimensions and "cogredience" in the Theory of Interpoints, which have a basic role in the technical development of his theories of space and time in all his later works. The meanings of these terms are generalized later on, to be sure; thus "cogredience" has no perceptual significance here, but is merely an abstract characteristic (suggested by the idea of points at infinity) possessed, under certain conditions, by the linear ultimates at

an instant of time. Nevertheless there is a recognizable similarity, which deserves to be investigated by anyone interested in Whitehead's physical theory.

5. The Theory of Interpoints, which explicitly defines points of space in terms of proposed ultimate material entities, should be looked upon as the first, unnamed beginning of Whitehead's "method of extensive abstraction." This method, as developed in his *Enquiry Concerning the Principles of Natural Knowledge,* grows out of two purposes: to define meanings for "point" and "instant," thus giving the relational theory of space and time the exact mathematical formulation that its other adherents had neglected to provide; and to answer the question (of central importance for an empirical science) as to how the space and time of physics are rooted in experience. Only the first is relevant to the 1905 memoir, all epistemological questions being excluded from an investigation that is purely mathematical. But the memoir does show Whitehead's interest in mathematical physics, and shows him applying a logical technique to define for it ideal entities such as points.

iii

"On Mathematical Concepts of the Material World" was written in a period when much work was being done on the axiomatic formulation of geometry. (One thinks of Hilbert, Veblen, and the Italians.) Two points concerning geometry remain to be noticed.

1. Whitehead, as we might expect from the author of the *Universal Algebra,* chooses for his concepts of the material world sets of axioms that contain no reference to metrical ideas. Such ideas, which must of course appear in any geometry that is to meet the demands of physics, are subsequently introduced by definition, "according to the well-known method of projective metrics."[12]

2. Non-Euclidean geometries are not used. Whitehead adopts the natural course of taking the geometry as Euclidean throughout, and remarks that non-Euclidean structures can be obtained, if desired, by appropriate alterations in the properties of the fundamental polyadic relation from which the propositions of geometry spring in each concept.[13] However, a novel analysis of Euclidean parallelism is an essential part of Concept V as elaborated with the aid of the Theory of Interpoints and the Theory of Dimensions. One wonders whether this fact has anything to do with Whitehead's later choosing of a Euclidean

structure for the geometry of space-time in his work on the Theory of Relativity.

Turning now to physics, we should remember that in the late nineteenth century the mathematical theory of vectors and the physical concept of a field provided the means for tracing and integrating the propagation of vibratory energy through space. In the latter chapters of *Universal Algebra* Whitehead made contributions to vector theory. In the memoir we are considering, the term "vector" is not used, for a vector requires an origin and a length. Notice, however, that in order to provide a foundation for all geometrical concepts and as many physical concepts as possible in a single-based set of axioms, Whitehead in Concepts IV and V chooses for his base-class lines considered as simple entities. He notes that in this way the controversy as to how physical forces can possibly act at a distance is resolved by the fact that two distant particles possess linear ultimates in common.

In this memoir each of the five "Concepts of the Material World" is shown to be adequate for Euclidean geometry and for expressing the motion of matter. But the existence of relatively permanent "corpuscles," that is, units of matter such as electrons, is also considered.* Generally speaking, in the linear concepts a corpuscle is defined as a volume whose permanence is the persistence of some peculiarity of motion of the linear ultimates "passing through" it. Here, as in his later books on the philosophy of physics and on metaphysics, Whitehead has in mind the scientific effort to conceive of that which to common sense is a solid, enduring body as in reality a particular result of the aggregate operation of vector forces.

iv

This magnificent paper had even less influence on the development of physics than did the *Universal Algebra* on the development of mathematics. The mainstreams of both sciences flowed around Whitehead's depositions. "On Mathematical Concepts of the Material World" is only of historical interest. In his Preface to it he wrote that his Concept V required "the discovery of some appropriate laws of motion before it can be applied to the ordinary service of physical science." On the last page of the memoir he offered a few general suggestions as to how physicists might fit existing theories into the

*Not, however, quanta of radiation (Planck, 1901).

structure of this Concept. The advantages to be hoped for from success in this enterprise might well have made the attempt worthwhile.

> What is wanted at this stage is some simple hypothesis concerning the motion of [the linear ultimates] . . . and correlating it with the motion of electric points and electrons. From such a hypothesis the whole electromagnetic and gravitational laws might follow with the utmost simplicity. The complete concept involves the assumption of only one class of entities as forming the universe. Properties of "space" become simply the properties of this single class of entities.[14]

Then geometry would be created rather than presupposed.

Whitehead's knowledge of the psychology of research warned him that this ideal would not be carried out in the near future.

> In physical research so much depends upon a trained imaginative intuition, that it seems most unlikely that existing physicists would, in general, gain any advantage from deserting familiar habits of thought.[15]

But the chief reason why the attempt was not made seems to me to have been the nonexistence (even in the leading scientific country, Germany) of a sufficient number of theoretical physicists who were both interested in the axiomatic method and willing to work through the highly complex *Principia* symbolism of Whitehead's memoir.

V

"On Mathematical Concepts of the Material World" has been noticed by philosophers interested in Whitehead, but only in the last quarter century.[16] Of course, he did not write it for philosophers. He said in his Preface that he was discussing its general problem "purely for the sake of its logical (i.e., mathematical) interest." He then noted that the general problem "has an indirect bearing on philosophy by disentangling the essentials of the idea of a material world from the accidents of one particular concept."

The writing of the memoir marks one step in Whitehead's "large scheme of work," which I described in section iii of Chapter VIII. Apart from this, its most obvious characteristic is that it was written by one of the authors of *Principia Mathematica,* who was fired with a vision of the possibilities of mathematical logic and was applying it to new regions, geometry and physics. The proofs, Whitehead says, were mostly elaborated in symbolic form, then translated into words.[17]

The fundamental relation, in terms of which geometry and dynamics are expressed in each Concept, is itself specified by logical ideas only, though it is immediately given a physical interpretation. Consequently, the paper may be said to broach the idea that physics is one application of a logical system; or, alternatively, to exhibit the logical component of physics. This interpretation carries some risk of misconceiving the memoir. It is accurate only when applied to a logician's endeavor to state the conceptual structure of physics *as physics stands,* the existing science being called upon to supply the entities and relationships among entities which, when substituted for the logical variables, give the conceptual structure its important manifestation. But it is of the essence of Whitehead's work to propose for physics new entities and new relationships among entities. Mathematical logic is only the great instrument, which by its generalizing symbolism makes these novel possibilities visible. Thus the work is more than a logician's analysis of the structure of physics. It is also an excursion into mathematical cosmology.

Concerning the classical concept of the material world, it is important to realize that logic does not provide the reason for its logical inelegance. The situation, as Whitehead describes it after setting forth a modified form of Veblen's axioms of geometry, is this:

> Nothing could be more beautiful than the above issue of the classical concept, if only we limit ourselves to the consideration of an unchanging world of space. Unfortunately, it is a changing world to which the complete concept must apply, and the intrusion at this stage into the classical concept of the necessity of providing for change can only spoil a harmonious and complete whole.[18]

The integral inclusion of geometry—the perfect static science—within the world of change: that, once the three volumes of *Principia* were published, was to be Whitehead's favorite problem.*

When Whitehead wrote this memoir, he and Russell had done some work on geometry together, for in a letter dated April 30, 1905,

---

*Concept V offered a possible common platform for geometry and electromagnetic physics. Whitehead enjoyed pushing toward a greater generality than that which prevailed, thereby lifting theories out of their mutual independence. His career as a whole shows a kinship between this interest of Whitehead as a mathematician and the philosopher's interest in gaining synoptic vision. *Principia Mathematica* itself is a colossal endeavor to replace two sciences, mathematics and logic, by one.

Whitehead refers to "the existing state of our Geometrical mss." (Those manuscripts were probably attempts to express in symbols some of the nonpolemical sections of Part VI, "Space," of Russell's *Principles of Mathematics.*) In the memoir itself Russell is frequently cited in footnotes, but there is no indication that he wrote any of it, or that anything in it, other than Concept II, was suggested by him. It is Whitehead's work, not a joint product.

CHAPTER XV

Last Years in Cambridge

i/ *The Axioms of Projective Geometry* and *The Axioms of Descriptive Geometry*.

ii/ Memories of Whitehead as a teacher at this time.

iii/ Keir Hardie's visit to Cambridge. Whitehead's support of women's suffrage.

iv/ Leaving Grantchester. Whitehead's reaction to the death of friends and the illnesses of his children.

v/ The Forsyth affair. Other motives for the Whiteheads' move from Cambridge to London.

Whitehead's next publications were two short companion books, *The Axioms of Projective Geometry* (1906) and *The Axioms of Descriptive Geometry* (1907)—Numbers 4 and 5, respectively, in the series "Cambridge Tracts in Mathematics and Mathematical Physics." Their interest is almost entirely mathematical. They offer no original set of axioms. Yet they show their authorship. The axioms of projective geometry are presented only after a short first chapter on the nature of the axiomatic method and an original definition of geometry in general. On the former subject, this chapter deserves recognition as a good compressed account, still worth reading. In it Whitehead follows Russell, saying:

> The axioms are not propositions at all: they are propositional functions. . . . The existence theorem for a set of [geometrical] axioms is the proposition that there are entities so interrelated, that the axioms become true propositions, when the points are determined to be these entities and the relations between points to be these interrelations.

Exhibition of an existence theorem, Whitehead notes, seems to be the only way of proving the consistency of a set of axioms. His comment on this:

> But the only rigid proofs of existence theorems are those which are deductions from the premises of formal Logic.* Thus there can be no formal proof of the consistency of the logical premises themselves. This is only one instance of the absolute distinction between the premises of Logic, which are necessary for reasoning itself, and the axioms of various mathematical subjects, such as Geometry.

That "absolute distinction" was, for Whitehead, a practically necessary addition to Russell's logicist thesis.

The definition of geometry that Whitehead offers is "the science of

*This was also Russell's position in *P of M,* Secs. 360, 413, 474. See Whitehead's *APG,* Sec. 43.

cross-classification," in contrast to the classification of things into mutually exclusive genera and mutually exclusive species. When he repeated this definition in his late philosophical book, *Adventures of Ideas,* he remarked that other sciences of cross-classification "have not been developed, partly because no obvious applications have obtruded themselves, and partly because the abstract interest of such sciences has not engaged the interest of any large group of mathematicians."[1] The remark is typical of Whitehead's outlook. His definition of geometry accords with the view that as pure mathematics it has no determinate subject matter in the usual sense of the term, but deals merely with types of relation, and applies to any entities whose interrelations satisfy the formal axioms. However, in the axioms of classification that are set out in this tract the undefined ideas are those of "a point" and "a [straight] line joining a point *A* with a point *B*."[2]

Whitehead's way of dividing geometries into "projective" and "descriptive" is rather unusual among mathematicians. Russell had used it in his *Principles of Mathematics,* where he defined geometry as "the study of series of two or more dimensions."[3] A geometry is called "projective" if two coplanar lines necessarily intersect, "descriptive" if they do not.[4] The subject matter of Whitehead's first tract is easily seen to coincide with "geometry of position," or "projective geometry" in the usual meaning of the phrase. The set of axioms used is mainly due to Pieri. Axioms of order follow those of classification. The latter part of the tract stresses the proof that numerical coordinates can be defined without introducing distance as a fundamental idea. "The establishment of this result," Whitehead says in his Preface, "is one of the triumphs of modern mathematical thought. It has been achieved by the development of one of the many brilliant geometrical conceptions which we owe to the genius of von Staudt." There he also refers the reader to Russell's book for "the various logical and philosophical enquiries" suggested by the principles of projective geometry.

The Axioms of Descriptive Geometry deals first with what others might call affine geometry; Veblen's axioms, for which the undefined ideas are "point" and "between," are mainly used, though others are discussed. The latter half of this tract is written round the theory of transformation groups, and emphasizes Sophus Lie's work on the analysis and definition of congruence, as superior to using congruence as an undefined idea.

Whitehead tries to fit the expositions in both tracts to the preparation of students, already introduced to the subject, for reading the detached writings on it. Accordingly, the effort is not to approximate exhausting the possibilities of a set of axioms, but to lead the reader to see when a fresh axiom is required if a particular proposition is to be proved. I am not sure that the tracts alone, without a teacher, accomplish their educational purpose—because of their extreme compression and the absence of external aids such as boldface type. The reviewer for the *Mathematical Gazette*[5] complained of this, and of Whitehead's frequent introduction of new nomenclature in place of what he expected.

Recall that in the *Treatise on Universal Algebra* Whitehead expressed his belief that a spatial interpretation will be possible for the symbolism of each of the new algebras. This interesting fact, he wrote there, "deserves investigation,"[6] as it makes a treatise on Universal Algebra "also to some extent a treatise on certain generalised ideas of space. In order to complete this subsidiary investigation, an appendix on a mode of arrangement of the axioms of geometry is given at the end of this volume."[7] The appendix is not in the volume, presumably because it had not been completely written. Should the pair of "Cambridge Tracts" be considered a greatly expanded substitute for it? I think not. Neither tract contains any reference to the *Treatise* or to these algebras. And their writing was arranged for independently.

A condensed account of both geometrical tracts, together with some interesting philosophical discussions, may be found in Whitehead's article "The Axioms of Geometry," which appeared in the eleventh edition of the *Encyclopaedia Britannica*. It is Division VII of the many-authored article "Geometry."[8] Division VI, "Non-Euclidean Geometry," was jointly written by Whitehead and Russell; it calls for no comment here.

"The Axioms of Geometry" begins by noticing the controversy between adherents of the traditional absolute theory and the relational theory of space. Whitehead defines the former as asserting, "It is not intrinsically unmeaning to say that any definite body occupies *this* part of space, and not *that* part of space, without reference to other bodies occupying space." After noting Newton's position and the fact that the impossibility of measuring any but relative motion "has led to the general adoption of the relational theory," Whitehead says, "But no decisive argument for either view has at present been elaborated." He classes Kant as an upholder of the absolute theory. In another respect he criticizes Kant: we ought not to say, with him, "Space is represented as an infinite given *quantity*,"[9] which implies that measurability is essential to the very idea of space. In the light of Lie's work, we ought now to say, rather, that "systems of quantities can be found [as properties of congruence-groups] in a space."[10]

ii

Descriptions of Whitehead's teaching can be found in *A Mathematician's Miscellany,* which was written by a pupil who became distinguished, J. E. Littlewood.[11] He recalls that when he entered Trinity College in 1903 Cambridge inspired awe, but that for one who had been well trained at school, the mathematical lectures given under the Tripos system were a waste of his time, with two exceptions that "touched enthusiasm." One of these was "a stimulating course in the first term by A. N. Whitehead on the foundations of mechanics."[12] In a 1965 interview, Professor Littlewood told me he had then "formed the view that however deeply you went into Foundations of anything you still ended in mush. It was a revelation to find that with the right technique it could all be clear-cut." He also remembered that Whitehead's stock lecture courses on statics and dynamics were "routine things done very well—straight lectures without frills." In the book Littlewood says, "His stock College lectures . . . were solid and unexciting affairs on applied mathematics: a mathematician can have the duty of being dull."[13] He narrates that in his fourth year there were few courses left for him to take. "What I did go to were courses by Whitehead on foundations of geometry and foundations of mathematics, given for the first time. There were three or four of us in the class and we found them very exciting."[14]

Another distinguished former pupil whom I interviewed was A. V. Hill—not professionally a mathematician but a physiologist; at an early age he won the Nobel Prize. Hill took the Mathematical Tripos two years after Littlewood. He remembered Whitehead's course of lectures on geometrical optics unfavorably. Whitehead seems to have deduced optical conclusions from abstract principles; Hill, a practical man, was not interested in that, since the real foundation of the conclusions was empirical. In another interview, the physicist Geoffrey Taylor, who heard the same lectures, disagreed with Hill; the subject, after all, was *geometrical* optics, and he found the lectures interesting. Taylor (and Littlewood) remembered Whitehead as making his own presentation, not following a textbook.

Hill frequently saw Whitehead outside the classroom and came to have great affection for him. Littlewood, who saw less of him, told me that he found Whitehead "a bit of an enigma," and could not analyze his complicated personality. "I never felt I understood him." Whitehead's mind, however, was another matter: his mind was "simple, in a good sense of 'simple,' the sense in which we say that Einstein was simple in respect to abstract intellectual questions."

My notion is that Whitehead's character was sweet and highly civilized; his universal helpfulness inspired universal affection. Moreover, he had a quality that is unique in my experience of men—an almost complete absence of egoism, of impulse to push himself forward; because of this, his actions were often not what ordinary mortals expected, so he appeared enigmatic. His mind, I think, was simple in the sense that Littlewood perceived. It was very precise, but more given to reorganizing and generalizing familiar ideas than to solving stock problems with them. So he was a stimulating teacher to young mathematicians, but not himself in the mainstream of mathematical research. Littlewood mentioned Whitehead's absorption in "that rather cranky subject of universal algebra."[15]

In appearance he was benign and smiling; some called him "the cherub." All of his former pupils that I talked to remember him, in the years 1905 to 1910, as already stooped, fairly bald, and looking venerable; one, I was told, referred to him as "the dear old man." He was not yet fifty. The habit of bending his head over sheets of paper on a table was responsible for stooped shoulders, and I dare say that the strain of constant self-discipline and of life with Evelyn had something to do with the premature look of age. The Whiteheads had a large silver coffeepot and a smaller silver teapot. The children called their mother the coffeepot, their father the teapot.

iii

The most dramatic event at Cambridge in this period was Keir Hardie's attempt to make a speech in the Guildhall on February 16, 1907; the Whiteheads were there.

I suppose that Hardie is unknown to my American readers, and that in Britain he is now a legendary figure, alive only to persons familiar with such books as the memoirs of Hugh Dalton. Hardie, a contemporary of Eugene Debs, was a coal miner who became a union organizer, then founder of the Scottish Labor Party and the first workers' representative elected to Parliament. In 1901 he had made the first no-private-ownership-of-land-or-capital declaration in the House of Commons. Now the fighting leader of the Socialist-Labour Group there, he was invited to speak to the public in Cambridge by the local

Fabian Society. Hardie, Dalton wrote, "was a red rag to all Tory bulls."[16] An attempt was made to prevent the speech by kidnapping him. When this was foiled by a disguise, an army of opponents, mostly undergraduates, stormed the Guildhall. Hardie tried to state the case for the new Labour Party, but the meeting was soon broken up by boos, catcalls, rotten eggs and oranges, and stink bombs (hydrogen sulfide).

The University Fabian Society was strong at Trinity and King's, but I doubt that Whitehead was a member. Never a Tory, he was a Liberal who wanted the new Party to have a fair hearing. Evelyn was more radical than he was; possibly he went to the meeting as much to support her as to support Hardie. In 1941 Whitehead wrote, "My wife and I were on the platform, sitting behind him, and there was a riotous undergraduate audience. The result was that any rotten oranges that missed Keir Hardie had a good chance of hitting one of us."[17] There is a good chance that Whitehead's memory confused this meeting with some political meeting in Grantchester or a neighboring village, where he spoke in parliamentary campaigns and was often pelted;* the account of the Hardie meeting in the Cambridge *Chronicle* mentions no senior people of the University except the Senior Proctor, and puts trade union people on the platform. But I do not doubt that the Whiteheads went to the meeting.

Another political cause attracted their unqualified support: votes for women. It had recently become a great issue. While Emmeline Pankhurst and her daughters captured headliness, many men and women, believing that the militant—not to say violent—tactics of these Suffragettes hurt the cause, worked peacefully for it. They called themselves Suffragists. On November 5, 1906, Whitehead spoke to the annual meeting of the Cambridge Women's Suffrage Association, which then published an extract from his speech as a pamphlet entitled *Liberty and the Enfranchisement of Women.*[18] The pamphlet is a vigorous, but careful, statement of the basic argument for women's suffrage and a conclusive answer to the various arguments (including Asquith's) then heard against it. On February 4, 1907, Bertrand Rus-

---

*See Chapter X, section v. To the account given there, it should be added that when Edwin Montagu first ran for Parliament, from West Cambridgeshire in the General Election of 1906, Whitehead spoke for him in the villages of the constituency. Jessie and North were in the audience when Montagu opened his campaign in Grantchester, and were appalled that he addressed the villagers as if they were on the intellectual level of the nursery rhyme "Old Mother Hubbard." Their father then coached him. Whitehead believed that his activities secured Montagu's election.

sell wrote to Lucy Donnelly,* "Whitehead has blossomed out into an orator, and is one of the speakers at a great Woman's Suffrage meeting in Exeter Hall this week." It was quite impossible—and Russell knew it was impossible—for Whitehead's platform manner ever to be that of an orator. You saw that he felt strongly about the question, but his manner was that of a man setting out fundamental facts and their logical consequences. He demolished opposing arguments, briefly and simply.

A national Men's League for Women's Suffrage was now founded; Whitehead became chairman of its Cambridge branch. In May 1907 Russell stood for Parliament in a by-election, not as a Liberal (which he was) but as the first candidate of the National Union of Women's Suffrage Societies. The National Union had asked him to do this just two weeks before election day, and as he believed strongly in women's suffrage he had agreed, though he knew he had no chance of winning. In July the Council of the National Union held its quarterly meeting in Cambridge; Evelyn gave a garden party for the members of the Council.[19]

iv

In April 1907 the Whiteheads left the old Mill House in Grantchester for a house in Cambridge. Cranmer Road was on the western edge of the town; it ran off the Grange Road and ended in fields. Russell wrote to a friend that all the family were glad of the move. Evelyn turned 11 Cranmer Road into a charming house, beginning as usual by knocking down a wall to turn two rooms into one, large enough for her to hold a salon and to have students in for Sunday teas.

In their last years in Cambridge several unhappy events took place. Some of them were of the kind that happen to almost everybody, but one was not. Since his happy years at Sherborne School, Whitehead loved the place. John Blanch, married to his sister, Shirley, had for many years been a successful housemaster and a good mathematics master there, perhaps Whitehead's teacher.† He was a good teacher, and a kind man who did many things to help people. As he approached old age he fell into depression, and one day in January 1907

*Lucy Donnelly was on the faculty of Bryn Mawr College; she had been Russell's friend since his first visit to the United States in 1896.

†See Chapter IV, section vii.

disappeared. Several days later his body, throat cut from ear to ear, was found in a wood.[20] At the inquest the verdict was suicide. Whitehead came down, was the chief support of the family, and tried to tend to affairs. (Henry was in India, and Charles could not be of much help.)

The previous summer, Whitehead had made a quick trip from Grantchester to Edinburgh to make arrangements for a corpse there. His neighbor Evelyn Shuckburgh, Fellow of Emmanuel College and one of his dearer friends,* had died of a heart attack on a train en route to Edinburgh.

In 1906 there was illness in Whitehead's own family, and his closest mathematical friend, Forsyth, was seriously ill for months. On March 26 Russell wrote to Margaret Llewelyn Davies,† "Forsyth is mending fast, and so are Tiny [Evelyn's nickname for Jessie] and Goo [Eric]. Forsyth . . . has been a close friend of Alfred ever since his undergraduate days. It must have been a very severe strain both for Alfred and for Evelyn."

In the spring of 1907 influenza struck the family. The most dangerous illness that any of them had in the course of their years in England occurred in the summer of 1908: North caught scarlet fever while on holiday in Switzerland; his recovery took over a year.

Whenever his wife or one of the children became ill, Whitehead was more upset than most men would be. Apart from that, Evelyn saw to it that Christmas was elaborately celebrated in the family. On December 10, 1908, Whitehead wrote to his collaborator, "You had better not come before the New Year. We are full of detail now in respect to the children and Xmas, etc. Also I must give them my spare time."

V

Early in 1910 something happened outside the family which was one of the causes of their quitting Cambridge for London. Cambridge was rocked by the Forsyth affair. Andrew Forsyth, that prim, correct man, in his early forties had fallen in love, and the lady with him. Amelia was his own age, and of excellent family. Unfortunately she was a married woman. Her husband, who had ignored her for years,

---

*See page 202 and footnote.

†Margaret was the sister of Crompton, Theodore, and Arthur Llewelyn Davies, and a good friend of the Whitehead family.

was the physicist C. V. Boys, F.R.S. The A's eloped,* and Boys started divorce proceedings. In the clerical-moral atmosphere that prevailed, it would surely have been demanded that Forsyth resign his professorship of pure mathematics. That required residence in Cambridge. His Trinity College fellowship was a different matter. Like Whitehead, he had held it long enough to be entitled to hold it for the rest of his life, wherever he chose to live. Forsyth was emotionally attached to the College, and had served it well. What should the poor, scrupulous man do? Whitehead urged him to sit tight. But Forsyth, rather precipitously, resigned both chair and fellowship, and prepared to leave Cambridge.

Whitehead was now again a member of the Council of Trinity. When Forsyth's letter of resignation was read on February 4, Council agreed to consider it a week later. At that fateful meeting, a motion to ask Dr. Forsyth to reconsider got nowhere; so did a motion to adjourn the question for a term. When the resignation was accepted by an 8-to-4 vote, Whitehead angrily left the meeting.[21] This was not like him, but he did have a flash point.

A protest followed: twenty-seven Fellows who were not on the Council told it that they "deeply regret the action of the Council in accepting immediately the resignation of his fellowship by Dr. Forsyth."[22] This did not change anything.

In late April the Trinity Council received a letter from Whitehead resigning his Lectureship as of September 30 next. This was the Senior Lectureship to which he had been appointed for a ten-year term in 1903. The College Statutes specified twenty-five years as the maximum period of service for a Lecturer, but also empowered the Council to extend the tenure for five years at a time, and in the past decade this had been done in several cases. Whitehead began to teach at Trinity, on a temporary basis, in March 1884, but the first annual appointment he received took effect in October 1885. Someone had asked the Council to determine how many years Whitehead had been a Lecturer within the meaning of the Statutes. At the meeting following its acceptance of Forsyth's resignation, Council decided, "continuously since 24th October, 1884."[23] (The grounds of this determination are obscure to me.) This meant that Whitehead had completed twenty-five years of service in October 1909 and nothing had been done about it. Nothing needed to be done now.†

---

*Subsequently they married. Adultery was then the only ground for divorce in England.

†The Council did, however, have to consider a proposal that Whitehead submitted to

The Whiteheads moved to London in the summer. He had not lined up a job there. Until he got one, he would be sacrificing his Trinity Lecturer's stipend, a little over £500 a year. He had not resigned his Fellowship, and kept it all his life. It was now giving him about £300. He probably had a little outside income, but I have been unable to discover any source of sizable amounts.

He and Evelyn were entirely in sympathy with Forsyth. Evelyn thereafter represented their move to London as made out of indignation. That made it dramatic. (Her stories always were.) There is evidence of other motives. In a hurried letter of application for a professorship which he wrote to the Provost of University College, London, on March 16, 1912, Whitehead said that he left Cambridge "chiefly from a desire to preserve by a change to the more varied life of London the necessary freshness for my work."[24] This sounds laid on, but I am sure that the desire mentioned was a weighty factor. When, in March 1941, long before I knew of this letter, I asked Whitehead for his explanation of the move, he answered, "I was in a groove in Cambridge."

Year after year he had been teaching the same courses in applied mathematics to the same sort of students. He was too smart not to know that whatever creative ability he had was not in applied mathematics. I think he dimly felt that he had a kind of unused genius which in Cambridge would go on being unused, stultified, never felt out and discovered for what it might really be. What he clearly felt was that he had some capabilities for which there was no scope in Cambridge. His chief interest, after mathematics, was education. He had already sketched out the *Introduction to Mathematics,* which he published the next year, a year in which he delivered the first of his addresses on what education in mathematics should be. More broadly speaking, London was in the twentieth century, Cambridge was not. Cambridge had shown this in its almost solitary refusal to grant degrees to women and in its shameful treatment of Keir Hardie. The future of England—something always on Whitehead's mind—depended on scrutinizing, then improving, the education of the masses in cities like London. In London direct opportunity to exert a guiding

---

it early in May. It was that Russell be elected to a research fellowship, with the duty of giving a course of lectures each term. Whitehead listed Russell's achievements in a lengthy memorandum, and laid out the evidence of his high reputation in logic and philosophy (especially, but not only, philosophy of mathematics). Whitehead made a strong case, but the Council appointed Russell only a Lecturer, not a Fellow. Russell believed that this was because a Fellow has a voice in governing the College, and he was a freethinker.

influence on this process was greatest. In London he would meet the real world, find out what it was at, and tell it some of the things about education that he had been feeling out, but in the worst way in the world to feel such things out—at a distance, without a real audience. Friends who lived in that real world, like Crompton Davies, urged him to move to London. He really wanted to.

Of course he talked about these things with dear Evelyn. She was sympathetic. Alfred habitually worried about money, but she had always had real courage about taking financial chances. And if worst came to worst, she could sponge on a friend. No one now alive knows whether Bertie's surreptitious help ended when the manuscript of *Principia Mathematica* was finished.* If it had ended she could try, not without hope of success, to get it renewed should that become necessary. In a pinch, she could sponge on some other friend, such as the wealthy Henry Beeton. Money matters should not deter Alfred from resigning his lectureship and moving to London.

The idea appealed to her taste for adventure. Besides, she was happiest when going to the theater. Think of the infinite superiority of the West End to Cambridge. How dull Cambridge was! She had put up with it for twenty years, but enough was enough.

Evelyn had not been much of a success in Cambridge society. According to her lights she had been well behaved, but she was out of her element, and had not been treated as "in." Unlike Mary Paley Marshall, she had not been invited to join the exclusive Ladies Dining Society founded by Mrs. Creighton. Evelyn was not the favorite friend of Lady Darwin or Lady Jebb, top of Cambridge society. She would have the Jebbs to dinner when her husband suggested it,[25] but she did not pay calls as dutifully as others did. She did not have the excuse of being an active intellectual, and would have been bored by the company of many don's wives, but it was acid in her soul to be the refused and not the refuser. In London there were real people.

In sum, for both Alfred and Evelyn the action of the Trinity Council on Forsyth's resignation, that so angered them, was primarily the catalyst in a mixture of feelings that had been growing in their minds.

A full expression of indignation would have occurred if Whitehead had resigned not only his Lectureship but his Fellowship. He, and Evelyn even more clearly, saw that this would have been unrealistic for their purpose, which was not so much to damn the Council as to leave Cambridge for new life and work in London.

---

*See Chapter XI, section v.

Notes

In these notes and the footnotes, the full-reference/shortened-reference system of citation has generally been used. Frequently cited works by Whitehead have been abbreviated as noted in the Bibliography. The frequently cited works of others have been abbreviated as follows:

Levy, *Moore*	Paul Levy, *Moore: G. E. Moore and the Cambridge Apostles* (London: Weidenfeld & Nicholson, 1979).
Price, *Dialogues*	Lucien Price, ed., *Dialogues of Alfred North Whitehead* (New York: New American Library, Mentor, 1956).
Russell, *MPD*	Bertrand Russell, *My Philosophical Development* (New York: Simon & Shuster, 1959).
Russell, *P of M*	Bertrand Russell, *Principles of Mathematics,* 2d ed. (unaltered) (Cambridge, 1903).
Russell, *Portraits*	Bertrand Russell, *Portraits from Memory and Other Essays* (New York: Simon & Shuster, 1963).
Russell, *Abiog.*	*The Autobiography of Bertrand Russell,* Vol. I, *1872–1914* (London: George Allen & Unwin, 1967).
Russell, W&*PM*	Bertrand Russell, "Whitehead and *Principia Mathematica,*" *Mind,* n.s., 57 (1948): 137–38.
Schilpp, *LLP-W*	Paul A. Schilpp, ed., *Library of Living Philosophers,* Vol. IV, *The Philosophy of Alfred North Whitehead* (Evanston: Northwestern University Press, 1941).
Thomson, *R&R*	J. J. Thomson, *Recollections and Reflections* (London: G. Bell & Sons, 1936).

I. Introduction

1. Russell, *Portraits,* p. 100. The passage is from his 1952 BBC talk on Whitehead.
2. C. D. Broad, in *Mind* 57 (April 1948): 145.
3. *PR* I i ii.
4. Ibid. V ii ii.
5. Victor Lowe, *Understanding Whitehead* (Baltimore: Johns Hopkins Press, 1962).

II. The Whiteheads

1. Abiog. Notes, in Schilpp, *LLP-W,* p. 3. Whitehead wrote this in 1941.

2. T. North Whitehead, "Now I Am an American: Habit and Change." This autobiography was written ca. 1960. It was copyrighted in 1966 but never published.

3. These recollections of Thomas Whitehead's son Henry, who was a pupil at Chatham House Academy, were reported by H. D. Rawnsley in *Henry Whitehead, 1825–1896: A Memorial Sketch* (Glasgow, 1898).

4. Ibid.

5. A photograph of the portrait was supplied to me by the Ramsgate Public Library (see p. ix above). It was taken just after the artist, W. M. Tweedie, finished his work. At the end of the Hitler war, in which all portraits were removed from the Town Hall (a building now replaced), this one could not be found.

6. Thomas C. Whitehead, *Village Sketches, descriptive of Club and School Festivals and other Village Gatherings and Institutions* (London, 1861).

7. Peter Burn, *Reminiscences of the late Rev. Henry Whitehead: A Noncomformist's tribute to a Churchman* (Carlisle, 1899).

8. Rawnsley, *Henry Whitehead.*

9. Ibid., p. 12.

10. Most of Henry Whitehead's studies were published in *Transactions of the Cumberland and Westmorland Antiquarian and Archeaological Society.*

11. A list of Uncle Henry's writings is published as an appendix in Rawnsley's *Henry Whitehead.* Another appendix in that volume prints verbatim Whitehead's farewell speech at the Rainbow Tavern in 1874, "The Experience of a London Curate."

12. Alfred North Whitehead, "England and the Narrow Seas" (1927), in *ESP,* p. 48.

13. This description .s from the first chapter of T. North Whitehead's unpublished autobiography, "Now I Am an American." Of course the words primarily convey what North heard from his parents. North was six years old when his grandfather died. Born and brought up in Cambridge, he saw much of his grandfather during Alfred North Whitehead's academic vacations.

14. Ibid.

15. Russell, *Portraits,* p. 99.

16. Whitehead, "England and the Narrow Seas," in *ESP,* p. 49.

17. Bishop Henry Whitehead to Alfred North Whitehead, September 13, 1927.

18. Whitehead, "England and the Narrow Seas," in *ESP,* pp. 48ff.

19. Ibid., p. 48.

20. *Times* (London), March 22, 1898, p. 10.

21. So Whitehead says twice in Price's *Dialogues:* in XXXVI (January 19, 1945) and in XXX (June 19, 1943). Either the numbering on Piccadilly has changed or the park was St. James's.

22. Ibid., XXXVI (January 19, 1945).

23. Ibid., XXX (June 19, 1943). Mrs. Buckmaster's family were gentry, not "in trade."

24. Herbert Buckmaster, *Buck's Book: Ventures, Adventures, and Misadventures* (London: Grayson & Grayson, 1937).
25. Ibid., p. 13.
26. Ibid., p. 126.
27. Ibid.
28. Ibid., p. 282.
29. Ibid., p. 82.

III. Childhood

1. Price, *Dialogues,* XIV (February 27, 1939).
2. Whitehead, "England and the Narrow Seas," in *ESP,* p. 41.
3. Ibid., pp. 41–42.
4. Ibid., pp. 44–45.
5. Ibid., pp. 20–21.
6. Report to Victor Lowe from Professor John V. Kelleher, a Junior Fellow in the early 1940s.
7. Abiog. Notes, in Schilpp, *LLP-W,* p. 5; or *ESP,* p. 5.

IV. Sherborne

1. H. D. Harper, headmaster of Sherborne, in a letter to the Secretary of the Royal Commission studying the public schools in 1868; quoted in L. V. Lester, *A Memoir of Hugo Daniel Harper, D. D.* (London, 1896), p. 87.
2. Ibid.; Harper's italics.
3. C. E. Pascoe, *A Practical Handbook to the Principal Schools of England* (London, 1877), p. 131.
4. Alfred North Whitehead, "The Education of an Englishman," *Atlantic Monthly* 138 (1926): 192–98; reprinted in *ESP.*
5. Whitehead, "The Education of an Englishman," in *ESP,* pp. 30ff.
6. Ibid., p. 31.
7. Vivian Ogilvie, *The English Public School* (London: Batsford, 1957), p. 8.
8. A. B. Gourlay, *A History of Sherborne School* (Winchester: Warren, 1951), p. 126.
9. Mrs. Ernest Stewart Roberts, *Sherborne, Oxford, and Cambridge: Recollections* (London: M. Hopkinson, 1934).
10. *ESP,* p. 37.
11. Price, *Dialogues,* XXIX (June 10, 1943).
12. William Ernest Hocking, "Whitehead as I knew Him," *Journal of Philosophy* 58 (1961): 512.
13. Ibid.
14. Ibid.
15. *ESP,* p. 38.
16. Gourlay, *Sherborne School,* p. 216.
17. Arthur Waugh, Preface to *A Cricket Eleven,* ed. R. H. Lowe (London: Gerald Howe, 1927), p. 10.
18. Ibid.
19. C. E. Pascoe, *Where Shall I Educate My Son?* (London, 1884), p. 50.
20. Lester, *Memoir of Hugo Daniel Harper,* p. 208.

21. "'No, No!' he used to say, 'keep them all out!'" Ibid., p. 221.

22. Further, see Arthur Waugh, *One Man's Road* (London: Chapman & Hall, 1931), pp. 88ff., 96ff.

23. G. G. Bradley, *Macmillan's Magazine,* March 1870, p. 406.

24. Dr. James W. Felt, S. J., to Victor Lowe, November 16, 1973.

V. Student at Cambridge

1. See, e.g., Thomson, *R&R,* p. 58.

2. Whitehead's copy was found by an American, Mr. Harlan L. Umansky.

3. Abiog. Notes, in Schilpp, *LLP-W,* p. 7; or *ESP,* p. 7.

4. Sorley was one of a group of nine who published *Essays in Philosophical Criticism* (London, 1883), a volume dedicated to T. H. Green.

5. Hermann Müller, *The Fertilization of Flowers,* trans. and ed. D'Arcy Wentworth Thompson, with a Preface by Charles Darwin (London, 1883).

6. *NL,* p. 24; or *MT,* p. 203.

7. Thomson, *R&R,* p. 54.

8. This is emphasized in the last third of Sheldon Rothblatt's book about Victorian Cambridge, *The Revolution of the Dons* (London: Faber & Faber, 1968).

9. The universality of chaperonage is beautifully illustrated in the delightful book *Period Piece,* by George Darwin's daughter, Gwen Raverat (London: Faber & Faber, 1952). It depicts the mode of life that was still *de rigueur* in the Cambridge of her childhood, ca. 1890.

VI. Mathematics at Cambridge

1. Thomson, *R&R,* pp. 56ff.; A. R. Forsyth, "Old Tripos Days at Cambridge," *Mathematical Gazette* 19 (July 1935): 162ff.; Karl Pearson, "Old Tripos Days at Cambridge as Seen from Another Viewpoint," ibid. 20 (February 1936): 27ff.

2. Thomson, *R&R,* p. 44.

3. Ibid., pp. 42–43.

4. J. L. (presumably Joseph Larmor), Obituary of W. D. Niven, *Proceedings of the London Mathematical Society* 16 (1917).

5. Lecture notes by Dr. Louise Heath, then a Radcliffe student.

6. W. W. Rouse Ball's *Short Account of the History of Mathematics* (London, 1888), is not very short.

7. W. W. Rouse Ball, *Mathematical Recreations and Essays* (London, 1892).

8. E. T. Whittaker's opinion in the obituary of W. W. Rouse Ball, *Mathematical Gazette* 12 (1924–25): 449–54. Ball did not neglect the present or the future of the College, to which he was devoted. He became the beneficent chairman of its important Education Committee. When he died Trinity received a large bequest.

9. Thomson, *R&R,* p. 48.

10. E. J. Routh, *Treatise on the Dynamics of a System of Rigid Bodies,* 4th ed. (London, 1882), p. vii.

11. *Times* (London), June 11, 1907, p. 10.

12. Whitehead, discussion at the meeting of the Physical Society of London, February 22, 1924, as reported in the Society's *Proceedings* 36 (1924): 193.

13. Thomson, *R&R,* p. 42.

14. Ibid., pp. 35–42; Pearson, "Old Tripos Days at Cambridge," pp. 28–29. There are also descriptions of Routh's coaching in Forsyth, "Old Tripos Days at Cambridge," pp. 73–74; and in the obituaries of Routh by Joseph Larmor (Senior Wrangler, 1880), *Proceedings of the Royal Society of London,* ser. A, 84 (1907): 94, and by H. H. Turner (Second Wrangler, 1882), *Monthly Notices of the Royal Astronomical Society* 68 (February 1908).

15. Thomson, *R&R,* pp. 40–41.

16. E. J. Routh, *A Treatise on the Stability of Motion, particularly Steady Motion* (London, 1877).

17. Leslie Stephen, "Early Impressions," *National Review,* September 1903, p. 131.

18. Pearson, "Old Tripos Days," p. 27.

19. Russell, *MPD,* p. 43.

20. This is most readily found in Whitehead's *AI,* Chapter XI. Section 16 of the address is entitled "The Flux of Energy."

21. J. J. Thomson, ed., *A History of the Cavendish Laboratory, 1871–1910* (Cambridge, 1910), p. 87.

22. Price, *Dialogues,* XL (June 19, 1945).

VII. The Cambridge Apostles

1. This seems to be true, at least with regard to Britain. Levy's remark is taken from his review of Peter Allen's book, *The Cambridge Apostles: The Early Years* (Cambridge University Press, 1979). The review is in *Books and Bookmen* 24, no. 6 (March 1979): 28.

2. Abiog. Notes, in Schilpp, *LLP-W,* p. 5; or *ESP,* p. 5.

3. Levy, *Moore.* I am particularly indebted to Levy for his Appendix, "The Apostles' Elections up to the First World War."

4. Arthur Sidgwick and Mrs. E. M. Sidgwick, *Henry Sidgwick: A Memoir* (London, 1906), p. 34.

5. *Autobiography of Dean Merivale,* ed. Judith Anne Merivale (London, 1889), p. 80. Charles Merivale was an Apostle at the same time as Tennyson and Hallam.

6. Abiog. Notes, in Schilpp, *LLP-W,* p. 8.

7. Sidgwick and Sidgwick, *Henry Sidgwick,* p. 29.

8. The title of this paper was "Can Faith remove mountains?" See W. R. Sorley, "James Ward," *Mind* 34 (July 1925): 277.

9. Abiog. Notes, in Schilpp, *LLP-W,* pp. 7–8; or *ESP,* p. 8.

10. Levy, *Moore,* p. 76.

11. H. H. Turner to Bertrand Russell, June 19, 1916.

12. For more on Raleigh's "Laughter from a Cloud," see the account in Levy's *Moore,* pp. 91–92.

13. Michael T. Holroyd, *Lytton Stachey,* 2 vols. (New York: Harcourt, Brace & World, 1968), 1:101.

14. E. M. Forster, *Goldsworthy Lowes Dickinson* (London: E. Arnold & Co., 1934), p. 66.

15. Levy, *Moore,* p. 95.

16. Forster, *Dickinson,* p. 104.

17. Lewis S. Feuer to Victor Lowe, May 5, 1970.
18. Mary Agnes Hamilton, *Remembering My Good Friends* (London: Jonathan Cape, 1944), pp. 40–41.
19. Forster, *Dickinson,* pp. 203–4.
20. *ESP,* p. 116.
21. See Avrom Fleishman, "Woolf and McTaggart," *ELH* 36, no. 4 (December 1969).
22. Virginia Woolf, *Roger Fry: A Biography* (New York: Harcourt, Brace & Co., 1940), p. 52.
23. *Letters of Roger Fry,* ed. Denys Sutton, 2 vols. (London: Chatto & Windus, 1972), 1:95.
24. Ibid., 1:115.
25. Woolf, *Roger Fry,* p. 259.
26. Ibid., p. 51.
27. Victor Lowe, "The Development of Whitehead's Philosophy," in Schilpp, *LLP-W.* See also Lowe, *Understanding Whitehead,* Pt. II.
28. Abiog. Notes, in Schilpp, *LLP-W,* p. 7; or *ESP,* p. 7.
29. Ibid.

VIII. The Young Mathematician

1. Abiog. Notes, in Schilpp, *LLP-W,* p. 8.
2. Price, *Dialogues,* XXX (June 19, 1943).
3. *Cambridge Review,* February 6, 1884, p. 167.
4. Minutes of the Council of Trinity College, January 23, 1885.
5. Rawnsley, *Henry Whitehead,* p. 7.
6. Hermann Günther Grassmann, *Die lineale Ausdehnungslehre, ein neuer Zweig der Mathematik* [The theory of linear extension, a new branch of mathematics] (Leipzig, 1844).
7. Abiog. Notes, in Schilpp, *LLP-W,* p. 9.
8. Clerk Maxwell, *Treatise on Electricity and Magnetism* (Oxford, 1873), Secs. 526, 687.
9. Hermann Günther Grassmann, "Neue Theorie der Elektrodynamik," *Annals of Physics* (Poggendorf) 64 (1845).
10. *UA,* pp. x, 346.
11. Homersham Cox, "On the Application of Quaternions and Grassmann's *Ausdehnungslehre* to Different Kinds of Uniform Space," *Transactions of the Cambridge Philosophical Society,* 1883.
12. W. K. Clifford, "Applications of Grassmann's Extensive Algebra," *American Journal of Mathematics* 1 (1878): 350–58.
13. Victor Lowe, "A. N. Whitehead on His Mathematical Goals: A Letter of 1912," *Annals of Science* 32 (1975): 85–101.
14. Further, see Michael J. Crowe, *A History of Vector Analysis* (Notre Dame, Ind.: University of Notre Dame Press, 1967), p. 139.
15. Volume I of *UA* was published by Cambridge University Press in 1898. No other volumes were published.
16. *Quarterly Journal of Pure and Applied Mathematics* 23 (1889): 78–93, 143–52. The journal was then edited at Cambridge (by Cayley, Glaisher, Forsyth, and N. M. Ferrers). The year of publication was wrongly given as 1888 in my

bibliography of Whitehead's writings, published in Schilpp, *LLP-W.* I am obliged to Professor Patrick J. Hurley for calling my attention to this error.

17. *PR,* p. 13; corrected ed. (1978), p. 9.

18. See C. W. Oseen, *Neuere Methoden und Ergebnisse in der Hydrodynamik* (Leipzig: Akademische Verlagsgesellschaft, 1927), pp. 162–65.

19. Ian Proudman and J. R. A. Pearson, "Expansions at Small Reynolds Numbers for the Flow Past a Sphere and a Circular Cylinder," *Journal of Fluid Mechanics* 2 (1957): 237–62.

20. This was done by S. Goldstein in "The Steady Flow of Viscous Fluid Past a Fixed Spherical Obstacle at Small Reynolds Numbers," *Proceedings of the Royal Society of London,* ser. A, 123 (1929): 225–35.

21. Russell, *Portraits,* p. 104.

22. Julia Bell to Victor Lowe, June 30, 1967.

23. *Obituary Notices of Fellows of the Royal Society* 6 (1948): 293.

24. Price, *Dialogues,* XXIX (June 10, 1943).

25. *ESP,* p. 41.

IX. Whitehead's Thirtieth Year

1. Russell, *Portraits,* p. 103. Russell gave this BBC talk on July 3, 1952.

2. Gerard Tracy, Librarian and Archivist at the Oratory in Edgbaston, to Victor Lowe, February 25, 1980.

3. Lewis S. Feuer to Victor Lowe, June 5, 1970.

4. Price, *Dialogues,* XLII (September 11, 1945).

5. Ibid., XXIX (June 10, 1943).

6. Reported to Victor Lowe by Charles Hartshorne, who was Whitehead's assistant in 1925–26.

X. The Married Mathematician

1. *The Letters of Frederick William Maitland,* ed. C. H. S. Fifoot (London: Selden Society, 1963), letter no. 321.

2. Whitehead discussed the crumbling of Newtonian physics in his lectures at Harvard and in Price, *Dialogues,* XXVIII (June 3, 1943) and XLII (September 11, 1945).

3. Olwen W. Campbell, *The Lighted Window* (Cambridge: Heffer, 1940), p. 47. The family is identified as that of "a Cambridge mathematician who is now a very distinguished idealist philosopher." Olwen's daughter, Clare Campbell, assured me that Whitehead was meant.

4. *UA,* p. viii.

5. Bertrand Russell to Victor Lowe, September 21, 1965.

6. *UA,* second paragraph of Preface and Sec. 22.

7. Ibid., p. vi.

8. Ibid., p. vii.

9. Ibid., p. viii.

10. C. J. Gerhardt, *Philosophische Schriften von Leibniz,* vol. VII (Berlin, 1890), p. 31; translated by C. I. Lewis and quoted in his *A Survey of Symbolic Logic* (Berkeley, 1918), p. 9.

11. *UA,* Sec. 6; F. H. Bradley, *The Principles of Logic* (London, 1883), Bk. II, Pt. I, Chap. III.

12. *UA,* Secs. 47, 1; G. F. Stout, "Thought and Language," *Mind* 16 (April 1891): 187.

13. *UA,* Sec. 7.

14. *SMW,* second page of Chap. VII; John Dewey, *Logic, the Theory of Inquiry* (New York: Henry Holt & Co., 1938), p. 391n.

15. *UA,* p. vii; see also Secs. 5, 8.

16. Ibid., Sec. 3.

17. Ibid.

18. Ibid.; italics in the original.

19. W. V. Quine, "Whitehead and the Rise of Modern Logic," in Schilpp, *LLP-W,* p. 128. I believe that Quine still adheres to this criticism.

20. *UA,* Sec. 14.

21. Russell, *MPD,* pp. 39, 43n.

22. Bertrand Russell to Victor Lowe, June 18, 1941.

23. *Obituary Notices of Fellows of the Royal Society* 6 (1948).

24. Hugh MacColl, "Review of A. N. Whitehead, *Universal Algebra,*" *Mind,* n.s., 8 (1899): 108–13.

25. See the references to MacColl in Lewis's *A Survey of Symbolic Logic.*

26. Abiog. Notes, in Schilpp, *LLP-W,* p. 9; or *ESP,* p. 9. Price (*Dialogues* [April 28, 1938]) reports that the Whiteheads showed him a picture of the house in the September 1936 issue of *National Geographic.*

27. T. North Whitehead, "Now I Am an American," Ch. I.

28. Price, *Dialogues,* XII (April 28, 1938).

29. Abiog. Notes, in Schilpp, *LLP-W,* p. 13; or *ESP,* p. 13.

30. Lady Whittaker to Victor Lowe, July 1965.

31. T. North Whitehead, "Now I Am an American," Ch. I.

32. From Bertrand Russell's 1952 BBC talk on Whitehead, as reprinted in his *Portraits,* p. 103.

33. Gregory Bateson to Professor William Coleman, December 1, 1966 (quoted by permission).

34. Ibid.

35. T. North Whitehead, "Now I Am an American," Ch. I.

36. Russell, *Abiog.,* Vol. I, p. 135.

37. Whitehead's pamphlet, along with one by R. T. Glazebrook and R. St. J. Parry which supported the building scheme, was printed by the Cambridge University Press; it was dated March 4, 1895. As it was not published (being in fact marked "strictly private and confidential"), it ought not to have been included in the bibliography of Whitehead's writings that Robert C. Baldwin and I compiled in 1941 (published in Schilpp, *LLP-W*). And that bibliographical entry contained two errors: in the title, "Proposed Extension of the Paddocks," "of" should have been "on"; and Parry should have been listed as an author in place of J. Ward.

38. Victor Lowe's conversation with T. C. Nicholas, July 1967. Nicholas was elected a Fellow in 1912.

39. Ibid. Whitehead had come from London for the College Meeting.

40. Minutes of the Council of Trinity College, June 3, 1903.

41. In Abiog. Notes (Schilpp, *LLP-W,* p. 12) Whitehead is incorrect on this point.

42. His speech was printed in the *Cambridge University Reporter,* March 26, 1897.

43. Fifoot, *Letters of Frederick William Maitland,* letter no. 191 (February 23, 1897).

44. Goldsworthy Lowes Dickinson, *J. McT. E. McTaggart* (Cambridge: Cambridge University Press, 1931). Dickinson's authority for Whitehead's membership, and Russell's later, was McTaggart's diary.

XI. Bertrand Russell

1. R. Crawshay-Williams, *Russell Remembered* (Oxford: Oxford University Press, 1970), p. 32.

2. Russell, *Abiog.,* Vol. I, p. 125.

3. Russell's *An Essay on the Foundations of Geometry* was reprinted by Dover Press in 1956, with a Foreword by Morris Kline. There is also a French translation (Gauthier-Villars, 1901), for which Russell made corrections and additions; Louis Couturat also had a hand in it.

4. Henry Sidgwick to James Ward, September 6, 1895.

5. See Russell's article "A Turning-Point in My life," in *The Saturday Book,* ed. Leonard Russell (London: Hutchinson, 1948), pp. 142–46.

6. Ibid., p. 146.

7. Crawshay-Williams, *Russell Remembered,* p. 43.

8. Whitehead's technical letters to Russell regarding the *Principia Mathematica* are now on file in the Bertrand Russell Archives at McMaster University, Hamilton, Ontario.

9. Giuseppe Peano, *Calcolo geometrico secondo l'Ausdehnungslehre di H. Grassmann* (Turin, 1888). Two chapters of it are included in *Selected Works of Giuseppe Peano,* trans. and ed. Hubert C. Kennedy (Toronto: University of Toronto Press, 1973).

10. Russell's translation of Gerhardt VII.401; see Bertrand Russell, *A Critical Examination of the Philosophy of Leibniz* (Cambridge, 1900), p. 13.

11. Russell, *Abiog.,* Vol. I, pp. 144–45.

12. Bertrand Russell, "Sur la logique des relations avec des applications à la theorie des series," *Rivista di matematica* 7 (1901): 115–48. An English translation by Robert C. Marsh, revised and corrected by Russell, may be found in Bertrand Russell, *Logic and Knowledge: Essays, 1901–1950,* ed. Robert C. Marsh (London: George Allen & Unwin, 1956).

13. Russell, *Logic and Knowledge,* pp. 3–4.

14. Russell, *Abiog.,* Vol. I, p. 146.

15. Ibid.

16. Bennett Simon, M. D., and Nancy Simon, R. N., "The Pacifist Turn: An Episode of Mystic Illumination in the Autobiography of Bertrand Russell," *Journal of the American Psychoanalytic Association* 20 (1972): 109–21. A slightly shorter version, addressed to a lay audience, appeared in *Russell* (journal of the Bertrand Russell Archives) 13 (Spring 1974): 11.

17. Letter no. 154 in the collection of letters from Bertrand Russell to Lady Ottoline Morrell, Humanities Research Center, University of Texas at Austin; quoted by Ronald W. Clark in *The Life of Bertrand Russell* (New York: Alfred A. Knopf, 1976), p. 85.

18. Russell, *Abiog.,* Vol. I, p. 150.

19. Bertrand Russell to Alys Russell, July 13, 1901, Bertrand Russell Archives, McMaster University, Hamilton, Ontario.

20. Russell, *Abiog.,* Vol. I, p. 150.

21. Bertrand Russell to Victor Lowe, June 3, 1967.

22. Russell, *Abiog.,* Vol. I, p. 150.

23. Barbara Strachey, *Remarkable Relations: The Story of the Pearsall Smith Family* (London: Gollancz, 1980).

24. Bertrand Russell, "The Free Man's Worship," *Independent Review* 1 (December 1903): 415–24. Everybody has read this essay. For those who have forgotten it, two long quotations will be sufficient reminders. The first (p. 416) is Russell's statement of his point of departure. The second is his closing paragraph.

> That Man is the product of causes which had no prevision of the end they were achieving; that his origin, his growth, his hopes and fears, his loves and his beliefs, are but the outcome of accidental collocations of atoms; that no fire, no heroism, no intensity of thought and feeling, can preserve an individual life beyond the grave, that all the labours of the ages, all the devotion, all the inspiration, all the noonday brightness of human genius, are destined to extinction in the vast death of the solar system, and that the whole temple of Man's achievement must inevitably be buried beneath the debris of a universe in ruins—all these things, if not quite beyond dispute, are yet so nearly certain, that no philosophy which rejects them can hope to stand. Only within the scaffolding of these truths, only on the firm foundation of unyielding despair, can the soul's habitation henceforth be safely built.

> Brief and powerless is Man's life; on him and all his race the slow, sure doom falls pitiless and dark. Blind to good and evil, reckless of destruction, omnipotent matter rolls on its relentless way; for Man, condemned today to lose his dearest, tomorrow himself to pass through the gate of darkness, it remains only to cherish, ere yet the blow falls, the lofty thoughts that ennoble his little day; disdaining the coward terrors of the slave of Fate, to worship at the shrine that his own hands have built; undismayed by the empire of chance, to preserve a mind free from the wanton tyranny that rules his outward life; proudly defiant of the irresistible forces that tolerate, for a moment, his knowledge and his condemnation, to sustain alone, a weary but unyielding Atlas, the world that his own ideals have fashioned despite the trampling march of unconscious power.

XII. *Principia Mathematica*

1. Clark, *Bertrand Russell,* p. 84.

2. Russell, *W&PM*, p. 137.

3. This has been published with commentary in I. Grattan-Guinness, ed., *Dear Russell–Dear Jourdain* (New York: Columbia University Press, 1977).

4. Grattan-Guinness, *Dear Russell–Dear Jourdain,* p. 134.

5. Letter no. 64 (December 5, 1911) in the collection of letters from

Bertrand Russell to Ottoline Morrell, Harry Ransom Humanities Research Center, University of Texas at Austin.

6. The application is printed in I. Grattan-Guinness, "The Royal Society's Financial Support of the Publication of Whitehead and Russell's *Principia Mathematica,*" *Notes and Records of the Royal Society of London* 30, no. 1 (July 1975): 89–104; see esp. p. 96.

7. Grattan-Guiness, *Dear Russell–Dear Jourdain,* letter of April 15, 1910.

8. Bertrand Russell, "My Mental Development," in *The Philosophy of Bertrand Russell,* ed. Paul A. Schilpp (Evanston: Northwestern University Press, 1944), pp. 12–13.

9. Abiog. Notes, in Schilpp, *LLP-W,* p. 10.

10. *Proceedings of the Royal Society of London* 64 (1898–99): 319–20.

11. Cf. *UA,* sec. 22; and Chapter X, section iii, above.

12. Part I of Whitehead's "Memoir on the Algebra of Symbolic Logic" was published in *American Journal of Mathematics* 23, no. 2 (April 1901); Part II in ibid. 23, no. 4 (October 1901).

13. A summary of this is given by W. V. Quine in "Whitehead and the Rise of Modern Logic." See Schilpp, *LLP-W,* pp. 131–32.

14. Alfred North Whitehead, "On Cardinal Numbers," *American Journal of Mathematics* 24, no. 4 (October 1902): 367–94.

15. Quine, "Whitehead and the Rise of Modern Logic," p. 157.

16. Alfred North Whitehead, "Theorems on Cardinal Numbers," *American Journal of Mathematics* 26, no. 1 (January 1904): 31–32.

17. Quine, "Whitehead and the Rise of Modern Logic," p. 158.

18. Alfred North Whitehead, "The Logic of Relations, Logical Substitution Groups, and Cardinal Numbers," *American Journal of Mathematics* 25, no. 2 (April 1903): 157–78.

19. Further, see Quine, "Whitehead and the Rise of Modern Logic," p. 133.

20. W. E. Johnson, "Report to the Sub-Committee on Higher Degrees of the Cambridge University Special Board on Mathematics," *Cambridge University Register* 28 (1905).

21. E. V. Huntington, in *Transactions of the American Mathematical Society* 35 (1933): 278n.

22. Russell, *Abiog.,* Vol. I, p. 151.

23. Russell, W&*PM*, p. 138.

24. Russell, *MPD,* p. 74.

25. Alfred North Whitehead to Bertrand Russell, April 30, 1905, Bertrand Russell Archives, McMaster University, Hamilton, Ontario.

26. Russell, *MPD,* p. 158.

27. Ibid., p. 157.

28. C. I. Lewis, in *Mind* 21 (1912): 522–31 and 23 (1914): 240–47.

29. Bertrand Russell, *Introduction to Mathematical Philosophy* (London and New York, 1919), pp. 153–54.

30. *PM,* Introduction, Chap. I. In the second edition, the quoted passage is on p. 14.

31. Russell, W&*PM,* p. 137. Russell's memory is not contradicted by Whitehead's remark in *PR* (n. 3 in I I iii), "These introductory discussions are practically due to Russell." The Introduction to *PM* comprises three chapters,

of which the second (on the theory of logical types) and third (on incomplete symbols) were in Russell's province.

32. Russell, *W&PM,* p. 138.

33. Russell, *MPD,* p. 95.

34. Russell, *W&PM,* p. 138.

35. Ibid.

36. *PM,* Preface to Vol. III.

XIII. *Principia Mathematica* (Continued)

1. *PM,* Vol. I, p. 68.

2. Bertrand Russell to Gottlob Frege, June 16, 1902, in Schilpp, *The Philosophy of Bertrand Russell,* p. 13. Russell's letter and Frege's reply have been published in John van Heijenoort, *From Frege to Gödel: A Source Book in Mathematical Logic* (Cambridge, Mass.: Harvard University Press, 1967).

3. Bertrand Russell, "Mathematical Logic as Based on the Theory of Types," *American Journal of Mathematics* 30 (1908): 22–62; reprinted in the Russell anthologies compiled by Robert C. Marsh and Douglas P. Lackey.

4. Russell, "Mathematical Logic," sec. III, penultimate paragraph.

5. Cf. *PM,* Vol. II, Summary of Pt. III; and Quine, "Whitehead and the Rise of Modern Logic," p. 152.

6. *OT,* p. 121; as reprinted in *AE,* p. 168.

7. Russell, *Introduction to Mathematical Philosophy,* 2d ed. (London: George Allen & Unwin; New York: Macmillan Co., 1920), p. 135.

8. *PM,* Vol. I, p. 173.

9. Ibid., *20.

10. *OT,* p. 122.

11. These archives also contain a dozen file boxes of rough drafts of sections of the *Principia,* omitted portions of the final manuscript (which was destroyed), miscellaneous notes by Whitehead on topics dealt with in the book, and copies of several hundred letters from Russell to his wife, Alys.

12. Douglas P. Lackey, "The Whitehead Correspondence," *Russell* 5 (Spring 1972): 14.

13. This is one of the two or three collaboration letters from Russell to Whitehead that somehow escaped destruction and are now in the Bertrand Russell Archives, McMaster University, Hamilton, Ontario.

14. C. J. Keyser, "Concerning the Axiom of Infinity and Mathematical Induction," *Bulletin of the American Mathematical Society* 9 (1904): 424–35.

15. Bertrand Russell, "The Axiom of Infinity," *Hibbert Journal* 2 (July 1904): 809–12; reprinted in Bertrand Russell, *Essays in Analysis,* ed. Douglas P. Lackey (New York: George Braziller, 1973).

16. Russell, *Introduction to Mathematical Philosophy,* p. 135.

17. Alfred North Whitehead, "The Philosophy of Mathematics," *Science Progress in the Twentieth Century* 5, no. 18 (October 1910): 234–39; the quotations are from p. 237. This article is nominally a review—the only review Whitehead ever published—of *Mysticism in Modern Mathematics,* by Hastings Berkeley.

18. See n. 3.

19. Arts. 2–4 in I. Grattan-Guinness, "The Russell Archives: Some New Light on Russell's Logicism," *Annals of Science* 31 (1974): 389–98, esp. 394.

20. Russell, *Portraits,* p. 56.
21. Bertrand Russell to Philip Jourdain, March 30, 1908, in Grattan-Guinness, *Dear Russell–Dear Jourdain.*
22. Russell, *Abiog.,* Vol. I, p. 152.
23. Bertrand Russell to Lucy Donnelly, October 18, 1909, Bertrand Russell Archives, McMaster University, Hamilton, Ontario. I have used part of a passage that has been printed in Clark, *Bertrand Russell,* p. 118.
24. Russell, *Abiog.,* Vol. I, p. 152.
25. Grattan-Guinness, "The Royal Society's Financial Support of Whitehead and Russell's *Principia Mathematica,*" pp. 94–95.
26. Russell, *Abiog.,* Vol. I, p. 152.
27. *Encyclopaedia Britannica,* 11th ed., s.v. "Mathematics"; reprinted under the title "Mathematics, Nature of," in the 14th ed.
28. Russell, *MPD,* p. 86.
29. *Sydney Chapman, Eighty* (Boulder: The National Center for Atmospheric Research, 1968), p. 187.
30. Russell, interview with Victor Lowe, June 4, 1965.
31. Whitehead, conversation with Victor Lowe, May 14, 1941.
32. Russell, *Portraits,* p. 100.
33. Abiog. Notes, in Schilpp, *LLP-W,* p. 11; or *ESP,* p. 11.
34. Russell, *Portraits,* pp. 39–40.

XIV. "On Mathematical Concepts of the Material World"

1. Whitehead, conversation with Victor Lowe, December 2, 1936.
2. F. S. C. Northrop and Mason W. Gross printed the memoir in its entirety as the opening selection in their splendid *Alfred North Whitehead: An Anthology* (New York: Macmillan Co., 1953).
3. MC, p. 473 (Northrop and Gross, *Whitehead,* p. 18).
4. Ibid., p. 466 (Northrop and Gross, *Whitehead,* p. 13). MC does not depend on the adoption of any *theory* of classes or types.
5. Russell, *P of M,* Chap. LIII. See MC, p. 479 (Northrop and Gross, *Whitehead,* p. 29). Russell's treatment of Rational Dynamics is an important part of the background of MC.
6. MC, p. 467 (Northrop and Gross, *Whitehead,* p. 14). See Bertrand Russell, "Is Position in Time and Space Absolute or Relative?" *Mind,* n.s., 10 (1901); and idem, *Philosophy of Leibniz,* p. 120.
7. Russell, *P of M,* Chap. LXXX.
8. MC, p. 466 (Northrop and Gross, *Whitehead,* p. 12). In each Concept of the Material World the relation from whose properties geometry issues is called "the essential relation." In Whitehead's Concept I the essential relation is triadic, and gives rise to geometry *only;* in Concept V it is pentadic. Relations other than the essential relation and the time-relation are called "extraneous." Cf. MC, pp. 468–69 (Northrop and Gross, *Whitehead,* pp. 15–16).
9. MC, pp. 467–68 (Northrop and Gross, *Whitehead,* pp. 13, 15); italics in the original.
10. Alfred North Whitehead to Bertrand Russell, September 3, 1911, Bertrand Russell Archives, McMaster University, Hamilton, Ontario.

11. MC, p. 484 (Northrop and Gross, *Whitehead,* p. 35).
12. MC, p. 477 (Northrop and Gross, *Whitehead,* p. 26).
13. MC, p. 476 (Northrop and Gross, *Whitehead,* p. 24).
14. MC, p. 525 (Northrop and Gross, *Whitehead,* p. 82).
15. Ibid.
16. Cf. W. Mays, "The Relevance of 'On Mathematical Concepts of the Material World' to Whitehead's Philosophy," in *The Relevance of Whitehead,* ed. Ivor Leclerc (New York: Macmillan Co., 1961).

F. S. C. Northrop has said it is "likely" that "one of the mathematical possibles listed in [this memoir] . . . is the metaphysical system which was stated in his [Whitehead's] technically modified English prose in *Process and Reality*" (Foreword to *A Whiteheadian Aesthetic,* by Donald W. Sherburne [New Haven: Yale University Press, 1961], p. xxii). This is not merely *un*likely, it is impossible. The impossibility is assured by many features of the later work, such as the complete demotion of the instant of time. *Process and Reality* introduces a different and much larger set of basic ideas in terms of which our *total experience* of the world is to be interpreted. One quotation from MC is enough to point up the great change of subject matter. Whitehead, on remarking that the material world, as described in Concept II, "labours under the defect that it can never be perceived," says, "But this is a philosophic question with which we have no concern" (MC, p. 480 [Northrop and Gross, *Whitehead,* pp. 11–12]). A few of the ideas of MC appear, in a different context, in *PR.* An important feature that is kept is the use of the axiomatic method.

An entirely adequate detailed account of MC takes up most of R. D. Bendall's 1973 Master's thesis for the Graduate Theological Union, Berkeley, Calif. ("'On Mathematical Concepts of the Material World' and the Development of Whitehead's Philosophy of Organism"). On p. 3 Bendall rightly says that "it is just as harmful to overestimate the significance of MC to the development of Whitehead's philosophy as it is detrimental to neglect it altogether."

17. MC, p. 471 (Northrop and Gross, *Whitehead,* p. 19).
18. MC, p. 479 (Northrop and Gross, *Whitehead,* p. 28).

XV. Last Years in Cambridge

1. *AI* VIII ix.
2. *APG,* Sec. 4.
3. Russell, *P of M,* Sec. 352.
4. *APG,* Sec. 3.
5. F. S. M., in *Mathematical Gazette* 4 (June 1908): 287–90.
6. *UA,* Sec. 22.
7. Ibid.
8. "Geometry" was not reprinted in the fourteenth or later editions of the *Encyclopaedia Britannica.* Whitehead's article was reprinted in *ESP.*
9. Immanuel Kant, *Critique of Pure Reason,* trans. Max Müller (New York: Macmillan Co.; London: Macmillan & Co., 1907), Trans. Aesth., sec. 1.
10. *Encyclopaedia Britannica,* 11th ed., s.v. "Geometry"; see also *ESP,* p. 264.

11. J. E. Littlewood, *A Mathematician's Miscellany* (London: Methuen, 1953).

12. Ibid., p. 70.

13. Ibid., p. 77.

14. Ibid. Among other things, Whitehead discussed Hilbert's axioms in his course on foundations of geometry.

15. J. E. Littlewood, interview with Victor Lowe, July 1965.

16. Hugh Dalton, *Call Back Yesterday: Memoirs, 1887–1931* (London: Muller, 1953), p. 46.

17. Abiog. Notes, in Schilpp, *LLP-W,* p. 13; or *ESP,* p. 14.

18. Copies of this pamphlet are very rare. After Professor John Slater discovered one, the pamphlet was reprinted in the American quarterly *Process Studies* 7, no. 1 (Spring 1977): 37–38.

19. *Women's Franchise* (published by the National Union of Women's Suffrage Societies), June 27, 1907, p. 4.

20. *Times* (London), January 11, 1907.

21. Minutes of the Council of Trinity College, February 11, 1910.

22. *Trinity College Report Book, 1906–1910,* p. 112. The protesting memorial statement was received by the Council March 11, 1910.

23. Minutes of the Council of Trinity College, February 18, 1910, Min. 18.

24. Professorial Board file 1884–1916, fol. 151, University College, London.

25. Price, *Dialogues,* XXIX (June 10, 1943).

Bibliography

This bibliography is believed to include everything written by Whitehead and published in his lifetime. It is arranged chronologically. No translations of his books are listed. Post-1947 printings of Whitehead's books are mentioned only if they are cited in the notes to this volume.

"A Celebrity at Home: The Clerk of the Weather." *Cambridge Review,* February 10, 1886, pp. 202–3.

"Davy Jones." *Cambridge Review,* May 12, 1886, pp. 311–12.

"A Visitation." *Cambridge Fortnightly,* March 6, 1888, pp. 81–83.

"On the Motion of Viscous Incompressible Fluids: A Method of Approximation." *Quarterly Journal of Pure and Applied Mathematics* 23 (1889): 78–93.

"Second Approximations to Viscous Fluid Motion: A Sphere Moving Steadily in a Straight Line." *Quarterly Journal of Pure and Applied Mathematics* 23 (1889): 143–52.

"On Ideals: With Reference to the Controversy Concerning the Admission of Women to Degrees in the University." *Cambridge Review,* May 14, 1896, pp. 310–11.

"The Geodesic Geometry of Surfaces in non-Euclidean Space." *Proceedings of the London Mathematical Society* 29 (1897–98): 275–324.

UA *A Treatise on Universal Algebra, with Applications,* Vol. I (no others published). Cambridge: Cambridge University Press, 1898.

"Sets of Operations in Relation to Groups of Finite Order." Abstract. *Proceedings of the Royal Society of London* 64 (1898–99): 319–20.

"Memoir on the Algebra of Symbolic Logic." *American Journal of Mathematics* 23 (1901): 139–65, 297–316.

"On Cardinal Numbers." *American Journal of Mathematics* 24 (1902): 367–94.

"The Logic of Relations, Logical Substitution Groups, and Cardinal Numbers." *American Journal of Mathematics* 25 (1903): 157–78.

"Theorems on Cardinal Numbers." *American Journal of Mathematics* 26 (1904): 31–32.

APG *The Axioms of Projective Geometry.* Cambridge Tracts in Mathematics and Mathematical Physics, No. 4. Cambridge:

Cambridge University Press, 1906. New York: Hafner Publishing Co., 1971.

Liberty and the Enfranchisement of Women. Cambridge: Cambridge Women's Suffrage Association, 1906.

MC "On Mathematical Concepts of the Material World." *Philosophical Transactions of the Royal Society of London,* ser. A, 205 (1906): 465–525.

ADG *The Axioms of Descriptive Geometry.* Cambridge Tracts in Mathematics and Mathematical Physics, No. 5. Cambridge: Cambridge University Press, 1907; New York: Hafner Publishing Co., 1971.

"The Philosophy of Mathematics." *Science Progress in the Twentieth Century* 5 (October 1910): 234–39.

"Axioms of Geometry." Div. VII of "Geometry." *Encyclopaedia Britannica,* 11th ed. (1910–11).

"Mathematics." *Encyclopaedia Britannica,* 11th ed. (1910–11).

"Non-Euclidean Geometry." With Bertrand Russell. Div. VI of "Geometry." *Encyclopaedia Britannica,* 11th ed. (1910–11).

PM *Principia Mathematica.* With Bertrand Russell. Vols. I–III. Cambridge: Cambridge University Press, 1910–13; 2d ed., 1925–27.

IM *An Introduction to Mathematics.* Home University Library of Modern Knowledge, No. 15. London: Williams & Norgate; New York: Henry Holt & Co., 1911.

"The Place of Mathematics in a Liberal Education." *Journal of the Association of Teachers of Mathematics for the Southeastern Part of England* 1 (1911). Reprinted in Pt. III of *ESP* (1947).

"The Principles of Mathematics in Relation to Elementary Teaching." *Proceedings of the Fifth International Congress of Mathematicians* (Cambridge, England, August 22–28, 1912), 2:449–54. Reprinted as Chap. V of *OT* (1917).

"Presidential Address to the London Branch of the Mathematical Association." *Mathematical Gazette* 7 (March 1913): 87–94. Reprinted, under the title "The Mathematical Curriculum," as Chap. IV of *OT* (1917) and Chap. VI of *AE* (1929).

"Space, Time, and Relativity." *Proceedings of the Aristotelian Society,* n.s., 16 (1915–16): 104–29. Reprinted as Chap. VIII of *OT* (1917) and Chap. X of *AE* (1929).

"The Aims of Education: A Plea for Reform." *Mathematical Gazette* 8 (January 1916): 191–203. Reprinted as Chap. I of *OT* (1917) and, with omissions, as Chap. I of *AE* (1929).

"The Organisation of Thought." *Report of the Eighty-sixth Meeting of the British Association for the Advancement of Science* (Newcastle-on-Tyne, 1916), 355–65. Presidential address to Section A

(Mathematical and Physical Science) of the Association. Reprinted as Chap. I of *OT* (1917) and Chap. VIII of *AE* (1929).

"La Théorie Relationniste de L'Espace." *Revue de Métphysique et de Morale* 23 (May 1916): 423–54.

OT *The Organisation of Thought, Educational and Scientific.* London: Williams & Norgate; Philadelphia: J. B. Lippincott Co., 1917.

"Technical Education and Its Relation to Science and Literature." *Technical Journal* 10 (January 1917): 59–74. Reprinted as Chap. II of *OT* (1917) and, with omissions, as Chap. IV of *AE* (1929).

"Graphical Solution for High-Angle Fire." *Proceedings of the Royal Society of London,* ser. A, 94 (1918): 301–7.

An Enquiry Concerning the Principles of Natural Knowledge. Cambridge: Cambridge University Press, 1919; 2d ed., 1925.

"Fundamental Principles in Education." *Report of the Eighty-seventh Meeting of the British Association for the Advancement of Science* (Bournemouth, 1919), 361. An abridgement of the address with which Whitehead opened a discussion of this subject before Section L (Education) of the Association.

"A Revolution in Science." *Nation,* November 15, 1919, 232–33. Reprinted in *Educational Review* 59 (1920): 148–53.

"Symposium: Time, Space, and Material: Are They, and If So in What Sense, the Ultimate Data of Science?" With Sir Oliver Lodge, J. W. Nicholson, Henry Head, Mrs. Adrian Stephen, and H. Wildon Carr. *Proceedings of the Aristotelian Society,* Suppl. 2 (1919), pp. 44–108. For Whitehead's contribution, see pp. 44–57.

The Concept of Nature. Cambridge: Cambridge University Press, 1920. Tarner Lectures, Trinity College, Cambridge.

"Einstein's Theory: An Alternative Suggestion." *Times* (London) *Educational Supplement,* February 12, 1920, p. 83.

"Report of Committee Appointed by the Prime Minister to Inquire into the Position of Classics in the Educational System of the United Kingdom." With the Marquis of Crewe (Chairman) and other members of the committee. London: His Majesty's Stationery Office, 1921.

"Science in General Education." *Proceedings of the Second Congress of Universities of the Empire,* pp. 31–39. London: G. Bell & Sons, 1921. Reprinted in Pt. III of *ESP* (1947).

"Discussion: The Idealistic Interpretation of Einstein's Theory." With H. Wildon Carr, T. P. Nunn, and Dorothy Wrinch. *Proceedings of the Aristotelian Society* 22 (1921–22): 123–38. For Whitehead's contribution, see pp. 130–34.

"The Philosophical Aspects of the Principle of Relativity." In *Proceedings of the Aristotelian Society* 22 (1921–22): 215–23.

R *The Principle of Relativity, with Applications to Physical Science.* Cambridge: Cambridge University Press, 1922.

The Rhythm of Education. London: Christophers, 1922. Reprinted as Chap. II of *AE* (1929).

"Uniformity and Contingency." *Proceedings of the Aristotelian Society* 23 (1922–23): 1–18. Reprinted in Pt. II of *ESP* (1947).

"The First Physical Synthesis." Chap. VI of *Science and Civilization,* edited by F. S. Marvin, pp. 161–78. London: Oxford University Press, 1923. Reprinted in Pt. IV of *ESP* (1947).

"Letter to the Editor." *New Statesman,* February 17, 1923, p. 568. A reply to C. P. Sanger's review of *The Principle of Relativity.*

"The Place of Classics in Education." *Hibbert Journal* 21 (1923): 248–61. Reprinted as Chap. V of *AE* (1929).

"The Rhythmic Claims of Freedom and Discipline." *Hibbert Journal* 21 (1923): 657–68. Reprinted as Chap. III of *AE* (1929).

"Symposium—The Problem of Simultaneity: Is There a Paradox in the Principle of Relativity in Regard to the Relation of Time Measured to Time Lived?" With H. Wildon Carr and R. A. Sampson. *Proceedings of the Aristotelian Society,* Suppl. 3 (1923), pp. 15–41. For Whitehead's contribution, see pp. 34–41.

"The Importance of Friendly Relations Between England and the United States." *Phillips Bulletin* 19, no. 3 (April 1925): 15–18. Address delivered at Phillips Academy, Andover, Mass.

"Religion and Science." *Atlantic Monthly* 136 (1925): 200–207. Reprinted as Chap. XII of *SMW* (1925).

SMW *Science and the Modern World.* New York: Macmillan Co., 1925; Cambridge: Cambridge University Press, 1926.

"The Education of an Englishman." *Atlantic Monthly* 138 (1926): 192–98. Reprinted in Pt. I of *ESP* (1947).

"Principia Mathematica." *Mind,* n.s., 35 (1926): 130. A note on the second edition of *Principia Mathematica.*

RM *Religion in the Making.* New York: Macmillan Co.; Cambridge: Cambridge University Press, 1926. Lowell Lectures, 1926.

"England and the Narrow Seas." *Atlantic Monthly* 139 (1927): 791–98. Reprinted in Pt. I of *ESP* (1947).

Symbolism: Its Meaning and Effect. New York: Macmillan Co., 1927; London: Cambridge University Press, 1928. Barbour-Page Lectures, University of Virginia, 1927.

"Time." *Proceedings of the Sixth International Congress of Philosophy,* pp. 59–64. New York and London: Longmans, Green & Co., 1927.

"Universities and Their Function." *Atlantic Monthly* 141 (1928):

638–44. Reprinted in *Harvard Business School Alumni Bulletin* 5, no. 1 (November 1, 1928): 9–14, and as Chap. VII of *AE* (1929).

AE *The Aims of Education and Other Essays.* New York: Macmillan Co.; London: Williams & Norgate, 1929.

FR *The Function of Reason.* Princeton: Princeton University Press, 1929. Louis Clark Vanuxem Foundation Lectures, Princeton University, March 1929.

PR *Process and Reality: An Essay in Cosmology.* New York: Macmillan Co.; Cambridge: Cambridge University Press, 1929. Corrected ed., edited by David Ray Griffin and Donald W. Sherburne, New York: Free Press, 1978. Gifford Lectures, University of Edinburgh, 1927–28.

"An Address." *Radcliffe Quarterly* 14, no. 1 (January 1930): 1–5. Delivered at the celebration of the fiftieth anniversary of the founding of Radcliffe College. Reprinted as "Historical Changes" in Pt. III of *ESP* (1947).

Prefatory Note to *The Practice of Philosophy,* by Susanne K. Langer. New York: Henry Holt & Co., 1930.

"On Foresight." Introduction to *Business Adrift,* by W. B. Donham. New York: McGraw-Hill; London: George Routledge & Sons, 1931. Reprinted, with additions, as Chap. VI of *AI* (1933).

"Objects and Subjects." *Proceedings and Addresses of the American Philosophical Association* 5 (1931): 130–46. Reprinted in *Philosophical Review* 41 (1932): 130–46 and as Chap. XI of *AI* (1933).

Symposium in Honor of the Seventieth Birthday of Alfred North Whitehead. Cambridge, Mass.: Harvard University Press, 1932. Printed for private circulation only. Partial publication of Whitehead's reply, under the title "Process and Reality," appears in Pt. II of *ESP* (1947).

AI *Adventures of Ideas.* New York: Macmillan Co.; Cambridge: Cambridge University Press, 1933.

"The Study of the Past: Its Uses and Its Dangers." *Harvard Business Review* 11 (1933): 436–44. Reprinted in *Educational Record* 14 (1933): 454–67 and in Pt. III of *ESP* (1947).

Foreword to *The Farther Shore: An Anthology of World Opinion on the Immortality of the Soul,* edited by Nathaniel Edward Griffin and Lawrence Hunt. Boston: Houghton Mifflin Co., 1934.

Foreword to *A System of Logistic,* by Willard Van Orman Quine. Cambridge, Mass.: Harvard University Press; London: Oxford University Press, 1934.

"Indication, Classes, Numbers, Validation." *Mind* 43 (1934): 281–97. For corrigenda to this article see ibid., p. 543.

NL *Nature and Life.* Chicago: University of Chicago Press; Cambridge: Cambridge University Press, 1934. Reprinted as Chaps. VII and VIII of *MT* (1938).

"The Aim of Philosophy." *Harvard Alumni Bulletin* 38 (1935): 234–35. Reprinted, with the omission of the first paragraph and the addition of a final paragraph, as the Epilogue to *MT* (1938).

"In Memoriam." In *Bernard Bosanquet and His Friends,* edited by J. H. Muirhead, p. 316. London: George Allen & Unwin, 1935.

"Harvard: The Future." *Atlantic Monthly* 158 (1936): 260–70. Reprinted in Pt. III of *ESP* (1947).

"Memories." *Atlantic Monthly* 157 (1936): 672–79. Reprinted in Pt. I of *ESP* (1947).

"Remarks." *Proceedings and Addresses of the American Philosophical Association* 10 (1936): 178–86. Reprinted in *Philosophical Review* 46 (1937): 178–86 and, without the first paragraph, under the title "Analysis of Meaning," in Pt. II of *ESP* (1947).

MT *Modes of Thought.* New York: Macmillan Co.; Cambridge: Cambridge University Press, 1938. Six lectures delivered at Wellesley College, Mass., and two lectures given at the University of Chicago.

"An Appeal to Sanity." *Atlantic Monthly* 163 (1939): 309–20.

"John Dewey and His Influence." In *The Philosophy of John Dewey,* edited by Paul A. Schilpp, pp. 477–78. Evanston: Northwestern University Press, 1939.

"The Issue: Freedom." *Boston Daily Globe,* December 24, 1940, p. 12. A letter to the editor.

Abiog. Notes "Autobiographical Notes." In *The Philosophy of Alfred North Whitehead,* edited by Paul A. Schilpp, pp. 1–14. Evanston: Northwestern University Press, 1941; Cambridge: Cambridge University Press, 1943; New York: Tudor Publishing Co., 1951. Reprinted in Pt. I of *ESP* (1947).

"Immortality." In *The Philosophy of Alfred North Whitehead,* edited by Paul A. Schilpp, pp. 682–700. Evanston: Northwestern University Press, 1941; Cambridge: Cambridge University Press, 1943; New York: Tudor Publishing Co., 1951. The Ingersoll Lecture for 1941, this sermon was originally delivered at the Harvard Memorial Church on April 22, 1941. Reprinted in *Harvard Divinity School Bulletin* 39 (1941–42): 5–21.

"Mathematics and the Good." In *The Philosophy of Alfred North Whitehead,* edited by Paul A. Schilpp, pp. 666–81. Evanston: Northwestern University Press, 1941; Cambridge: Cambridge University Press, 1943; New York: Tudor Publishing Co., 1951.

"Statesmanship and Specialized Learning." *Proceedings of the American Academy of Arts and Sciences* 75 (1942): 1–5. A shortened version, "The Problem of Reconstruction," was published in *Atlantic Monthly* 169 (1942): 172–75.

Preface to "The Organization of a Story and a Tale," by William Morgan. *Journal of American Folklore* 58 (1945): 169.

ESP *Essays in Science and Philosophy*. New York: Philosophical Library, 1947, London: Rider & Co., 1948.

Index

Entries for the writings of Whitehead which receive more than passing mention in the text are included, but quotations per se *are not indexed. Within entries,* Whitehead *is shortened to* W.

Victor Lowe is widely recognized as the preeminent authority on Whitehead's life and work. He was one of Whitehead's students at Harvard, is the author of *Understanding Whitehead,* and wrote the Whitehead article in the *Dictionary of American Biography*. His research for this biography has spanned two decades. Lowe is Professor Emeritus of Philosophy at the Johns Hopkins University, where he has been on the faculty since 1947.

The Johns Hopkins University Press
Alfred North Whitehead: The Man and His Work, Vol. I

This book was composed in Bembo text and display by The Composing Room of Michigan, Inc., from a design by Chris L. Smith. It was printed on S. D. Warren's 50-lb. Sebago Eggshell Cream paper and bound in Holliston Roxite A by The Maple Press Company.